ELECTRONICS

(Circuits and Devices)

Aldo Karlmann

© **Copyright 2016 by Aldo Karlmann**

This book may not be duplicated in any way without the express written consent of the publisher, except in the form of brief excerpts or quotations for the purpose of review. The information contained herein is for the personal use of the reader and may not be incorporated in any commercial programs, other books, database, or any kind of software without written consent of the publisher. Making copies of this book or any portion for purpose other than your own is a violation of copyright laws.

Limit of Liability/Disclaimer of Warranty:
The author and publisher make no representations or warranties with respect to the accuracy or completeness of the contents of this work and specifically disclaim all warranties, including without limitation warranties of fitness for a particular purpose. The advice and strategies contained herein may not be suitable for every situation. Neither the publisher nor the author shall be liable for damages arising here from.

Trademarks:
All brand names and product names used in this book are trademarks, registered trademarks, or trade names of their respective holders. The author and publisher are not associated with any product or vendor mentioned in this book

ACKNOWLEDGMENT

I remain grateful to my family members who supported me with their prayers and bore the burden of my absence during the writing of this book.

DEDICATION

This work is dedicated

to

You

the reader.

PREFACE

The main objective of writing this book has been to produce a good textbook from the student's point of view. The author has been teaching courses in Electronics at undergraduate level for several years. His experience as instrument engineer in the industry also helped to introduce and explain terminology and definitions as needed, placing emphasis on the topics considered most appropriate. The materials are arranged for ease of understanding.

The first three chapters deal with the fundamentals of Electronics and will be useful to those taking their first course in the subject. The remaining chapters contain what is required at undergraduate level. Topics like integrated circuits, microprocessors, digital memories etc. are adequately treated. The author believes that the book will also be useful to teachers in the field of Electronics Engineering.

TABLE OF CONTENTS

CHAPTER ONE 1
1.0 THE CONCEPT OF THERMIONIC EMISSION 1
1.1 INTRODUCTION 1
1.2 THE VACUUM DIODE: 3
1.3 STATIC CHARACTERISTICS OF A VACUUM DIODE:- 4
1.4 D.C AND A.C RESISTANCE OF VACUUM DIODE. 5
1.5 THE DIODE AS A CIRCUIT ELEMENT 7
1.6 THE LOAD LINE: 7
1.7 THE DYNAMIC CHARACTERISTICS OR CURVE: 8
1.8 DIODE APPLICATIONS: 9
1.9 FULL-WAVE RECTIFICATION WITH VACUUM DIODE. 10
1.10 VACUUM TRIODE: 11
1.11 DETERMINATION OF THE STATIC CHARACTERISTICS OF A TRIODE: 13
1.12 AMPLIFICATION FACTOR, MUTUAL CONDUCTANCE AND RESISTANCE 15
1.13 AMPLIFICATION FACTOR: 16
1.14 AMPLIFICATION BY MEANS OF A VACUUM TRIODE 18
1.15 EQUIVALENT CIRCUIT OF A VACUUM TRIODE 21
1.16: INTER-ELECTRODE CAPACITANCE OF A TRIODE VALVE 22
1.17 TETRODE AND PENTODE 23
1.18 NEGATIVE RESISTANCE EFFECT OF THE TETRODE 28

CHAPTER TWO 29
2.0 ENERGY LEVELS IN MATERIALS 29
2.1 THE NATURE OF ATOMS 29
2.2 THE BOHR ATOM 31
2.3 ATOMIC ENERGY LEVELS 32
2.4 COLLISIONS OF ELECTRONS WITH ATOMS 34
2.5 THE PHOTON NATURE OF LIGHT 35
2.6 THE ENERGY BAND THEORY OF CRYSTALS 36
2.7 METALS, SEMICONDUCTORS AND INSULATORS FERMI ENERGY LEVELS . 39
2.8 SEMICONDUCTORS 42
2.9 CONDUCTION IN SEMICONDUCTOR 43
2.10 THE EFFECT OF TEMPERATURE UPON INTRINSIC CONDUCTION 45
2.11 QUANTITATIVE THEORY OF THE PN JUNCTION 46
2.12 BIASING OF PN JUNCTION 48
2.13 STATIC CHARACTERISTICS OF THE P-N DIODE GERMANIUM 50
2.14 IMPORTANT TERMS 51
2.15 CAPACITANCE OF P-N JUNCTION 53
2.16 ZENER DIODE 54
2.17 APPLICATION OF ZENER DIODE 55
2.18 SILICON DIODE 56
2.19 AVALANCHE EFFECT 56
2.20 TUNNEL DIODE 57
2.21 CHARACTERISTICS OF A TUNNEL DIODE 58
2.22 SEMICONDUCTOR PHOTODIODE 59

	2.23	THERMISTORS 61

CHAPTER THREE 62

3.0	PN DIODE RECTIFIERS	62
3.1	HALF WAVE RECTIFIER	62
3.2	EFFICIENCY OF HALF WAVE RECTIFIER	64
3.3	FULL-WAVE RECTIFIERS:	69
3.4	EFFICIENCY OF FULL-WAVE RECTIFIER: Fig. 3.4.1	73
3.5	RIPPLE FACTOR:	79
3.6	FILTER CIRCUITS:	82
3.7	CAPACITOR FILTER CIRCUIT	83
3.8	CHOKE INPUT FILTER:	84
3.9	CAPACITOR INPUT FILTER OR π-FILTER:	85

CHAPTER FOUR 88

4.1	BIPOLAR TRANSISTORS	88
4.2	CURRENTS IN EMITTER, COLLECTOR, BASE CIRCUIT OF PNP	90
4.3	COMMON BASE CONFIGURATION	90
4.4	COMMON EMITTER CONFIGURATION:	91
4.5	COMMON COLLECTOR CONFIGURATION	92
4.6	STATIC CHARACTERISTICS FOR A COMMON BASE CIRCUIT	92
4.7	STATIC CHARACTERISTICS FOR A COMMON EMITTER CIRCUIT	95
4.8	RELATIONSHIP BETWEEN ∝ AND β	97
4.9	LOAD LINE FOR TRANSISTORS AND CURRENT AMPLIFICATION	97
4.10	AMPLIFIERS	100
4.11	TRANSISTOR BIASING	100
4.12	CAPACITIVE COUPLING	101
4.13	STATIC AND DYNAMIC LOAD LINES	102
4.14	THE FIXED BIAS CIRCUIT	103
4.15	TYPICAL TRANSISTOR JUNCTION VOLTAGE VALUES	103
4.16	SILICON CONTROLLED RECTIFIERS - THYRISTORS	106
4.17	THE MULTILAYER DIODE	107
4.18	OPERATION OF SILICON CONTROLLED RECTIFIER	108
4.19	THE SILICON CONTROLLED RECTIFIER (SCR)	108

CHAPTER FIVE 109

5.0	TRANSISTOR BIASING	109
5.1	INTRODUCTION	109
5.2	PROPER ZERO SIGNAL COLLECTOR CURRENT	110
5.3	PROPER MINIMUM BASE-EMITTER VOLTAGE	113
5.4	PROPER MINIMUM COLLECTOR-EMITTER VOLTAGE (V_{CE})	114
5.5	TRANSISTOR BIASING	115
5.6	INHERENT VARIATION OF TRANSISTOR PARAMETERS	118
5.7	STABILISATION	119

CHAPTER SIX 143

6.0	ELECTRONIC ENGINEERING CIRCUITS	143
6.1	INTRODUCTION	143
6.2	ELECTRONIC DEVICES AS NETWORK OR CIRCUIT ELEMENTS	144
6.3	EQUIVALENT T CIRCUITS	150

6.4	THE HYBRID π EQUIVALENT CIRCUIT	165
6.5	FREQUENCY RESPONSE PLOTTING WITH LOG CO-ORDINATE	166
6.6	FIELD EFFECT TRANSISTORS (F.E.T)	172
6.7	CHARACTERISTICS AND OPERATION	174
6.8	PARAMETERS	175

CHAPTER SEVEN 179

7.0	POWER AMPLIFIERS	179
7.1	INTRODUCTION	179
7.2	SINGLE ENDED AMPLIFIERS	180
7.3	TUNED CIRCUITS AND AMPLIFIERS	183
7.4	F.E.T TUNED AMPLIFIERS	187
7.5	TRANSFORMER COUPLED LOAD – CLASS A AMPLIFIER	191
7.6	CLASS A PUSH PULL	195
7.7	CLASS B PUSH PULL	197
7.8	CLASS AB PUSH PULL	201
7.9	OTHER OUTPUT CONFIGURATIONS-CLASS B	203
7.10	TRANSISTOR TUNED AMPLIFIERS	205
7.11	DOUBLE TUNED AMPLIFIERS	208
7.12	FEEDBACK AND ITS EFFECTS	211
7.13	THE IMPORTANCE OF NEGATIVE FEEDBACK	211
7.14	VOLTAGE GAIN OF NEGATIVE FEEDBACK AMPLIFIERS	212
7.15	PROPERTIES OF NEGATIVE FEEDBACK AMPLIFIERS	214
7.16	METHODS OF APPLYING NEGATIVE FEEDBACK	216
7.17	EFFECT OF NEGATIVE FEEDBACK ON INPUT AND OUTPUT IMPEDANCE	219

CHAPTER EIGHT 221

8.0	STABILITY OF FEEDBACK AMPLIFIERS	221
8.1	INTRODUCTION	221
8.2	NYQUIST CRITERION OF STABILITY	222
8.3	STABILISATION OF FEEDBACK AMPLIFIERS	225
8.4	OSCILLATORS	226
8.5	PROCEDURE FOR ACHIEVING OSCILLATION	227
8.6	R-C SINE WAVE OSCILLATORS	231

CHAPTER NINE 235

9.0	PRINCIPLES OF COMMUNICATION SYSTEMS	235
9.1	INTRODUCTION	235
9.2	MODULATION AND FREQUENCY TRANSLATION	236
9.3	THE FUNDAMENTAL MODULATION PROCESSES	238
9.4	AMPLITUDE MODULATION (AM)	238
9.5	POWER CONSIDERATIONS	241
9.6	SIDEBAND OPERATION	242
9.7	VESTIGAL SIDEBAND TRANSMISSION (VSB)	243
9.8	FREQUENCY MODULATION F.M	244
9.9	COMPARISONS OF AM AND FM	247

CHAPTER TEN 250

10.0	DIGITAL ELECTRONICS	250

10.1	BINARY ARITHMETIC & BOOLEAN ALGEBRA	250
10.2	BOOLEAN ALGEBRA	258
10.3	KARNAUGHS MAPS;	261
10.4	EXCLUSIVE OR CIRCUIT	265
10.5	HALF ADDER CIRCUIT	268
10.6	FULL ADDER CIRCUIT.	269
10.7	BINARY CODES	272
10.8	PARITY ERROR DETECTION IN NUMBERS	274
10.9	MULTIPLE BIT SERIAL ADDER	276

CHAPTER ELEVEN 279

11.1	ELECTRONIC SWITCHING	279
11.2	ELECTRONIC SWITCH	279
11.3	ADVANTAGES OF ELECTRONIC SWITCHES	282
11.4	SWITCHING ACTION OF A TRANSISTOR	284
11.5	MULTIVIBRATORS	287
11.6	TYPES OF MULTIVIBRATORS	288
11.7	ASTABLE MULTIVIBRATOR CIRCUIT:	289
11.8	MONOSTABLE MULTIVIBRATOR CIRCUIT	293
11.9	BISTABLE MULTIVIBRATOR CIRCUIT	295
11.10	DIFFERENTIATING CIRCUIT	297

CHAPTER TWELVE 307

12.0	INTEGRATED CIRCUITS	307
12.1	TECHNOLOGY REVIEW	307
12.2	LINEAR INTEGRATED CIRCUIT	309
12.3	NEGATIVE FEEDBACK	311
12.4	VOLTAGE DERIVED N.F.B	314
12.5	CURRENT DERIVED N.F.B	315
12.6	PINOUT DIAGRAM FOR TTL INTEGRATED CIRCUITS	318
12.7	DIGITAL INTEGRATED CIRCUITS- BASIC GATES	323

CHAPTER THIRTEEN 327

13.1	BASIC CONCEPT OF MICROPROCESSORS	327
13.2	HEXADECIMAL CODE	329
13.3	REGISTERS	331
13.4	INTERRUPT ROUTINES	335
13.5	MICROPROCESSOR PROGRAMMING	336

CHAPTER FOURTEEN 340

14.0	DIGITAL MEMORIES	340
14.1	INTRODUCTION:	340
14.2	THE MAGNETIC CORE MEMORY	340
14.3	SEMI-CONDUCTOR MEMORIES	345
14.4	RAM (Random Access Memories)	348
14.5	THE DYNAMIC CELL	353
14.6	ARRANGEMENT OF MEMORIES	355
REFERENCES		380

INDEX 381

CHAPTER ONE

1.0 THE CONCEPT OF THERMIONIC EMISSION

1.1 INTRODUCTION

An electrical conductor contains a large number of mobile or free electrons that are not attached to any particular atom of the material, but move randomly from one atom to the other within the boundary of the conductor; and the higher the temperature of the conductor, the greater the velocity attained by these electrons. In the diagram shown in fig 1.1.1 a metal cylinder A surrounds an incandescent filament C in an evacuated glass bulb G. A battery B is connected in series with a milliameter D between the cylinder and the negative end of the filament.

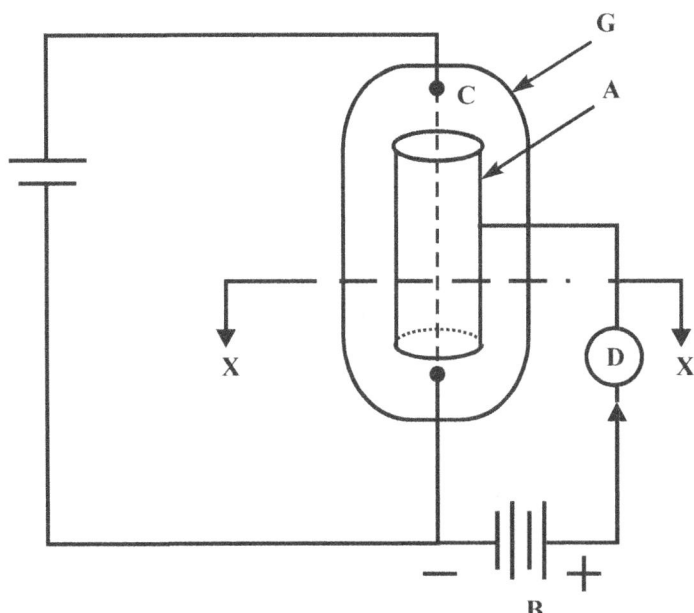

Fig. 1.1.1 A Vacuum diode

It is found that an electric current flows through the milliameter when the cylinder is made positive relative to the filament. But when the terminals of the battery are inter-changed, there is no current flow through D.

In this case, these free electrons may acquire sufficient energy to overcome the forces tending to hold them within the boundary of the filament. Consequently they escape outwards; but if there is no potential difference between the filament and the surrounding cylinder, the electrons emitted from the filament form a negatively-charged CLOUD or SPACE CHARGE around the wire, the wire being now left positively charged. The electrons near the surface of the filament experience a force urging them to re-enter the filament. This condition of equilibrium is established in which electrons re-enter the surface at nearly the same rate as they are being omitted. Consider the cross-section X-X of the glass bulb G, see Fig 1.1.2, electrons are attracted outwards from the space charge as indicated by the dotted. Line-cross-section X-X radial lines in the diagram with the application of positive potential to the glass tube. The number of electrons reaching the cylinder increases with increase in the positive potential of the cylinder.

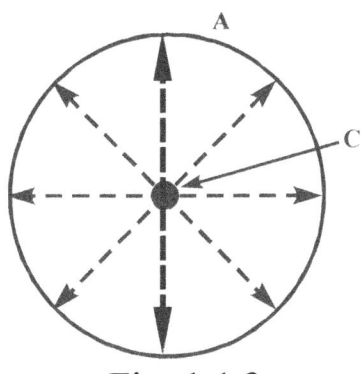

Fig. 1.1.2

The rate of flow of electrons is referred to as the saturation current. If the cylinder is made negative relative to the filament, the electrons of the space charge are repelled toward the filament and the reading of the milliameter will be zero.

ELECTRONICS

1.2 THE VACUUM DIODE:

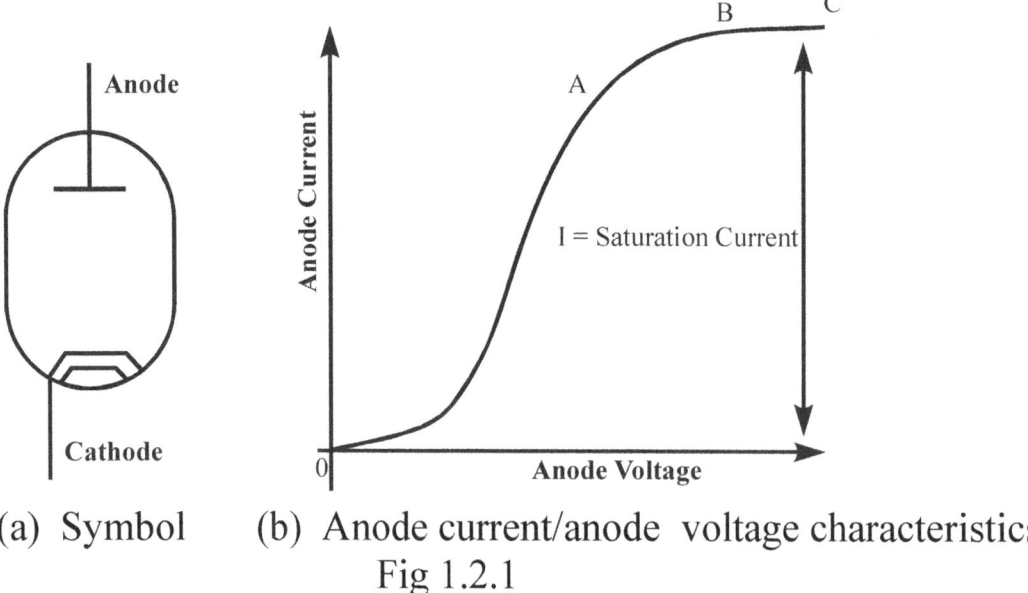

(a) Symbol   (b) Anode current/anode voltage characteristics
Fig 1.2.1

The vacuum diode is sometimes referred to as thermionic valve. The liberation of electrons from an electrode by virtue of its temperature is referred to as thermionic emission. The potential difference (p.d) between the anode and the cathode is termed the anode voltage, and the rate of flow of electrons from cathode to anode constitutes the anode current. The conventional direction of the anode current is from the anode to the cathode. It is important however to realize that the anode current is actually a movement of electrons in the reverse direction.

The thermionic emission current is given by

$$I = AT^2 e^{-b/T} \; A/m^2 \; [By \; richardson \; equ]$$

T is the absolute temperature in K

e the Napierian base = 2.718

and A and b are constants for a given material.

Fig 1.2.1 shows the symbol and characteristics of the vacuum diode. The saturation current is referred to as the temperature limited current (see

BC in the diagram of characteristics). Over range OB, the same numbers of electrons are being emitted from the cathode, but the number which reaches the anode depends upon the combined effect of the space charge and the anode voltage. The lower the anode voltage, the more effective is the opposition of the space charge to the movement of electrons from the cathode to the anode. The value of this space-charge limited current over range OA is approximately given by $i_A = K V_A^{1.5}$ *(K = a constant for a given valve)*

1.3 STATIC CHARACTERISTICS OF A VACUUM DIODE:-

The static characteristics gives the relationship between the anode current and the anode voltage for a given filament voltage and therefore for a given filament temperature.

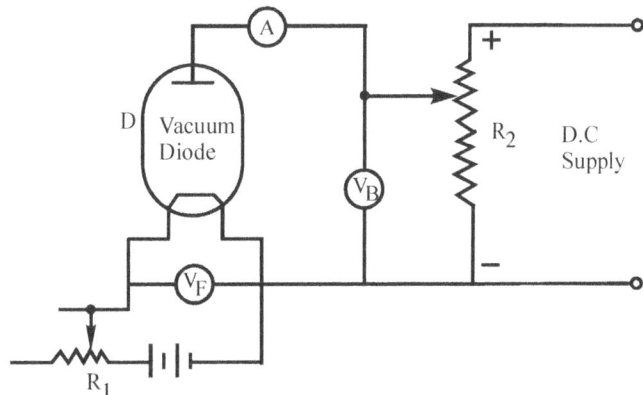

Fig 1.3.1 For the determination of the static characteristics of the vacuum diode

The circuit of fig 1.3.1 shows the arrangement for the determination of the static characteristics. The filament voltage is given by voltmeter VF and its value can be varied by R1. The anode voltage can be varied from **zero to any desired maximum by means of R_2**

For VF =2.5V, 3V and 4V the values of the current are plotted against the anode voltage as represented in fig 1.3.2. It could be observed that with a relatively large anode voltage, a further increase of anode current i.e a state of saturation is reached when practically all the electrons emitted by the cathode are being attracted to the anode. The smaller the filament voltage and therefore the lower the cathode temperature, the smaller is the saturation current.

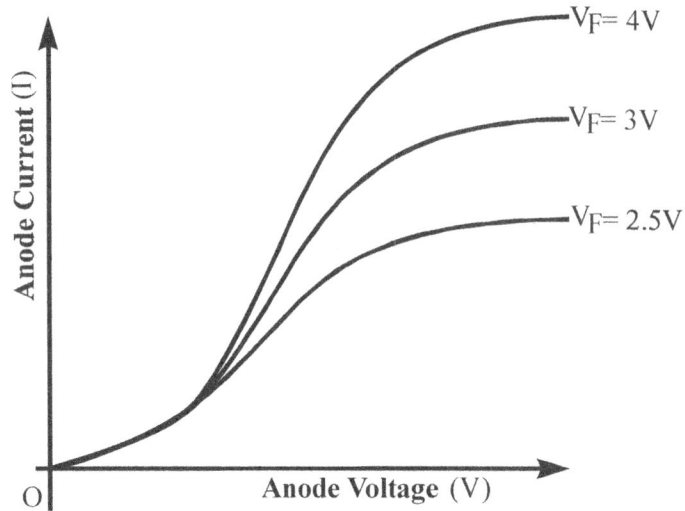

Fig. 1.3.2   Static Characteristics of a Vacuum Diode

1.4   D.C AND A.C RESISTANCE OF VACUUM DIODE.

Let us consider the characteristics of a vacuum diode shown in fig 1.4.1; the anode d.c resistance is given by,

$$\frac{OC}{CA} = \frac{Anode\ voltage}{Anode\ current}$$

(*for an anode voltage of OC*)

$$= \frac{1}{\tan AOC} \quad ----------(1)$$

$$\frac{OD}{DB} = \frac{1}{\tan BOD} \quad (\text{for an anode votage of OD})$$

The d.c resistance varies with anode voltage. It is high with low value of V_B and decreasing with increased voltage but at saturation, the value increases with increased voltage.

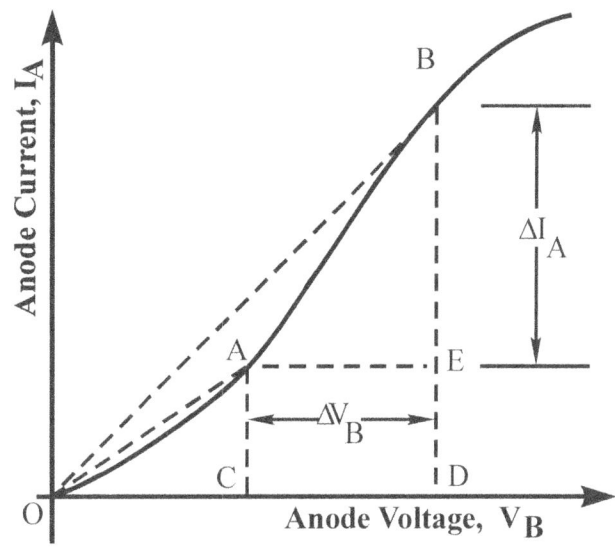

Fig. 1.4.1 Static Characteristics of a Vacuum Diode

If the diode is operated on the linear part of the characteristics i.e line AB of the characteristics, the ratio of the change of the anode current to a change in anode voltage remains constant.

The ratio $\frac{\Delta V_B}{\Delta I_A}$ is the anode a.c resistance or the anode slope resistance and is represented by r_a.

$$r_a = \frac{\Delta V_B}{\Delta I_A}$$

TUTORIAL EXERCISE

The following table gives the corresponding values of anode voltage and current for a certain diode:

V_A (v)	0	10	20	30	40	50	60
I_A (mA)	0	0.2	1	2	3	4	5

A 10KΩ load resistor R is connected in series with the anode. Determine

(a) the anode d.c resistance for an anode voltage of 50V and (b) the anode a.c resistance of the diode.

1.5 THE DIODE AS A CIRCUIT ELEMENT

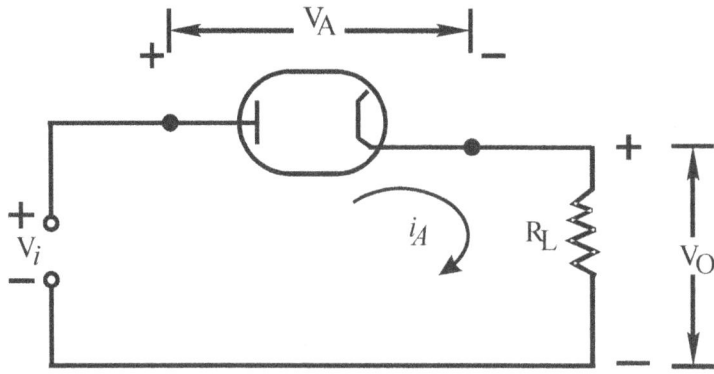

Fig. 1.5.1 : The basic diode circuit.

The circuit consists of a vacuum tube in series with a resistance R_L and an input signal source vi see fig 1.5.1

Since the heater plays no part in the analysis of the circuit it has been omitted from the diagram

This circuit can now be analysed to find the instantaneous anode current i_A and the instantaneous voltage across the diode V_A when the instantaneous input voltage is V_i.

1.6 THE LOAD LINE:

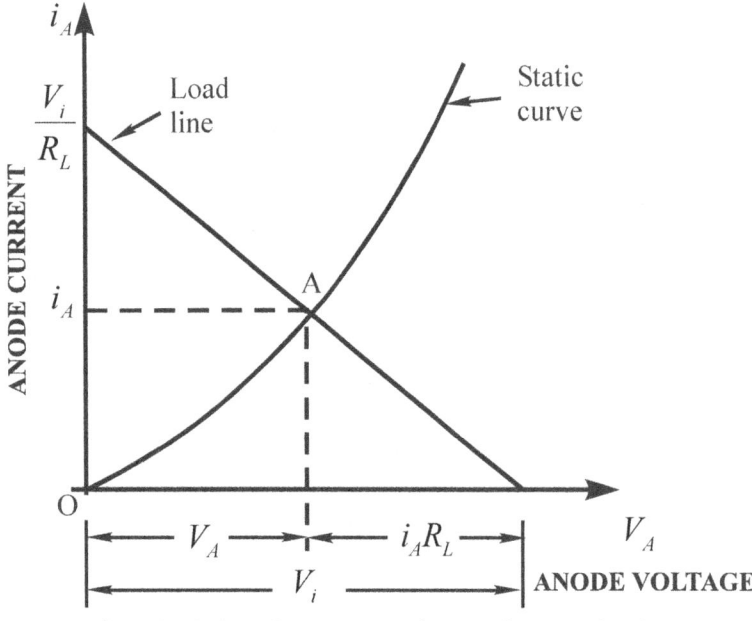

Fig. 1.6.1: Construction of Load Line

From Kirchoff's voltage law, $V_A = V_i - i_A R_L$. The load line passes through points $i_A = 0; V_A = V_i$ and $V_A = 0; i_A = \frac{V_i}{R_L}$. The slope of this line is determined by R_L. Fig 1.6.1 shows a typical load line.

1.7 THE DYNAMIC CHARACTERISTICS OR CURVE:

This is the plot of the anode current V_s the input voltage. It can be derived from the static characteristics as shown in the diagram below. (fig 1.7.1)

This is the plot of anode current vs the input voltage. Fig. 14.7.1 shows the construction of dynamic characteristics from static curve. The current i_p is plotted vertically above V_i at point B. As V_i changes, the slope of the load line does not change since RL is fixed. Thus when the applied potential has the value V_i', the corresponding current is i_p'. This current is plotted vertically above $V_i`$ at B`. The resulting curve OB'B that is generated as Vi varies is the dynamic characteristics.

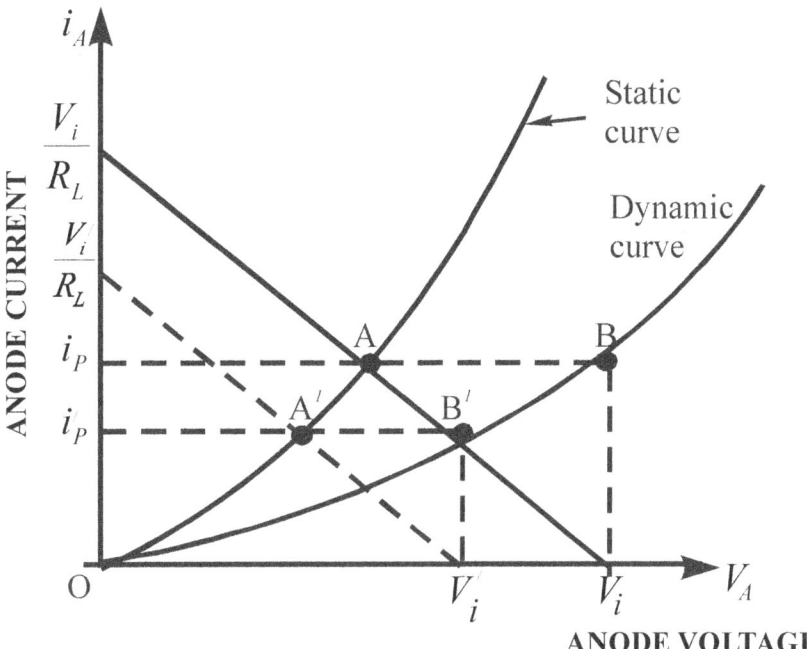

Fig. 1.7.1 Construction of the dynamic curve from static curve

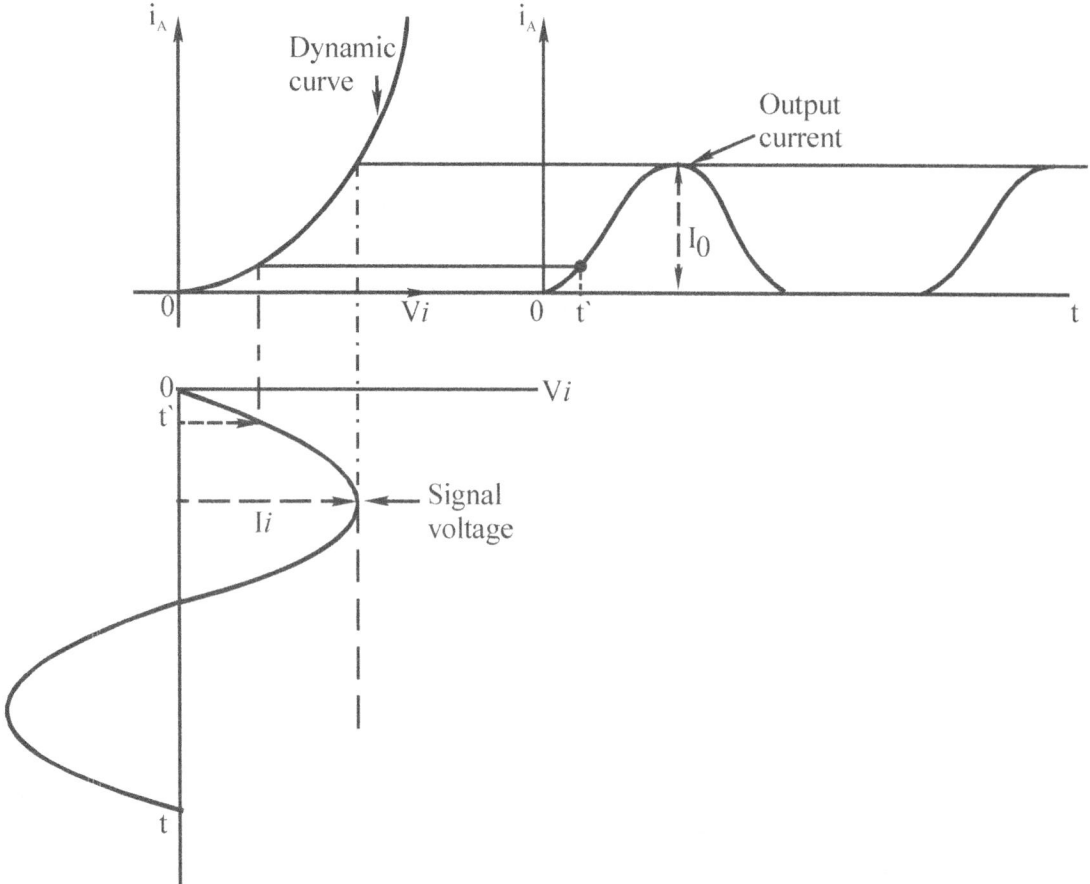

Fig 1.7.2 the out-put-current waveform from the dynamic curve for a given input voltage waveform.

It should be noted that regardless of the shape of the static characteristics or the waveform of the input voltage, the resulting waveform of the current in the output circuit can always be found graphically from the dynamic characteristics. The input signal waveform (not necessarily sinusoidal) is drawn with its time axis vertically downward. (see diagram of out-put current waveform as represented in fig.1.7.2).

1.8 DIODE APPLICATIONS:

From the above diagram it could be verified that with linear dynamic curve, the out-put $V_o = i_A R_L$, is an exact replica of the input voltage Vi except that the negative portion is missing. The diode acts as a clipper in this application. The positive portion of the waveform is clipped if the polarity of the diode is reversed. The clipping level need not be zero. For example if reference battery Vr is added in series with RL in the circuit with the negative battery terminal at ground, signal voltages smaller than Vr will be clipped. One of the most important applications of the diode is rectification. Diodes also find extensive use in digital computers and in circuits used to detect radio frequency signals.

1.9 FULL-WAVE RECTIFICATION WITH VACUUM DIODE.

A double diode is a valve containing within one envelope two anodes and a common cathode as shown in fig 1.9.1. The cathode is an indirectly heated type and the heater is supplied from a low voltage winding M. the high voltage winding (secondary) PQ has a mid point tapping N connected to one end of the cathode resistor R, the other end of which is connected to the cathode. The two anodes A and B, are connected to ends P and Q respectively of the high voltage winding.

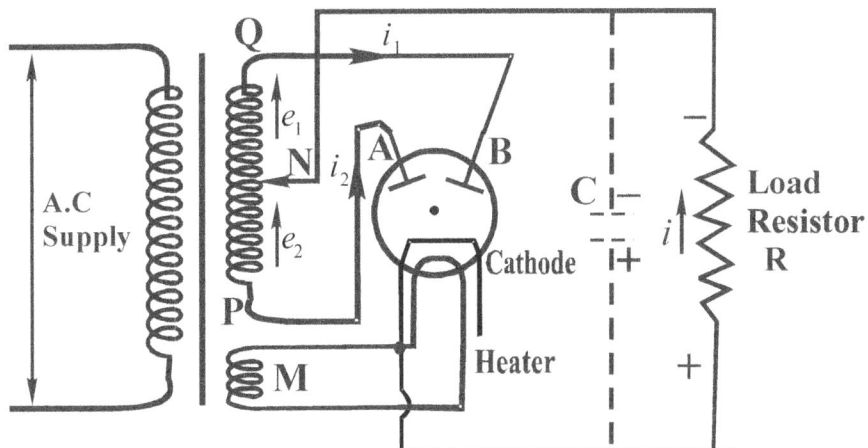

Fig. 1.9.1 Full wave ratification with vacuum diode

During the half cycle OC (see waveform in fig 1.9.2), the e.m.f induced in PQ is assumed to act in the direction of the arrow e_1 thereby making anode B positive and anode A negative with respect to the cathode.

Consequently current flows via A and none Via B. During the second half cycle CD, the emf in PQ is in the direction shown by arrow e2 so that the current i_2 flows via A and none via B. Hence the current through the load R is unidirectional as represented in the diagram. One method of reducing the fluctuation of the output current and voltage is to connect a capacitor C in parallel with the load. This helps to smoothen the ripple in the output waveform.

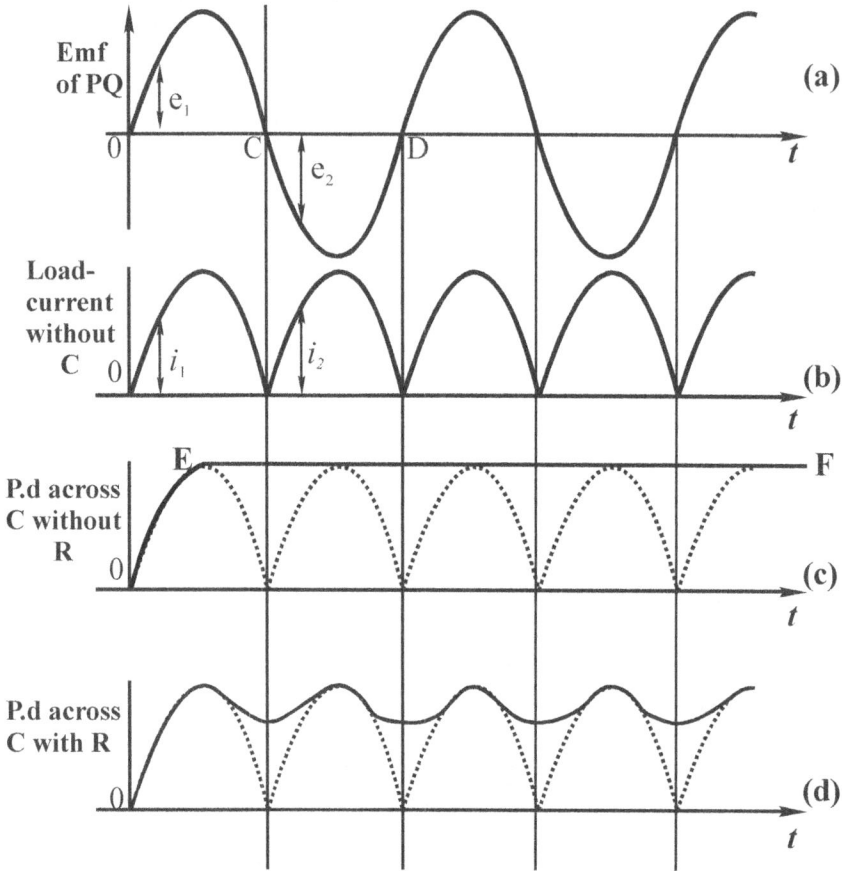

Fig.1.9.2 Wave form for a full-wave rectifier

1.10 VACUUM TRIODE:

Figure 1.10.1 shows the general arrangement of a triode having a directly heated cathode C. The anode cylinder A is shown cut to depict more clearly the internal construction. Grid G is usually a wire helix attached to one or two supporting rods. The pitch of this helix and the distance between the helix and the cathode are the main factors that determine the characteristics of the triode.

In a triode, the potential of the anode A is always positive with respect to the filament, so that electrons tend to be attracted towards A from the space charge surrounding cathode C. The effect of making the grid G positive with respect to C is to attract more electrons from the space charge. Most of these electrons pass through the gaps between the grid wires, but some of the m are caught by the grid as shown in fig 1.10.2a, and return to the cathode via the grid circuit. On the other hand the effect

The Three Electrode Vacuum Valve Or Triode.

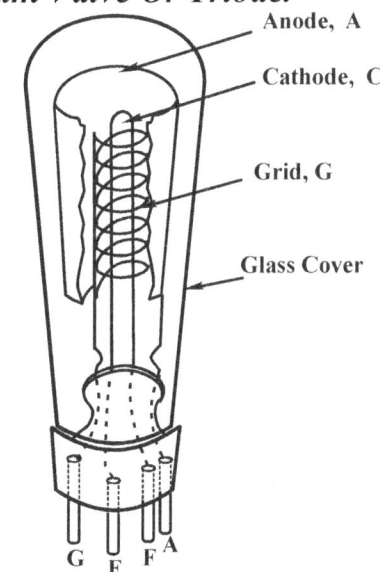

Fig. 1.10.1 A Vacuum Triode

of making the grip negative is to neutralize, partially or wholly, the effect of the positive potential of the anode. Thus fewer electrons reach the anode, the path of these electrons being as shown in fig 1.10.2b. No electrons are now reaching the grid i.e there is no grid current when the grid is negative by more than about 1V with respect to the cathode. The path of the electrons which are repelled back from the space charge into the cathode are not indicated in the diagram.

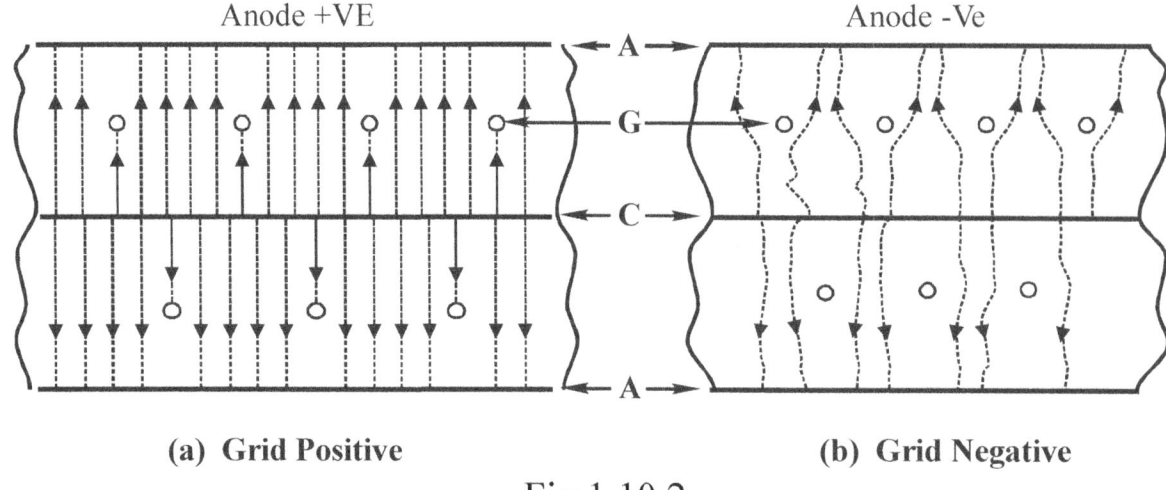

(a) Grid Positive   (b) Grid Negative

Fig 1.10.2

From the above, we can see that the magnitude of the anode current can be controlled by varying the potential difference between the grid and cathode: and since the grid is in close proximity to the space charge surrounding the cathode, a variation of say, 1V in the grid potential produces a far greater change of anode current than that due to 1V variation of anode potential. The relationship between the anode current and the grid voltage for a given anode voltage is termed the transfer or mutual characteristics of the triode and that between the anode current and the anode voltage for a given grid voltage is termed the anode or output characteristics.

1.11 DETERMINATION OF THE STATIC CHARACTERISTICS OF A TRIODE: (a) I_A/V_G CHARACTERISTICS.

The triode is connected as shown in fig 1.11.1. The resistor R1 is adjusted to give a constant voltage of 130V on voltmeter V_A. The readings on ammeter A are taken for various positive and negative values of V_A, the polarity being reversed by means of switch RS. The test is repeated for VA =100v and VA =70V.

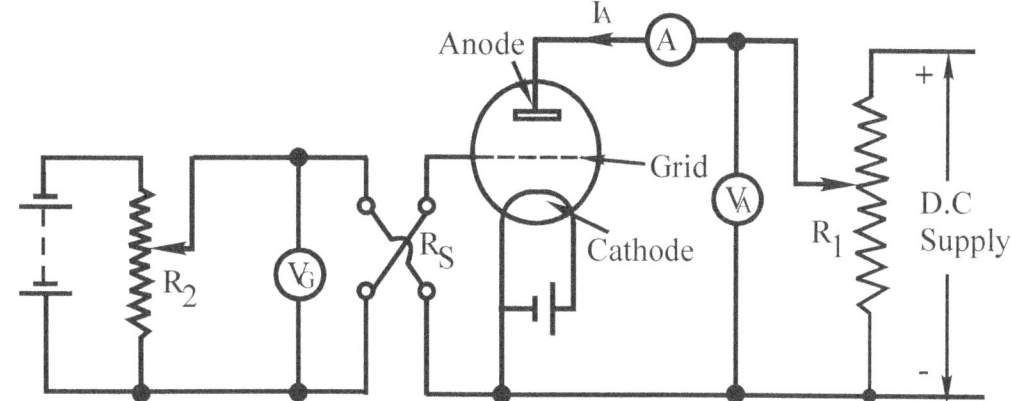

Fig1.11.1 Circuit for Determination of the Vacuum Triode Static Characteristics.

Graph of fig 1.11.2a shows the transfer or mutual characteristics of the triode valve.

NOTE: (a) that the graphs are practically linear for a considerable portion of their length and (b) that the graphs for equal differences of grid voltage are approximately equally spaced.

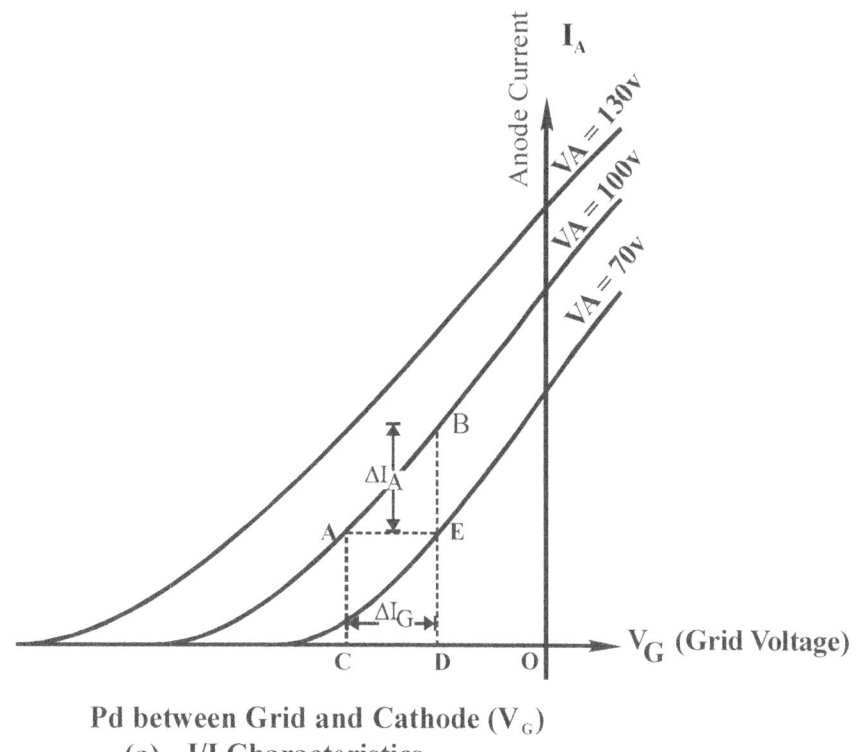

Pd between Grid and Cathode (V_G)
(a) $I_A I_G$ Characteristics

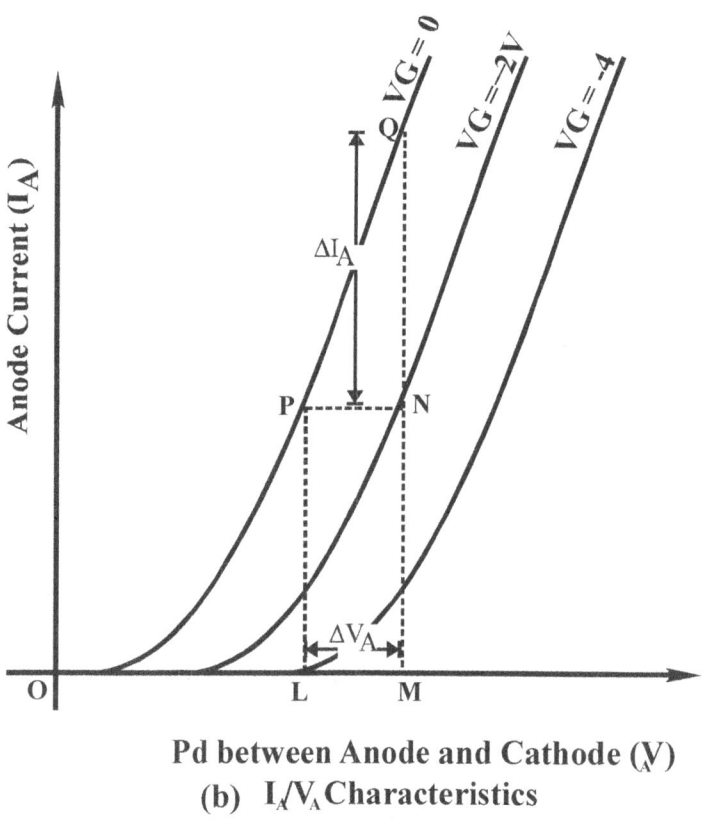

Pd between Anode and Cathode (V_A)
(b) I_A/V_A Characteristics
Fig. 1.11.2

b. IA/VA CHARACTERISTICS

The slider R2 is set for zero voltage on the voltmeter V_G and the anode current is noted for various values of the anode voltage. The test is repeated for $V_G = -2V$ and then $V_G = -4v$. The graph of fig 1.11.2b shows the I_A/V_G characteristics or the anode transfer characteristics. It will be seen (a) that the graphs are practically linear over a considerable portion of their length. (b). that graphs for equal differences of grid voltage are approximately equally spaced.

1.12 AMPLIFICATION FACTOR, MUTUAL CONDUCTANCE AND ANODE A.C. OR SLOPE RESISTANCE

Considering the I_A/V_G characteristics of the vacuum triode discussed in the last section, it will be seen that for the linear portion of the characteristics, the change of anode current is proportional to the change in the p.d between the grid and cathode, the anode voltage being constant.

The ratio $\Delta I_A / \Delta V_G$ is termed the mutual conductance of the triode and is denoted by the symbol g_m.

Therefore Mutual Conductance $= g_m = \dfrac{\Delta I_A}{\Delta V_G} = \dfrac{EB}{AE}$ in fig a (for constant V_A)

$\qquad$ = slope of I_A/V_G characteristics

Similarly, it is seen that for the linear portion of the I_A/V_A characteristics, the change of anode current is proportional to the change of potential difference between anode and cathode, the grid voltage constant. Again the ratio $\Delta V_A / \Delta I_A$ is the anode a.c resistance or the anode slope resistance.

$$\text{Anode slope or a.c resistance} = r_a = \dfrac{\Delta V_A}{\Delta I_A} \text{ for constant } V_G$$

$$= \dfrac{PN}{NQ} = \dfrac{1}{NQ/PN} = \dfrac{1}{\text{Slope of } I_A/V_A \text{ characteristics}}$$

1.13 AMPLIFICATION FACTOR:

From the above expressions, we have

$\qquad \Delta I_A = \Delta V_G \bullet g_m$ and

$\qquad \Delta I_A = \dfrac{\Delta V_A}{r_a}$

Hence if both V_G and V_A are varied simultaneously and if the triode is operated on the linear portions of the characteristics,

$$\Delta I_A = \Delta V_G \cdot g_m + \dfrac{\Delta V_A}{r_a};$$

if the variation of V_G and V_A be such that the anode current is constant, i.e. $\Delta I_A = 0$

$\qquad$ Then $0 = \Delta V_G \cdot g_m + \dfrac{\Delta V_A}{r_a}$

ELECTRONICS

$$\frac{\Delta V_A}{\Delta V_G} = -g_m r_A = -\mu$$

Therefore Amplification factor = $\mu = g_m \cdot r_a$ (for constant I_A)

The negative sign in the expression is due to the fact that if the anode voltage is increased, the grid potential has to be made more negative in order to make the anode current constant. For example in figure 1.11.2a if V_A = 70v and V_G = OD, the anode current is DE. If the anode voltage is increased to 100v without any change in the grid potential, the anode current increases to DB. This current can be reduced to its original value DE (=CA) by making the grid more negative by an amount represented by DC.

If OD = –1v and OC = -3v, $\Delta V_G = -2V$

Therefore, Amplification Factor; $= \mu = \dfrac{-100+70}{-2} = 15$

Similarly for fig 14.11.2b, if the anode voltage is increased from OL to OM, with VG =O, the anode current increases from LP to MQ the p.d between grid and cathode has to be changed from 0 to –2v.

Hence if LM =-30v,

Amplification factor; $\mu = \dfrac{-30}{(-2)} = 15$

Example 1.13:1

The following readings were obtained from the linear portion of the static characteristics of a vacuum triode

V_A (Volts)	120	120	80
V_G [volts]	−1.3	−3.8	−1.3
I_A (milliamperes)	10	4	6.3

Calculate:

a. The anode a.c. resistance

b. The mutual conductance

c. The amplification factor.

Solution:

a. With a grid voltage of −1.3v, a reduction of V_A from 120v to 80v is accompanied by a reduction of IA from 10 to 6.2 mA; i.e

$$\Delta VA = 40V \text{ and } \Delta I_A = 3.8mA$$

$$\therefore Anode\ a.c\ resistance = r_a = \frac{40 \times 10^{+3}}{3.8} = 10.50\ K\Omega$$

b. With a VA constant at 120V, I_A is increased from 4mA by changing the grid voltage from $-3.8\ to\ -1.3V$ i.e.

$$\Delta IA = 6mA \text{ and } \Delta VG = 2.5$$

$$\therefore mutual\ conductance = gm = \frac{6}{2.5} = 2.4\ mA/V$$

c. Amplification factor $= \mu = gm\ ra.$

$$= \frac{2.4}{1000} \times 10530 = 25.3$$

1.14 AMPLIFICATION BY MEANS OF A VACUUM TRIODE

Shown in fig 1.14.1 is a triode with a non reactive resistor R connected in series with the valve across a d.c source having a terminal voltage V_{HT}. Between the cathode and the grid a battery having a terminal voltage V_B is connected in series with an a.c source S. The function of V_B is to give the grid a negative bias. If V_B is greater than the peak value of the alternating voltage from S, the grid is never positive with respect to the cathode and therefore there is no grid current. It should be noted that considerable distortion of the output-voltage waveform can be caused by grid current if there is appreciable impedance in the grid circuit.

Fig 1.14.2 shows the dynamic transfer or mutual characteristic of the triode

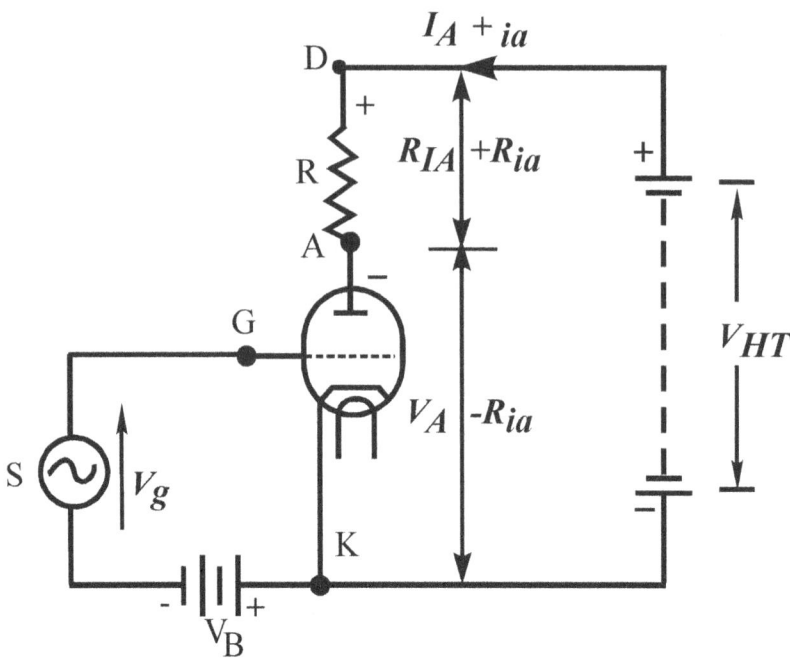

Fig. 1.14.1: A triode with resistive load

When the alternating voltage applied to the grid is zero, the anode current I_A is represented by LM as in fig 1.14.2. The corresponding p.d across the load resistor is RI_A and that between the anode and cathode is V_A.

$$\therefore \quad V_A + RI_A = V_{HT}$$

Let us consider the effect of applying a sinusoidal alternating voltage from source S. At the time when this alternating waveform is Vg the anode current has increased to $(I_A + i_a)$. The corresponding p.d across the load resistor is increased to the value $(RI_A + Ri_a)$ while the p.d between anode and cathode has decreased to $(V_A - Ria)$

ELECTRONICS

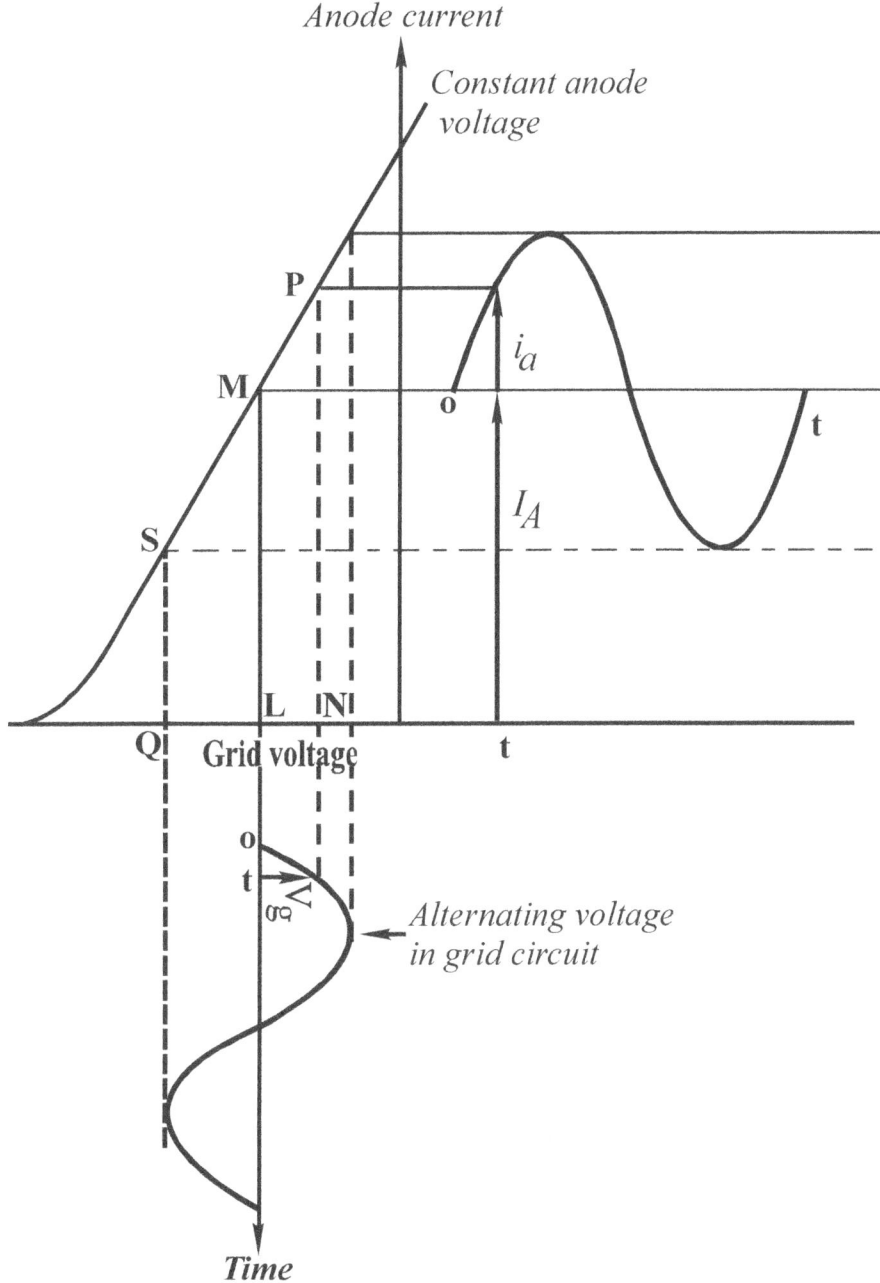

Fig. 1.14.2: Grid voltage and anode current.

Assuming the alternating waveform from the grid circuit varies as shown in fig. 1.14.2 the anode current then varies between a maximum NP and a minimum QS. The anode current can therefore be derived from the characteristics shown. It is evident that if the portion PS of the dynamic characteristics is linear, the waveform of the anode current and therefore of the potential difference across R are exactly the same as the wave of the alternating voltage applied to the grid.

It will be noted that when the alternating voltage on the grid is positive, the p.d between anode and cathode falls below V_A. The alternating voltage is assumed to be positive when the grid signal goes more negative. When the alternating voltage is negative, the anode-cathode p.d is exactly V_A. Hence the variation of the anode-cathode p.d is in antiphase to the alternating voltage applied to the grid. The ratio of the alternating component of the anode-cathode p.d to the alternating voltage applied to the grid is termed the voltage amplification and it is represented by the symbol A.

For a non-reactive load, $A = \dfrac{-Ri_a}{V_g}$. For a non-reactive, resistive load in which the amplification is distortionless, the instantaneous values i_a and V_g can be replaced by the r.m.s values of I_A and V_G

Therefore, Voltage Amplification = $A = \dfrac{-RI_A}{V_G}$ .. The condition of voltage amplification described above is referred to as Class A valve amplification since the valve is used in such a way that the anode current does not fall below the linear region of the I_A/V_G dynamic characteristics and the grid never goes positive with respect to the cathode so that there is no grid current.

1.15 EQUIVALENT CIRCUIT OF A VACUUM TRIODE

The d.c sources of fig 1.14.1 have negligible impedance to alternating current so that no alternating voltage appears across them. Consequently, as far as the alternating voltages are concerned, the terminals of the d.c sources can be short-circuited and the equivalent circuit of the vacuum triode amplifier can be shown as represented in the circuit, fig. 1.15.1.

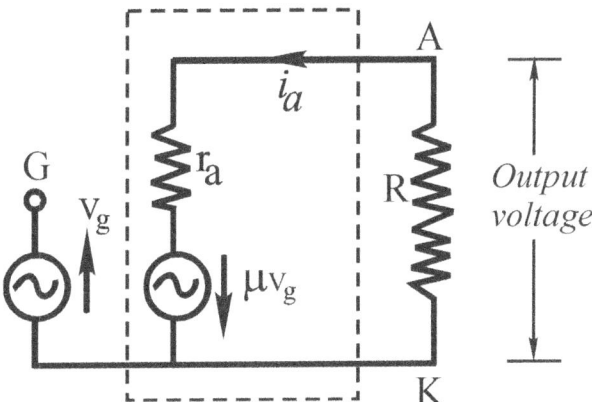

Fig. 1.15.1: Equivalent circuit of the vacuum triode amplifier

1.16: INTER-ELECTRODE CAPACITANCE OF A TRIODE VALVE

A triode has three inter-electrode capacitances namely

    a.   Grid cathode capacitance, Cgk

    b.   Grid anode capacitance, Cga

    c.   Anode cathode capacitance, C_{ak}

These capacitances are for convenience, shown as capacitances external to the triode as represented in fig 1.16.1. The values of Cgk and Cga are each of the other of 5pF but the value of Cak is much smaller.

It can be shown that the larger the voltage amplification A, the greater is the input capacitance and the lower the input reactance of the triode amplifier for a given frequency. This increase of the input capacitance due to voltage amplification is known as the *Miller effect.*

If Cgk = Cga = 5pF and A = 10

Input capacitance = C_{in} = 5 + (I + 10) × 5 = 60 pF

where C_{in} = Cgk + (1 + A) Cga.

Therefore at 100Hz, input reactance = $10^{12} / (2\pi \times 100 \times 60)$

$$= 26.5 M\Omega$$

At 1 MHz input reactance = 2650Ω

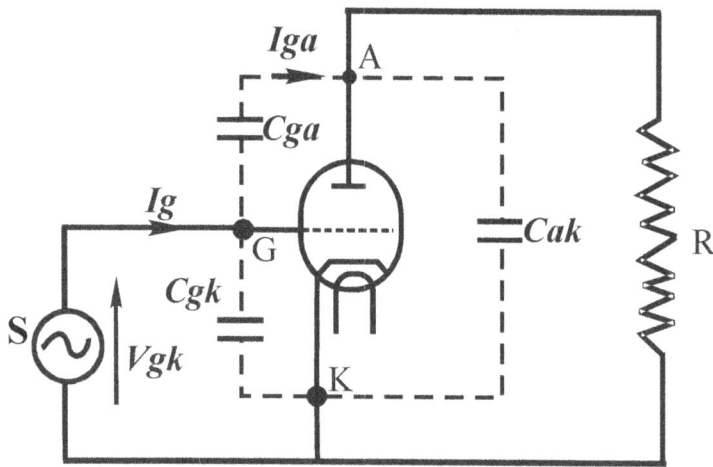

Fig. 1.16.1: Inter electrode capacitances

From these values it will be seen that at low frequencies the input reactance is very high but at radio frequency, the input reactance may be so low that the grid current may be sufficient to cause considerable reduction of the grid-cathode voltage if the source has a high internal impedance.

Another disadvantage of grid-anode capacitance is that it provides a coupling between the anode and grid; and when the load is inductive this feedback may produce instability, i.e. the triode may act as an oscillator.

It was to reduce the effect of the grid anode capacitance at radio frequency that the tetrode and subsequently the pentode were developed.

1.17 TETRODE AND PENTODE

The tetrode is a four-electrode valve containing a cathode, anode, control grid and an additional grid, which is referred to as a screen grid and is situated between the control grid and the anode. This is shown in the sectional elevation and plan of fig. 1.17.1.

Fig. 1.17.2 shows the conventional method of representing the tetrode.

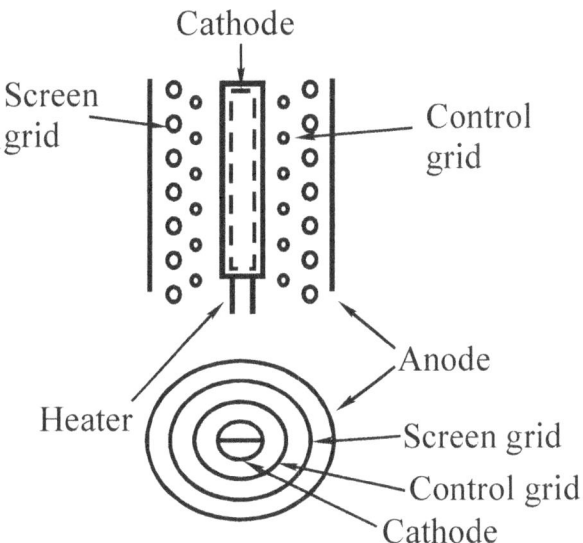

Fig. 1.17.1: A Tetrode

The main function of the screen grid is to provide an electrostatic shield between the anode and the control grid and thereby minimizing the feedback effect of the inter-electrode capacitances.

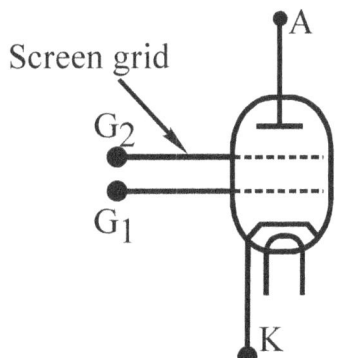

Fig. 1.17.2: Conventional representation of a tetrode

This electrostatic shielding is so effective that the inter-electrode capacitance between the anode and control grid is reduced to a value of the order of 0.01pF. With the value of capacitance and voltage amplification of say 100 and with the same interelectrode capacitance of 5pF between cathode and control grid assumed for the triode in the last section.

C_1 = input capacitance of tetrode = $5 + (101 \times 0.01) = 6$pF.

Therefore, the miller effect in a tetrode is practically negligible. An additional function of the screen grid is to reduce the effect of the anode voltage on the space charge at the cathode. Consequently the anode current remains nearly constant over a large variation of anode voltage.

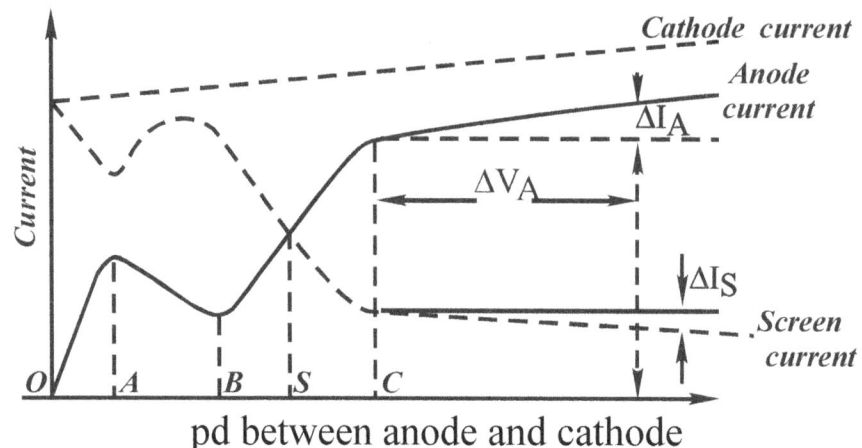

Fig. 1.17.3: Static Characteristics of the Vacuum Tetrode

The graph of fig. 1.17.3 shows the static characteristics of the vacuum tetrode. It shows the variation of anode and screen currents with variation of anode cathode potential difference.

In the graph OS represents the constant pd of about 100V between the screen and the cathode. The chain dotted graph represents the sum of the anode and screen currents, namely the cathode currents and indicates that the current remains approximately constant over the whole range of anode voltage. An increase of anode voltage from zero to OA is accompanied by a rapid increase of anode current, but as the anode voltage is increased to OB the anode current decreases. This is due to secondary emission from the anode due to the bombardment of the anode by the primary electrons emitted from the cathode.

In a tetrode with an anode voltage OA and a screen voltage OS of about 100V, the potential of screen is higher than that of the anode so that the secondary electrons are attracted to the screen grid. When the anode voltage is AB, the number of secondary electrons is high thereby reducing the net anode current. This accounts for the slope of the I_A/V_A characteristic over region AB. As the anode voltage is increased above OB, the secondary electrons tend to be attracted back to the anode and the current increases rapidly with the increase of anode voltage in the region of OS. When the anode voltage exceeds OC, the characteristic is practically linear and the increase of anode current is very small for a considerable increase of anode voltage.

For the characteristics, ΔI_A is the increase in anode current for ΔV_A increase in anode-cathode p.d

Therefore anode a.c or slope resistance = ra

$$= \frac{\Delta V_A}{\Delta I_A}$$

It should be noted that for the linear or working region of the characteristics, the slope is small and the value of ra is therefore very high-usually of the order of 0.25MΩ.

The screen grid is usually fed from the HT supply through a resistor having the resistance necessary to give the correct screen-cathode voltage. A capacitor is usually connected across the resistor to bypass from the resistor any alternating component of the screen current. In this way, the potential of the screen is maintained practically constant.

The presence of the kink in the I_A/V_A characteristics of the tetrode reduces considerably the working range available.;

The tetrode in the form described is now obsolete and is replaced by the Pentode in which a suppressor grid (G_3) is connected to the cathode either internally or externally as shown in fig. 1.17.4a.

ELECTRONICS

Its function is to prevent the secondary electrons, released from the anode, reaching the screen grid even when the anode-cathode pd is small. The kink of the tetrode is eliminated in this way. The suppressor grid also provides still further electrostatic shielding of the anode from the control grid G_1, so that suitable operation at still higher frequencies is possible.

Fig. 1.17.4b shows a typical I_A/V_A static characteristics of a Pentode for various p.ds between the control grid G_1 and the cathode K. The uniformly dotted graph shows the screen grid current for zero p.d between the control grid and the cathode. The chain dotted graph T gives the sum of the anode and screen currents, namely the cathode current, for this condition. The value of the anode a.c resistance for a control grid potential of –2 volt in the graph is given by:

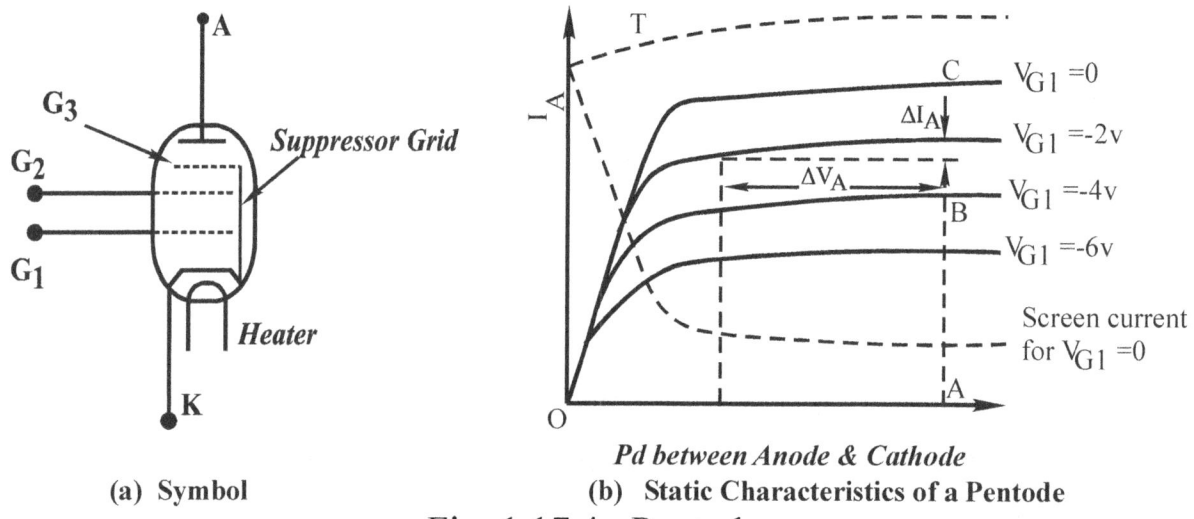

(a) Symbol (b) Static Characteristics of a Pentode

Fig. 1.17.4: Pentode

$$r_a = \frac{\Delta V_A}{\Delta I_A}$$ and is usually of the order of 1MΩ

1.18 NEGATIVE RESISTANCE EFFECT OF THE TETRODE

The static characteristics of the tetrode shows also the variation of the screen grid with the anode-cathode potential difference. From the characteristics we see that the screen current varies inversely as the anode current, reducing with increased anode-cathode potential difference. This clearly shows the negative resistance effect of the tetrode.

Therefore negative resistance $= \dfrac{\Delta V_A}{\Delta I_S}$

CHAPTER TWO

2.0 ENERGY LEVELS IN MATERIALS

2.1 THE NATURE OF ATOMS

In order to explain many phenomena associated with conduction in gases, metals and semiconductors, and the emission of electrons from the surface of a metal, it is necessary to understand that the atom has loosely bound electrons which can be torn away from it.

It has been found that the atom consists of a nucleus of positive charge, which contains nearly all the mass of the atom. Surrounding this central positive core are negatively charged particles or electrons.

Let us again use the hydrogen atom as our model of illustration, see sketch of fig 2.1.1. This atom consists of a positively charged nucleus (a proton) and a single electron. The charge on the proton is positive and equal in magnitude to the charge on the electron, thereby making atom as a whole electrically neutral. The proton carries practically all the mass of the atom and therefore remains substantially immobile whereas the electron moves about it in a closed orbit. The force of attraction between the electron and proton follows Coulomb's Law. And the resultant closed path is in a cycle or ellipse under the action of such a force.

This motion is analogous to that of the planets about the sun because in both cases, it has been found, that the force varies inversely as the square of the distance between the particles.

It has been proved that the force of attraction between the nucleus and the electron is equal to $\dfrac{e^2}{4\pi\varepsilon_O r^2}$, where

e is the electronic charge in coulomb

r is the separation between the two particles in meters

ε_O is the permittivity of free space.

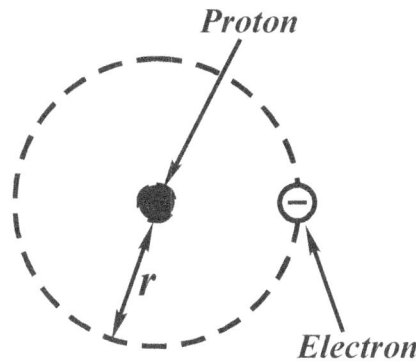

Fig. 2.1.1 Hydrogen Atom

By Newton's Law (Second) of motion, this force of attraction must be equal to the product of the electronic mass of the electron m in kg and the acceleration $\left(\dfrac{V^2}{r}\right)$ towards the nucleus

v is the speed of the electron in its circular path in m/s

Therefore $\dfrac{e^2}{4\pi\varepsilon_O r^2} = \dfrac{mV^2}{r}$ ----------------------------------(1)

m is the mass of the electron.

Furthermore, the potential energy of the electron at a distance r from the nucleus is $\left(-\dfrac{e^2}{4\pi\varepsilon_O r}\right)$ and its kinetic energy is $\dfrac{1}{2}mv^2$

According to the law of conservation of energy, the total energy of the electron,

$$W = \dfrac{1}{2}mV^2 - \dfrac{e^2}{4\pi\varepsilon_O r^2} \quad\text{------------------------(2)}$$

(the sum of the kinetic and potential energies)

From equation (1) above, $mV^2 = -\dfrac{e^2}{4\pi\varepsilon_O r},$

$$W = \dfrac{1}{2} \times \dfrac{e^2}{4\pi\varepsilon_O r} - \dfrac{e^2}{4\pi\varepsilon_O r}$$

$$= -\frac{e^2}{8\pi\varepsilon_0 r} \quad\text{---(3)}$$

Equation (3) gives the desired relationship between the radius and the energy of the electron. The above expression shows that the total energy of the electron is negative. It also shows that the energy of the electron becomes smaller as it approaches closer to the nucleus.

Let it be noted that an accelerated charge must radiate energy in accordance with the laws of electromagnetism. It has been found that if the charge is performing oscillations of frequency f the radiated energy will also be of this frequency

Therefore ***frequency of radiated energy = frequency of oscillation.***

From equation (3) above, the radius of electron radiating energy decreases as it loses its total energy. Also the frequency of oscillation depends upon the size of the circular orbit. Therefore, the energy of an orbiting electron would be of a gradually changing frequency. This conclusion is incompatible with the sharply defined frequencies of spectral lines.

2.2 THE BOHR ATOM

In order to surmount the above difficulties, Bohr postulated the following three fundamental laws;

1. Not all energies as given by classical mechanics are possible, but atom can possess only certain discrete energies i.e. energies exist in quanta. While in states corresponding to these discrete energies, the electron does not emit radiation and the electron is said to be in a stationary or non radiating state.

2. In a transition from one stationary state corresponding to a definite energy W_2 to another stationary state, with an associated energy W_1, radiation will be emitted.

The frequency of this radiant energy is given by:

$$f = \frac{W_2 - W_1}{h} \quad \text{...........................(4)}$$

where h is Planks constant in J-sec; the W's are expressed in joules and *f* is in cycles per second or hertz

3. A steady state is determined by the condition that the angular momentum of the electron in this state is quantized and must be an integral multiple of $h/2\pi$

$$\therefore mvr = \frac{nh}{2\pi} \quad \text{where n is an integral and h is Plank's Constant....(5)}$$

From this it can be proved that the energy level in joules of each state is given by;

$$Wn = \frac{-me^4}{8h^2\varepsilon_0^2} \cdot \frac{1}{n^2} \quad \text{............................(6)}$$

using equation (1) and (5) above. The exact frequency may be found by using equation (4)

2.3 ATOMIC ENERGY LEVELS

For each integral value of n in equation (6) above, a horizontal line is drawn. These lines are arranged vertically in accordance with the numerical values calculated from the equation $Wn = \frac{-me^4}{8h^2\varepsilon_0^2 n^2}$. Such a pictorial representation is called an energy level diagram and is as represented in fig. 2.3.1 for hydrogen atom. The number to the left of each line gives the

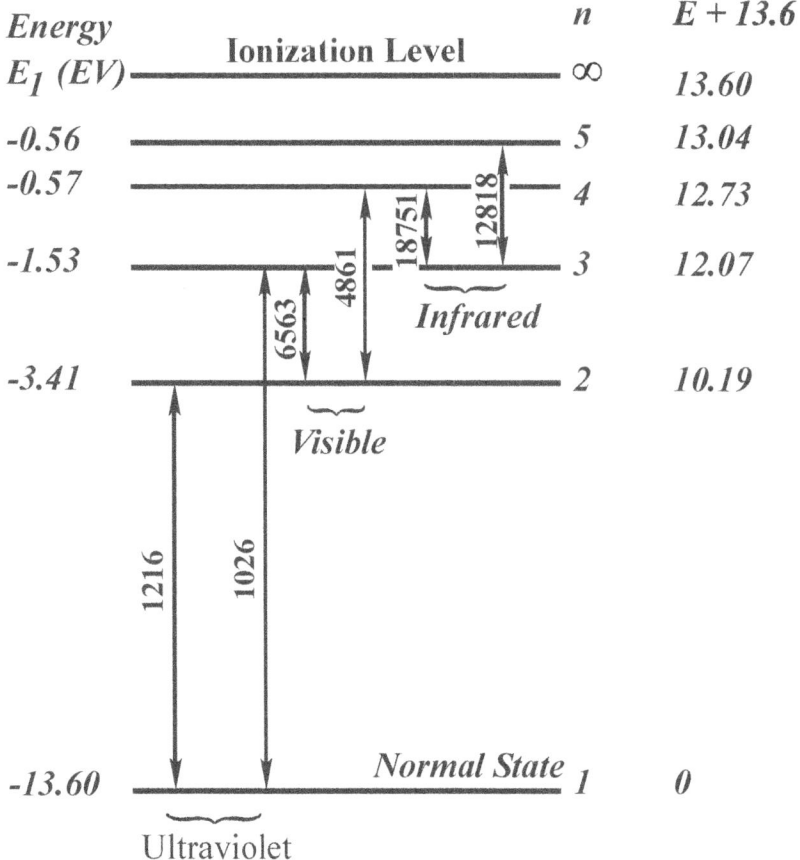

Fig. 2.3.1: The lowest five energy levels and the ionisation level of hydrogen- (the spectral lines are in Angstrom Units).

To the left of each line gives the energy of this level in electron volts. The number immediately to the right of a line is the value of n. Theoretically, an infinite number of levels exist for each atom, but only the first five and the level for n = ∞ are indicated in the diagram.

The horizontal axis has no significance here but in extending such energy level diagram to solids, the x-axis is used to represent the separation of atoms within a crystal or the distance within a solid. In such cases, the energy levels are not constant but rather are functions of x.

It is customary to express the energy value of the stationary state in electron volts E rather than in joules (W). It is also more common to specify the emitted radiation by the wavelength (λ) in angstroms rather than by its frequency f in hertz. In these units equation (4) becomes

$$\lambda = \frac{h}{E_2 - E_1} \text{ or } \frac{12,400}{E_2 - E_1} \quad \text{(where h = 12,400 J-Sec)}$$

It is customary and convenient to choose the lowest energy state as the zero level. The lowest energy state is called the normal or ground level and the other stationary states of the atom are called radiation, critical or resonance levels.

As the electron is given more and more energy, it moves into stationary states, which are further and further away from the nucleus. When its energy is large enough to move it completely out of the field of influence of the ion, it becomes detached from it. The energy required for this to occur is called the ionisation potential and is represented as the highest state in the energy level diagram:13.60 for hydrogen.

2.4 COLLISIONS OF ELECTRONS WITH ATOMS

The above discussions show that energy must be supplied to an atom in order to excite or ionize the atom. One of the ways to supply this energy is by electron impact. Suppose that an electron is accelerated by the potential applied to a discharged tube, the energy gained from the field may then be transferred to an atom when the electron collides with the atom. If the bombarding electron has gained more than the requisite energy from the discharge to raise the atom from its normal state to a particular resonance level, the amount of energy in excess of that required for excitation will be retained by the incident electron as kinetic energy after collision.

If an impinging electron possesses an amount of energy at least equal to the ionisation potential of the gas, it may deliver this energy to an electron of the atom and completely remove it from the parent atom. Three charged particles results from such an ionizing collision two electrons and a positive ion.

2.5 THE PHOTON NATURE OF LIGHT

Assume that an atom has been raised from the ground state to an excited level by electron bombardment. The mean life of excited state ranges from 10^{-7} to 10^{-10}s, the excited electron returning to its previous state after the lapse of this time. In this transition the atom must lose an amount of energy equal to the difference in energy between the two states that it has successfully occupied, this energy appearing in the form of radiation. According to Bohr, this energy is emitted in the form of photon of light, the frequency of this radiation being given by

$$f = \frac{W_2 - W_1}{h} \text{ or the wavelength } = \lambda = \frac{h}{E_2 - E_1}$$

The term photon denotes an amount of radiant energy equal to the constant h times the frequency. This quantized nature of an electromagnetic wave was first introduced by Plank in 1901. It must be noted that the atom radiates only when it makes a transition from one energy level to a lower energy state. In this transition, it emits a definite amount of energy of one particular frequency i.e. one photon, *hf* of light. Of course when a luminous discharge is observed, the discontinuous nature of radiation is not suspected because of the enormous number of atoms that are radiating energy and correspondingly, because of the immense number of photon that are emitted in unit time.

2.6 THE ENERGY BAND THEORY OF CRYSTALS

A crystal consists of a space array of atoms or molecules built up by regular repetition in three dimensions of some fundamental structural unit. The electron energy levels discussed for a single free atom (as in gas), do not apply to the same atom in a crystal. This is so because the potential characterizing the crystalline structure is now a periodic function in space whose value at any point is the result of contributions from every atom. When atoms form crystals, it is found that the energy levels of the inner-shell electrons are not affected appreciably by the presence of the neighbouring atoms.

The properties of any solid material including semiconductors depend on the nature of the constituent atoms and upon the way in which the atoms are grouped together. That is, the properties are a function of both the atomic structure of the atoms and the crystalline structure of the solid. Experiments have shown that an atom consists of a positively charge nucleus surrounded by electrons located in discrete orbits. It has been shown earlier that electrons can exist in stable orbits near the nucleus only for certain discrete values of energy called energy levels of the atom. Consider a simplified energy level diagram of the atom shown in fig 2.6.1 below.

ENERGY LEVELS IN MATERIALS

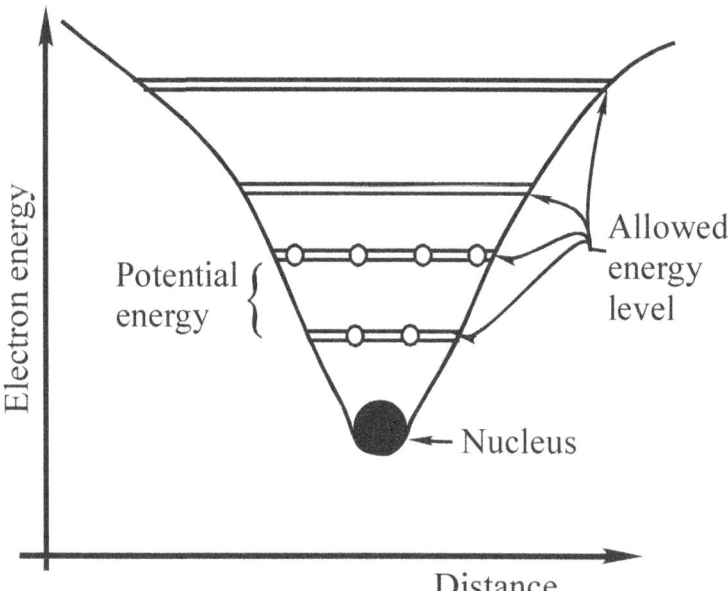

Fig. 2.6.1: Simplified Energy Level Diagram of the Atom.

The horizontal lines depict the allowed energies of electrons in an atom while the curved lines represent the potential energy of an electron near the nucleus as given by Coulomb's Law. As a consequence of Pauli exclusive principle only a certain maximum number of electrons can occupy a given energy level. The result is that in any atom, electrons fill up the lowest possible levels first.

When atoms come close together to form a solid crystal, electrons in the upper energy levels of adjacent atoms interact to bind the atoms together. Because of the strong interactions between these outer or VALENCE ELECTRONS, the upper energy levels are drastically altered. This is illustrated in an energy level diagram for the entire crystal.

Consider two isolated atoms, each with energy level diagram pertaining to the outer electrons as shown in fig 2.6.2.

When these are brought close together, the valence electrons in each atom are attracted together by both nuclei. The result is that the energy

ENERGY LEVELS IN MATERIALS

required to remove an electron from one nucleus and place it on the other is reduced.

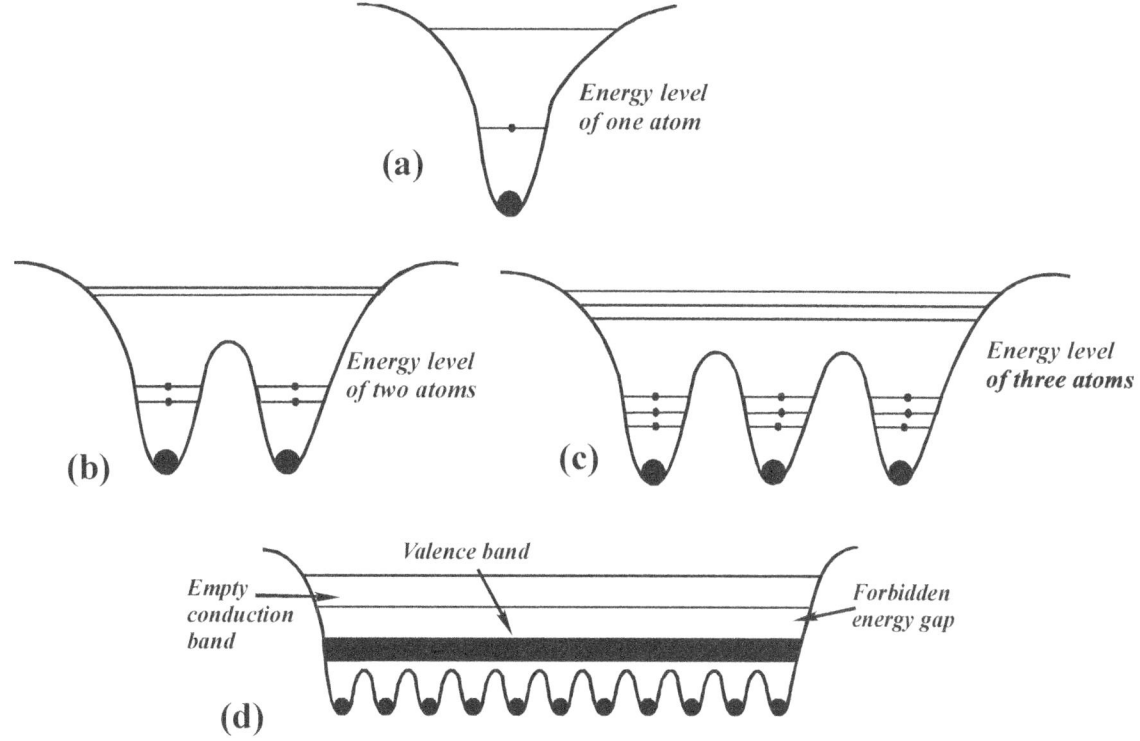

fig. 2.6.2: Energy levels of Electrons in crystals

Thus the outer electrons are no longer under the influence of one particular nucleus. The appropriate energy level diagram for the combination of two atoms has two energy levels near each atom core (see fig 2.6.2b. The higher, unoccupied levels are similarly split indicating that these levels, too can each contain two electrons. When three atoms are brought together, as shown in the fig 2.6.2c, above, the outer electrons of all three atoms can be associated with any of the three nuclei. Consequently three energy levels are available.

Even the tiniest crystal contains many hundreds of millions of atoms, so that many energy levels are associated with each nucleus. Thus, the energy level diagram appropriate for the entire crystal has a band of levels. The lowest energy band called the VALENCE BAND (see diagram) is filled with electron for there is one electron for each of the available energy level. The upper and unoccupied band is called the CONDUCTION BAND. The energy region between the valence and conduction band is called the FORBIDDEN ENERGY GAP, since no electrons with such energies exist in the crystal. The picture of the electron energy levels in a crystal is known as the energy model of crystal. It is very useful in determining the electrical properties of any solid since it shows how electrons can move in the crystal.

2.7 METALS, SEMICONDUCTORS AND INSULATORS FERMI-ENERGY LEVELS

The atomic and crystal structure of metals are such that the valence and the conduction bands overlap, as indicated in the conventional energy band model for a metal shown in fig. 2.7.1

Since there is no forbidden energy gap in a metal crystal, any of the many valence electrons are free to move throughout the solid and to move in response to an electric field. Therefore metals are excellent electric conductors. Electrons in the band are distributed in accordance with the Pauli exclusion principle. At absolute zero, all electrons fill up the lowest levels in the energy band model. The highest filled level is termed the FERMI LEVEL. At normal temperatures, some electrons at the highest energies are excited to levels slightly above the Fermi level by virtue of the heat energy in the crystal.

ENERGY LEVELS IN MATERIALS

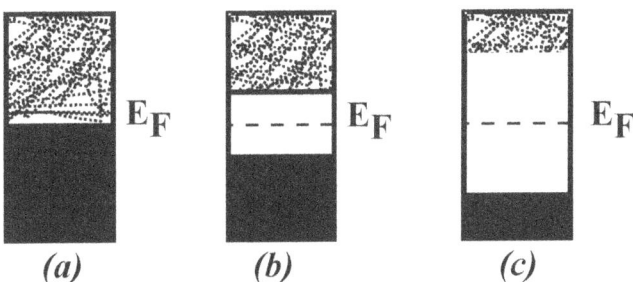

fig. 2.7.1: Energy band model for
(a) metal (b) semiconductor (c) insulator. E_F is the Ferm Level

Consequently, a few energy levels below the Fermi level are empty and a few energy above the Fermi level are filled. In this situation the Fermi level represents the energy at which the levels are half-filled and half-empty. Fig. 2.7.1 a, b, and c above indicate the position of the Fermi level (E_F) in metals, semiconductors and insulators.

An insulator has a wide forbidden energy gap as shown in the diagram c above. The valence band is completely filled with electrons and the conduction band is completely empty. Therefore the upper band cannot contribute to electric conductivity since no electron are present to act as carriers. It must be noted also that the electrons in the completely filled valence band cannot conduct electricity.

The energy band model of a semiconductor (see fig 2.7.1b) above is similar to that of an insulator except that the forbidden gap is comparatively narrow. A few electrons can be promoted from the valence band of the conduction band across the forbidden energy gap by virtue of the thermal energy of the crystal at room temperature. Electrons promoted to the conduction band can conduct electricity. Since the number of carriers is much fewer than in the case of metals, semiconductors are poorer conductors than metals but better than insulators. Thermal energies at very low temperatures are insufficient to excite electrons across the forbidden energy gap. Conversely, at sufficiently high temperature even insulators conduct electricity because some electrons can be promoted from the valence band to the conduction band. Width of the forbidden energy gap of semiconductors is of the order of 1eV.

In general, materials with a wide forbidden energy gap are desirable for semiconductor devices. The number of electrons promoted to the conduction band at high temperature is small and the change in device characteristics with temperature is less severe when the forbidden gap is wide. For this reason silicon crystals are now widely used than germanium crystals even though the latter are easier to prepare and less expensive.

Intrinsic and Extrinsic Semiconductors

The energy band model of fig. 2.7.1b refers to a perfect crystal structure which contain no chemical impurities and in which no atoms are displaced from their proper sites. The properties of the solid are therefore characteristic of an ideal structure and the crystal is called an intrinsic semiconductor.

But the electrical properties of a semiconductor are drastically altered when foreign or impurity atoms are incorporated into the crystal. Since the properties are now dependent upon the impurity content, the solid is called extrinsic semiconductor.

2.8 SEMICONDUCTORS

It has been explained that the electrical conductivity of semiconductors lies between that of metals and insulators (conductor, 10^{-6} Ω/m, semiconductor 10Ω/m, insulator -10^{6}Ω/m).

The graph of fig. 2.8.1 shows the variation of resistance with temperature for metals, semiconductors and insulators.

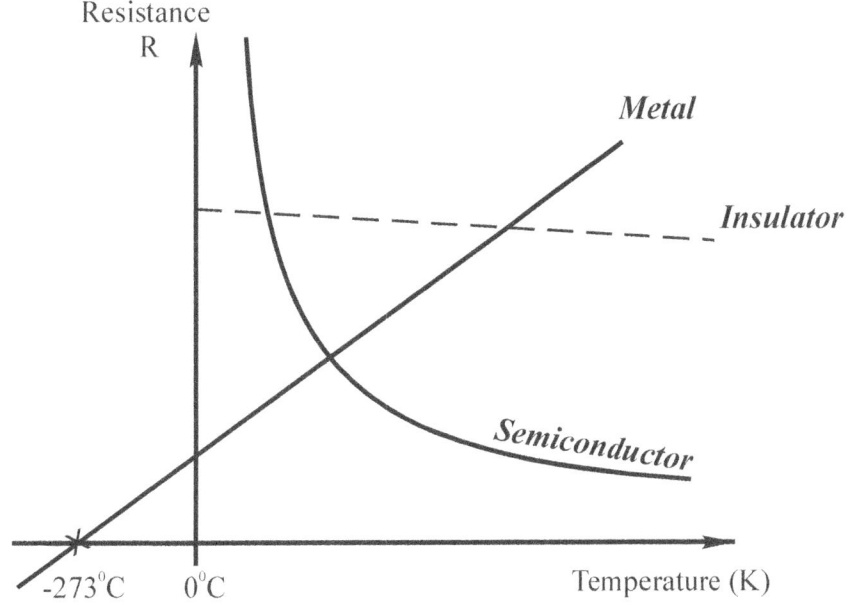

Fig. 2.8.1: Relationship between temperature and resistance for metal, semiconductor and insulator.

Pure semiconductors such as germanium and silicon are rather poor conductors and are of little technical interest. The interesting properties used in diodes and transistors are obtained by doping the intrinsic materials.

2.9 CONDUCTION IN SEMICONDUCTOR

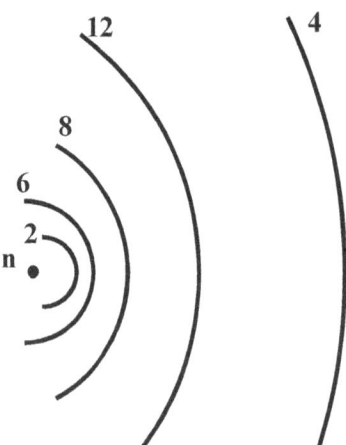

Fig. 2.9.1: Pure crystalline germanium with atomic number 32

Consider a piece of pure crystalline germanium which has are atomic number

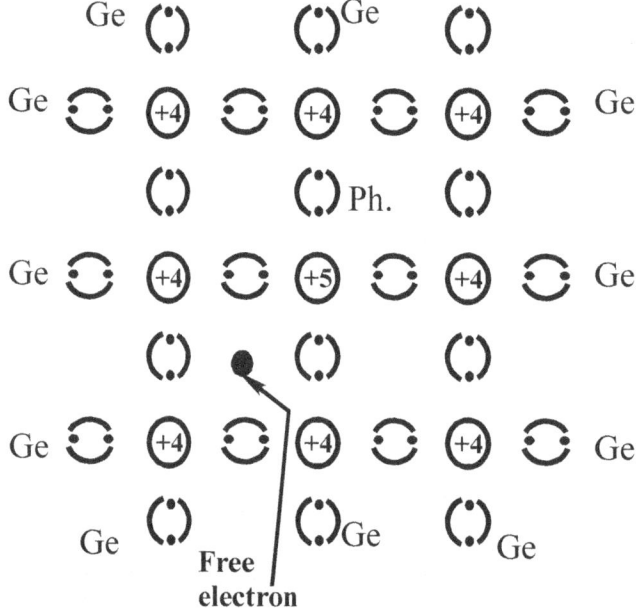

Fig. 2.9.3: Germanium atom replaced by phosphorous atom

32 and 4 electrons in the outer orbit, valency 4 as represented in fig 2.9.1. If we add a small impurity, that is a small quantity of an element which has a valency of 5 (pentavalent, e.g. phosphorus, arsenic or antimony in an atom of this material, it could be made to form a bond with the germanium leaving one electron free for conduction. This material so

doped is called an N-type semiconductor. The material used to dope the intrinsic semiconductor is called a donor impurity (phosphorous, arsenic or antimony). The resulting materials where 1 part in 10^8 of the crystal of germanium is replaced by say phosphorous, obeys ohms law.

If a trivalent impurity (boron, gallium or indium) is used for doping, only 3 electrons in the outer cell will form the bond thus leaving a hole in the intrinsic material. This is illustrated in the diagram of fig. 2.9.3. This type of impurity make available positive carriers because they create holes which can accept electrons.

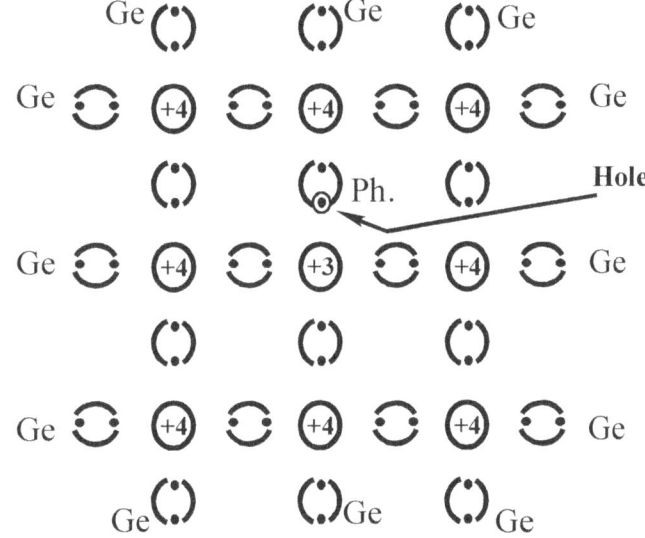

Fig. 2.9.3 Germanium atom replaced by boron.

These impurities are known as Acceptor or P-type impurities. If the impurity is added to the extent of 1 part in 10^8, the conductivity of germanium at 30^0C is multiplied by a factor of 12; thus by doping an intrinsic semiconductor we do not only increase the conduction but also produce a conductor in which the electric carriers are either predominantly holes or electrons. In an N-type semiconductor, the electrons are the majority carriers and holes the minority carriers. But in the P-type semiconductor the holes are the majority carriers while the electrons are the minority carriers.

2.10 THE EFFECT OF TEMPERATURE UPON INTRINSIC CONDUCTION

With each hole-electron pair created two charge carrying particles are formed. One is negative (the free electron) of mobility μn and the other positive (the hole) of mobility μp. They move in opposite directions in an electric field.

The current density J will be,

$$J = (n\mu n + p\mu_p)e\varepsilon$$

where   n   =   magnitude of free electron concentration

P   =   magnitude of hole (+ve) concentration

σ =   conductivity

Therefore $\sigma = (n\mu n + p\mu_p)e$.

For the intrinsic semiconductor (pure) consider n = p = ni where ni is the intrinsic concentration.

In pure germanium at room temperature there is about one hole-electron pair for every 2×10^9 germanium atoms. With increased temperature, the density of hole electron pair increases and as a result the conductivity increases.

It is found that the intrinsic concentration ni varies with temperature according to the following relationship

$$n_i = A_o T^3 e^{-E_{GO}/KT}$$

where E_{GO} and A_o are constants.

The constant E_{GO}, μn, μp are impurity physical quantities. The conductivity of germanium (silicon) is found from the above equation to increase approximately 6-8% per degree increase in temperature. This large change in conductivity with temperature places a limitation upon the use of semiconductor devices in some circuits. On the other hands, for some applications, it is exactly this property of semiconductors that is used to advantage. A semiconductor used in this manner is called a ***thermistor***.

2.11 QUANTITATIVE THEORY OF THE PN JUNCTION

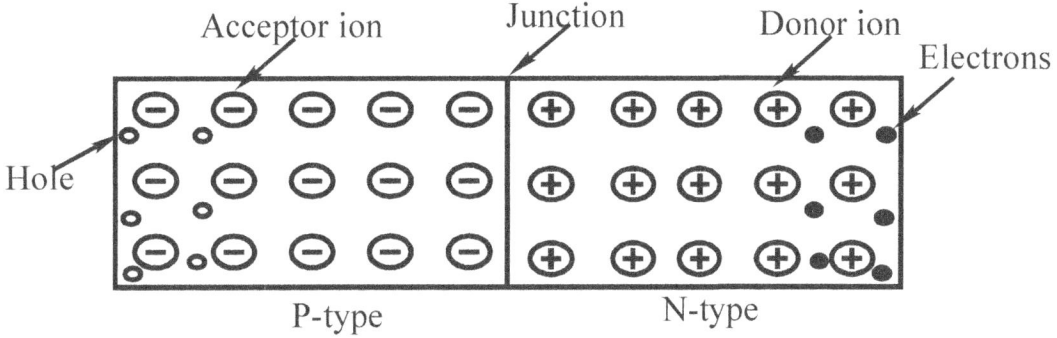

Fig. 2.11.1: P-N Junction

If a donor impurity is introduced into one side and acceptor into the other side of a single crystal of a semiconductor, a P-N junction is formed. In fig. 2.11.1, an acceptor impurity is introduced to the left while a donor impurity is to the right of the crystal.

Initially there are only P-type carriers to the left. But because of the density gradient across the junction, hole will diffuse to the right across the junction and electrons to the left.

Let us consider a crystal one half of which is doped with p-type impurity and the other half with N-type impurity. Initially the P-type semiconductor has mobile holes and the same number of fixed negative ions carrying exactly the same total charge as the total positive charge represented by the holes. Similarly the N-type semiconductor has mobile electrons and the same number of fixed positive ions carrying the same total charges as the total negative charge, the mobile electrons. Hence each region is initially neutral.

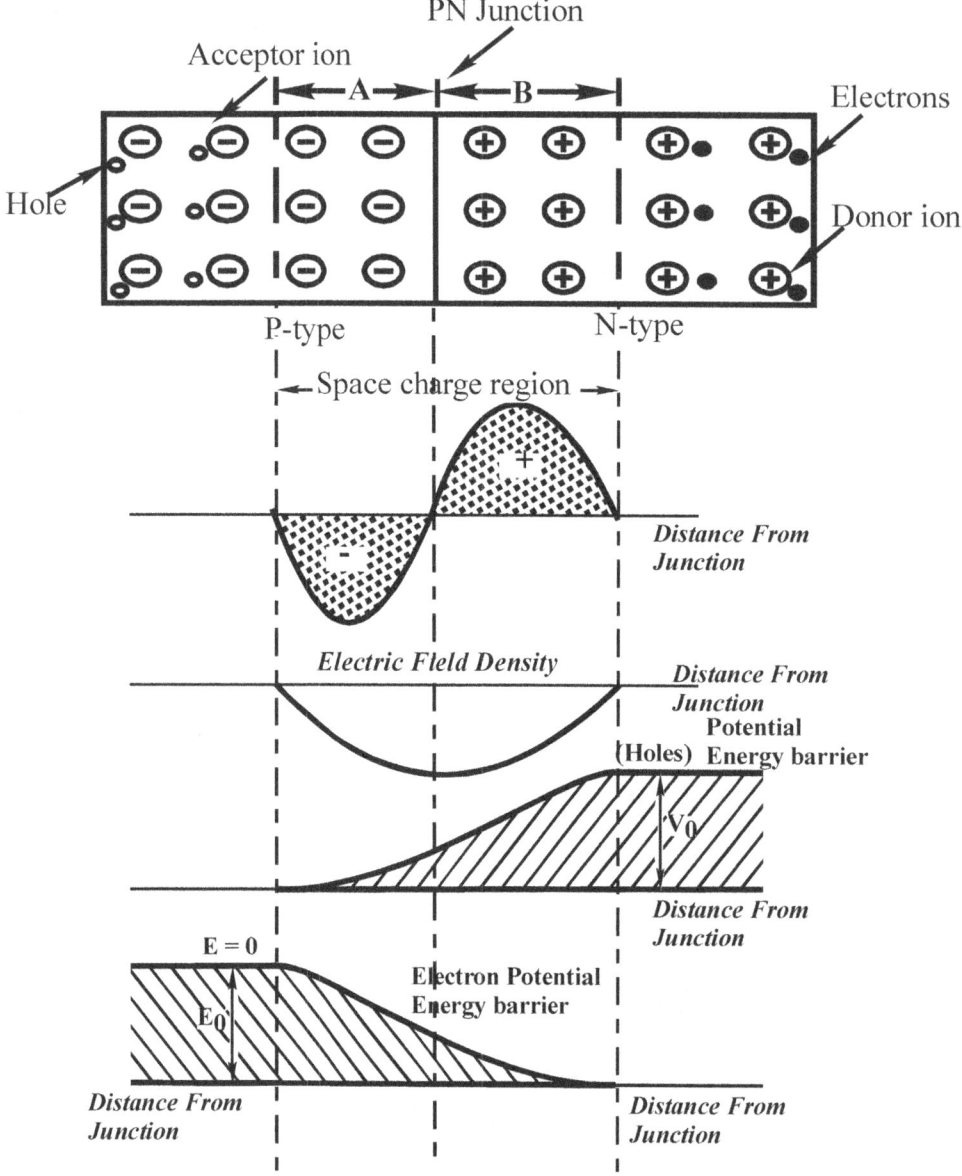

Fig. 2.11.2: Schematic diagram of a P-N Junction

Owing to their random movements, some of the holes will diffuse across the boundary into the N-type semiconductor and some of the free electrons will similarly diffuse into the P-type semiconductor as represented in fig. 2.11.2 above. Consequently region A acquires an excess negative charge, which repels any more electrons trying to migrate from the N-type into P-type region. Similarly, region B acquires a surplus of positive charge, which prevents any further migration of holes across the boundary. These positive and negative charges are concentrated near the junction as shown in the diagram, and therefore form a potential barrier between the two regions. The potential barrier is known as the space charge region.

2.12 BIASING OF PN JUNCTION

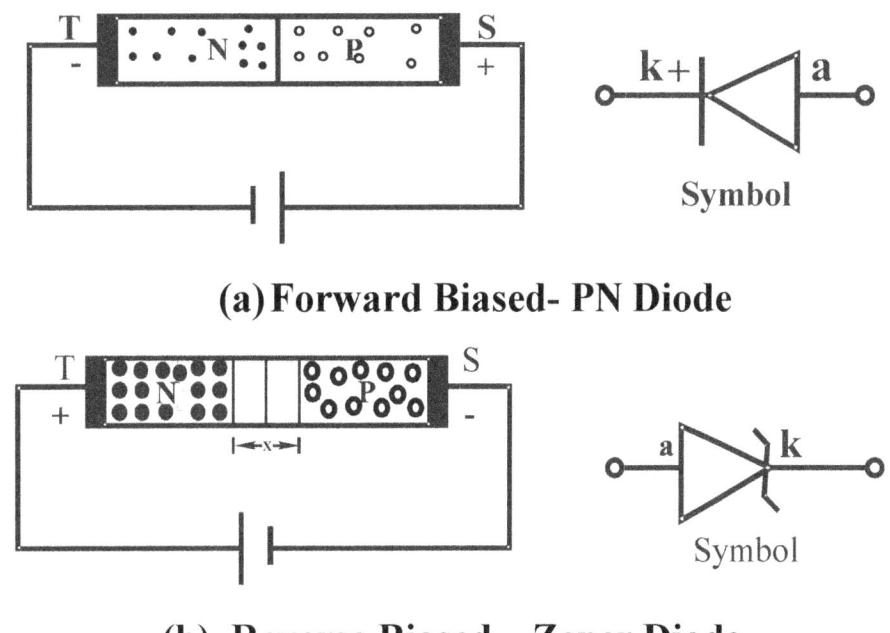

(a) Forward Biased- PN Diode

(b) Reverse Biased – Zener Diode

Fig. 2.12.1: Biasing of PN Junction

FORWARD BIAS:- Let us consider the effect of applying a p.d across S and T, S being positive relative to T as shown in fig 2.12.1a. the direction of the electric field is such as to produce a drift of holes towards the left in the P-type semiconductor and of a free electrons towards the right in the N-type semiconductor. In the region of the junction, free electrons and holes combine i.e. free electrons fill the vacancies represented by the holes. For each combination, an electron is liberated from a covalent bond in the region near positive plate S and enters the plate, thereby creating a new hole which moves through the p-type material towards the junction. Simultaneously, an electron enters the N-region from the negative plate T and moves through the N-type semiconductor towards the junction. The current in the diode is therefore due to the hole flow in the P-region, the electron flow in the N-region and a combination of the two in the vicinity of the junction. The electric symbol of the pn diode is represented in fig. 2.12.1a.

REVERSE BIAS: When the polarity of the applied voltage is reversed, the holes are attracted towards the negative electrode and the free electrons towards the positive electrode T, see fig. 2.12.1b. This leaves a region x known as a depletion layer in which there are no holes or free electrons, i.e there are no charge carriers in this region apart from the relatively few electrons produced spontaneously by thermionic agitation. Consequently the junction behaves as an insulator.

In practice, there is a small current due to the fact that at room temperature thermal agitation or vibration of atoms takes place in the crystal and some of the valence electrons acquire sufficient velocity to breakaway from their atoms, thereby producing electron-hole pair. The generation and recombination of electron-hole pair is a continuous process and is a function of the temperature, the higher the temperature, the greater is the rate of generation and recombination of electron-hole pair and therefore the lower the intrinsic resistance of a crystal pure germanium or silicon. The property of a reverse bias diode is used in the construction of a zener diode which is represented as shown in fig. 2.12.1b.

2.13　STATIC CHARACTERISTICS OF THE p-n DIODE GERMANIUM

When a germanium junction diode is biased in the reverse direction, the current remains for a bias varying between 0.1V and the breakdown voltage. The constant value is referred to as the Saturation Current and is represented by I_0 in the curve of fig. 2.13.1

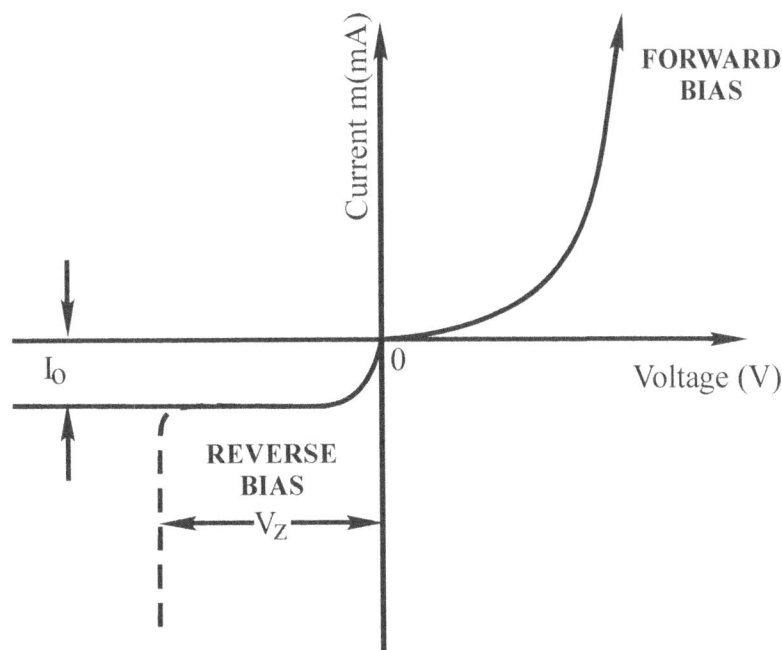

Fig. 2.13.1: Static Characteristics of the PN Junction.

Note: That in practice, the reverse current increases with increase in bias, this increase being due mainly to surface leakage.

ENERGY LEVELS IN MATERIALS

Considering the germanium junction diode in which the surface leakage is negligible, the current is given by;

$$i = I_0 \left[e^{Q_e V / KT} - 1 \right]$$

where I_0 = Saturation current with negative bias

Q_e = Charge on electron (1.6×10^{-19} C)

V = p.d in volts across junction

k = Botzmann's Constant (1.38×10^{-23} J/K)

and T = Absolute temperature (K)

Assuming a saturation current of say 10µA; then for a temperature of 300K (- 27°C), we have from expression above,

$i = 10(e^{38.6V} - 1) \mu A$. Values of current i is calculated for various values of V and plotted to give the static characteristics of the PN diode as represented in fig 2.13.1

2.14 IMPORTANT TERMS

Two important terms often used with PN junction (i.e. crystal diode) are breakdown voltage and knee voltage.

i. *Breakdown Voltage:*

It is the minimum reverse voltage at which PN junction breaks down with sudden rise in reverse current.

Under normal reverse voltage, a very little reverse current flows through a PN junction. However, if the reverse voltage attains a high value, the junction may break down with sudden rise in reverse current. Even at room temperature, some hole-electron pairs (majority carriers) are produced in the depletion layer. With reverse bias, the electrons move towards the positive terminal of supply. At large reverse voltage, these electrons acquire high enough velocities to dislodge valence electrons from semiconductor atoms.

The newly liberated electrons in turn free other valence electrons. In this way, we get an avalanche of free electrons. Therefore, the PN junction conducts a very large reverse current. Once the breakdown voltage is reached, the high reverse current may damage the junction. Therefore, care should be taken that reverse voltage across a PN junction is always less than the breakdown voltage.

ii. *Knee Voltage:*

It is the forward voltage at which the current through the junction starts to increase rapidly.

When a diode is forward biased, it conducts current very slowly until we overcome the potential barrier. For silicon PN junction, potential barriers is 0.7V whereas it is 0.3v for germanium diode.

Once the applied forward voltage exceeds the knee voltage, the current starts increasing rapidly. It may be added here that in order to get useful current through a PN junction, the applied voltage must be more than the knee voltage. It is clear from fig. 2.14.1 that knee voltage for silicon diode is 0.7v and 0.3v for germanium diode.

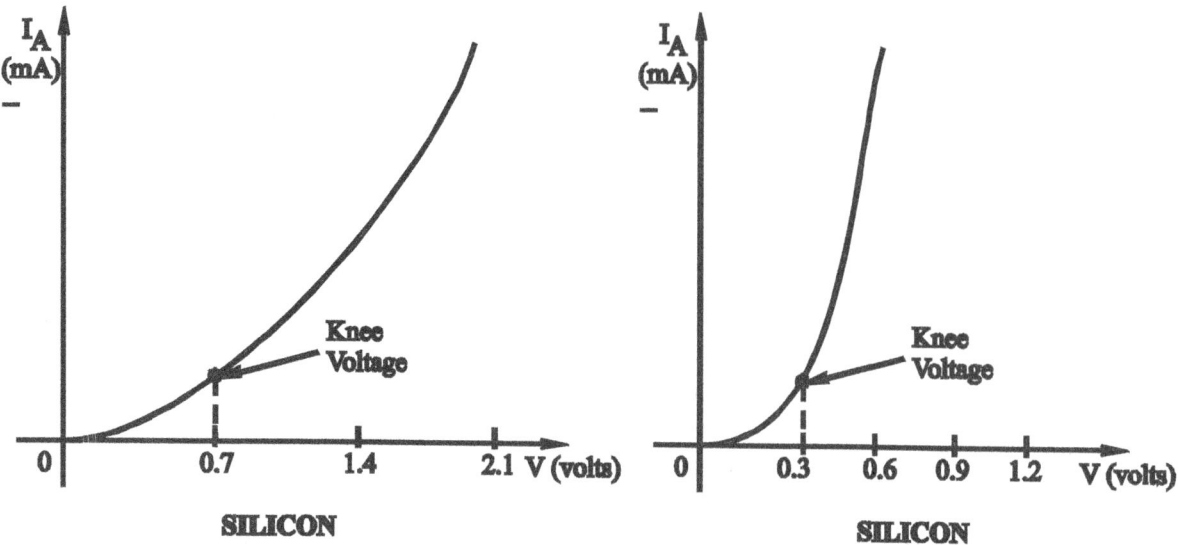

Fig. 2.14.1: Knee voltage for silicon and germanium

It should be noted that the potential barrier voltage is also known as turn-on voltage. This is obtained by taking the straight line portion of the forward characteristics and extending it back to the horizontal axis.

2.15 CAPACITANCE OF P-N JUNCTION

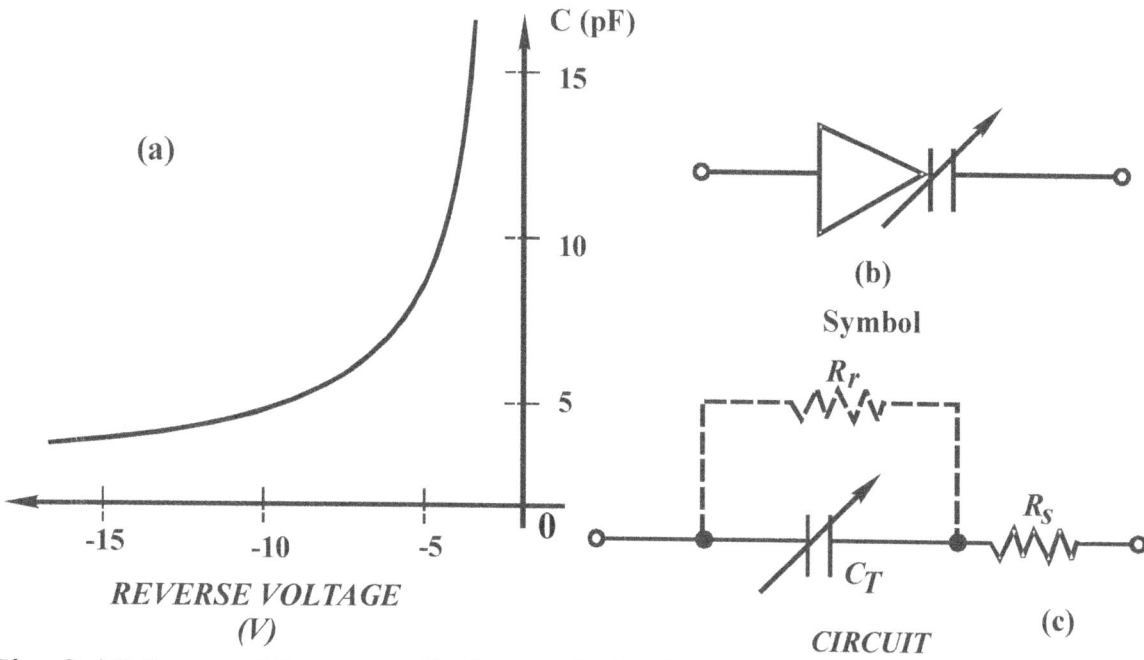

Fig. 2.15.1: Varactor diode symbol, circuit and reverse voltage/capacitance relationship

The application of a reverse bias to PN junction increases the width of the carrier free zone in the vicinity of the junction. This results in a decrease in the capacitance. The variation provides a use in a device known as VARACTOR DIODE which enables circuits to be tuned with the application of reverse voltage. Other applications include self-balancing bridge circuit and special types of amplifiers called parametric amplifiers.

The ability of a diode to have a constant potential drop over a range of reverse current enables it to be used as a voltage stabilizer. In the varactor diode circuit of fig. 2.15.1c the reverse diode resistance R_r shunting C_T is large > 1MΩ and hence is usually negligible. R_s is the body (ohmic) series resistance, Rr is the reverse diode resistance and C_T is the transition capacitance. Typical values of Rr and C_T are 20μF and 8.5Ω respectively at a reverse voltage of (-4v). Fig. 2.15.1a shows the relationship between the capacitance and the reverse voltage of a varactor diode. The symbol is as represented in fig. 2.15.1b

2.16 ZENER DIODE

Zener diodes are available at a range of voltages and of allowable power. The smallest are rated at about 30mW i.e. 6V, 50mA and the largest about 8W. Fig. 2.16.1 shows the ratings of two types of zener diodes, one 6.85V,10mA and the other 6.9V, 30mA.

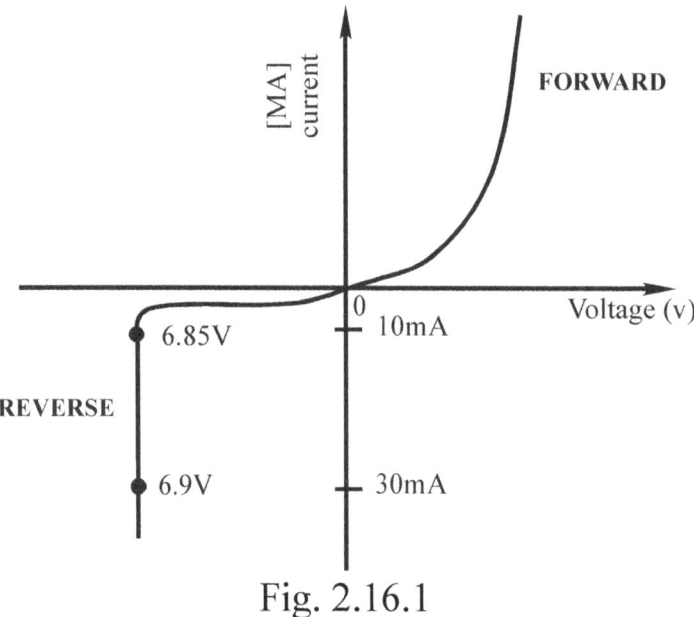

Fig. 2.16.1

2.17   APPLICATION OF ZENER DIODE

The zener diode is used as a voltage stabilizer. Fig. 2.17.1 shows a typical circuit where the zener diode stabilizes power across a load.

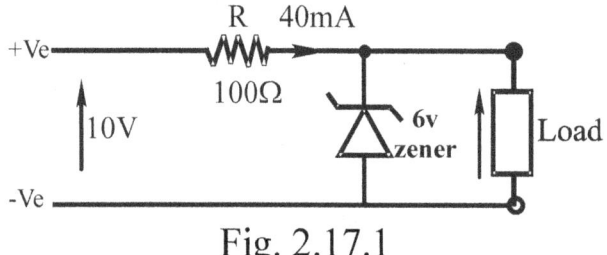

Fig. 2.17.1

In engineering, we are required to show over full operating range that the current with the zener is neither too small, giving a poor slope resistance nor too large so that power rating of the zener is exceeded.

Supply changes are reduced by the value $\frac{r}{R+r}$ where r is the slope resistance and R is the control resistance.

Assume r = 2Ω and R = 100Ω

$$\frac{2}{100+2} = \frac{1}{51} \text{ (approximately 2\% ripple left)}$$

The value of r tends to get less as the current is increased.

2.18 SILICON DIODE

The silicon junction diode is similar in appearance to the germanium diode and the static characteristics are also similar. The properties of silicon junction diodes differ from those of germanium junction diodes in the following respects.

a. The forward voltage drop is roughly double that of the corresponding germanium diode.
b. The reverse current at a given temperature and voltage is approximately a hundredth of that of the corresponding germanium diode but there is little sign of current saturation as is the case with germanium-infact the reverse current of a s silicon diode is roughly proportional to the square root of the voltage until breakdown is approached.
c. It can withstand a much higher reverse voltage and can operate at temperatures up to about $150^0 - 200^0 C$ compared with about 75^0 $90^0 C$ for germanium.
d. The reverse current of silicon diode, for a given voltage, practically doubles for every $8^0 C$ rise in temperature compared with $10^0 C$ for germanium.

2.19 AVALANCHE EFFECT

If the reverse voltage across a PN junction is gradually increased, a point is reached where the energy of the current carriers is sufficient to dislodge additional carriers. These carriers in turn, dislodge more carriers and the junction goes into a form of avalanche breakdown characterized by a rapid increase in current. The power due to a relatively large increase if maintained for an appreciable time can easily ruin the device.

2.20 TUNNEL DIODE

The PN junction diode so far discussed has an impurity concentration of about 1 part in 10^8. With this amount of doping, the width of the depletion layer, which constitutes a potential barrier at the junction, is small. This potential barrier restrains the flow of carriers from the side of the junction where there is a concentration of majority carriers to the side where there is a concentration of minority carriers.

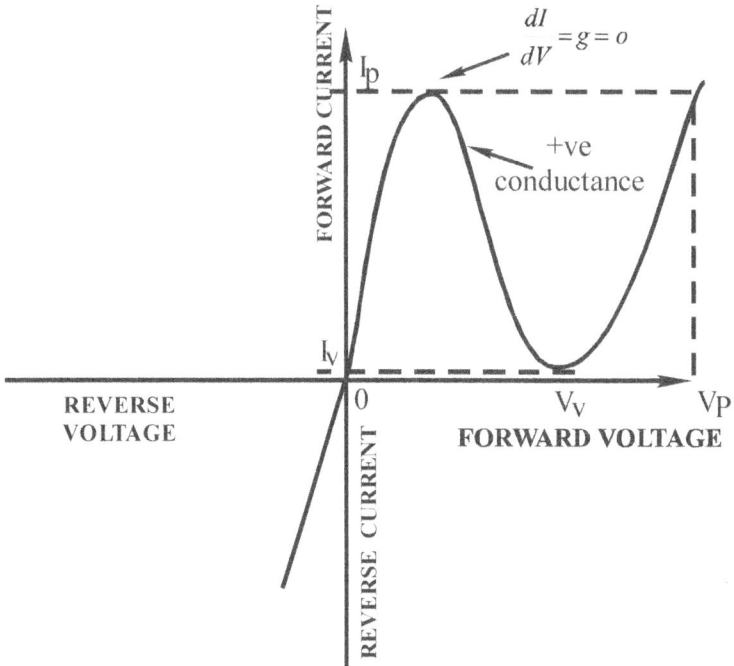

Fig. 2.20.1: Volt ampere characteristics of tunnel diode

If the concentration of impurity atoms is greatly increased say 1 part in 10^3, the device characteristics are completely changed and this results in the use of the PN junction in the construction of the tunnel diode.

The width of the junction barrier varies inversely as the square root of impurity concentration and therefore is reduced to less than 10^{-6} cm.

This thickness is only about one fifth the wavelength of visible light. For the barrier as thin as those estimated above in the ESAKI DIODE (as it is usually referred to), the Schrödinger equation indicates that there is a large probability that an electron will penetrate through the barrier. This quantum mechanical behaviour is referred to as tunnelling and hence these high impurity density PN junction devices are called tunnel diodes.

2.21 CHARACTERISTICS OF A TUNNEL DIODE

From the characteristics of the tunnel diode shown in fig. 15.20.1 we see that the tunnel diode is an excellent conductor in the reverse direction. Also for small forward voltage (up to 50mV for G_e), the resistance remains small (of the order of 5Ω). At the peak current corresponding to the voltage V_p, the slope of the characteristics is zero. If V is increase beyond Vp, the current decreases. As a consequence the dynamic conductance $g = \dfrac{dI}{dV}$ is negative. The tunnel diode exhibits a negative resistance characteristics between peak current Ip and the minimum value called the valley current, I_V At the valley voltage Vv at which I = Iv, the conductance is again zero and beyond this point the conductance becomes and remains positive. At the so-called peak forward voltage Vp the current again reaches the value Ip. For larger voltages the current increases beyond this value. For current whose values are between Iv and Ip, the curve is triple-valued as seen in the characteristic curve. It is this multi valued features which makes the tunnel diode useful in pulse and digital circuitry. The representation of fig. 2.21.1 shows the symbol of the tunnel diode.

Fig. 15.21.1 Symbol

2.22 SEMICONDUCTOR PHOTODIODE

If a reverse biased PN junction is illuminated the current varies almost linearly with the light flux. This effect is exploited in the semiconductor photodiode. The device consists of a PN junction embedded in a clear plastic as shown in fig. 2.22.1. Radiation is allowed to fall upon one surface across the junction. The remaining sides of the plastic are either painted black or enclosed in a metallic case.

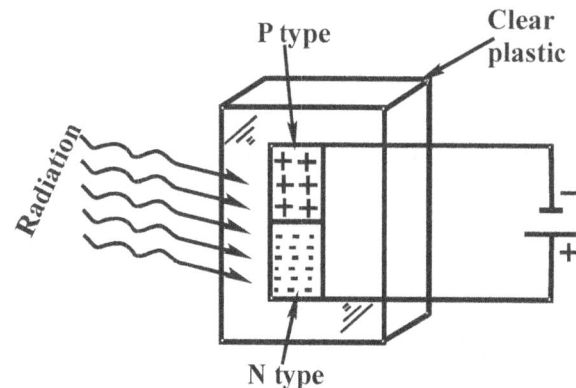

Fig. 2.22.1: The construction of semiconductor photodiode

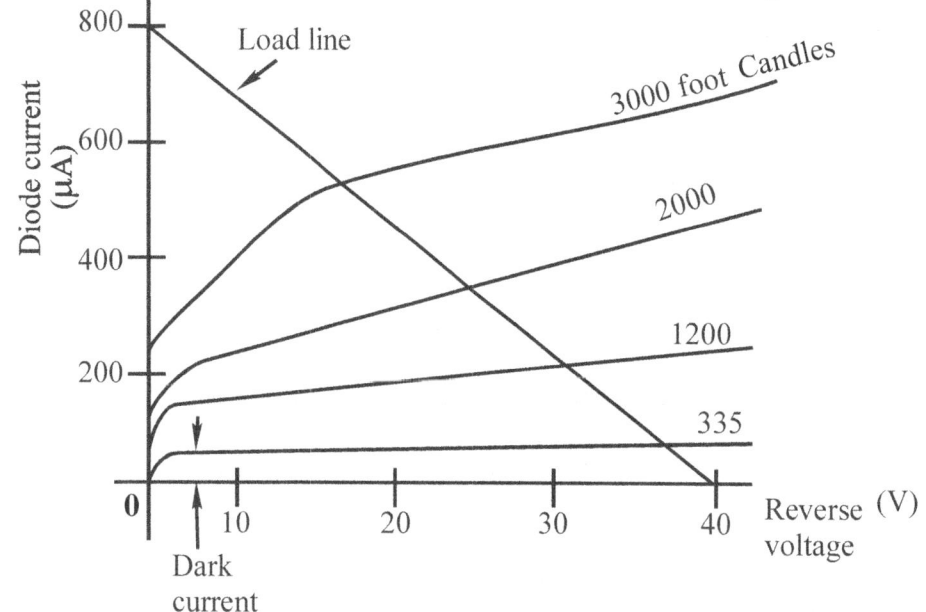

Fig. 2.22.2: Volt ampere characteristics

If reverse voltages are applied across the PN junction an almost constant current (independent of the magnitude of the reverse bias) is obtained.

The dark current shown on the current/reverse voltage characteristics of fig 2.22.2 corresponds to the reverse saturation current due to the thermally generated minority carriers. Now if light falls upon the surface, additional electron-hole pairs are formed. It should be noted that it is justifiable to consider the radiation solely as a minority carrier injector.

Fig. 2.22.3; Semiconductor photodiode

These injected minority carriers (for example, electron in the P-side) diffuse to the junction, cross it and contribute to the flows of electric current. The saturation current Io in a PN diode is proportional to the concentration Pno and Npo of diode in the N and P regions, respectively. If a reverse biased PN junction is illuminated, the number of new hole-electron pairs is proportional to the number of incident photons. Pno and Npo are the positive ion and electron concentration.

Hence, the current under large reserve bias is I = Io + Is.

Where Is, the short circuit current is proportional to the light intensity.

$$\therefore I = Is + Io\left(e^{Q_eV/KT} - 1\right)$$

where I_0 = Saturation current with negative bias

Q_e = Charge on electron (1.6×10^{-19} C)

V = p.d in volts across junction

K = Botzmann's Constant (1.38×10^{-23} J/K)

T = Absolute temperature (K)

It should be noted that the characteristics with exception of the dark-current curve do not pass through the origin.

The photodiode finds extensive use in light speed reading of computer punch cards, and tapes, light detection systems, reading of films and tracks, light operated switches, production line counting of objects which interrupt a light beam etc. The diagram of fig. 2.22.3 shows the symbol of a semiconductor photodiode.

2.23   THERMISTORS

The conductivity of germanium (silicon) is found to increase approximately 6(8) percent per degree increase in temperature. This property of the semiconductor is an advantage. A semiconductor used in this manner is called a thermistors. Such a device finds extensive application in thermometry, in measurement of microwave frequency power, as a thermal relay and in control devices actuated by changes in temperature.

CHAPTER THREE

3.0 PN DIODE RECTIFIERS

For reasons associated with economics of generation, transmission and distribution, electric power in available is usually an a.c supply. The supply voltage varies sinusoidal and has a frequency of 50Hz. It is used for lighting, heating and supplying electric motors. But there are many applications where d.c supply is needed. When such d.c supply is required, the mains a.c supply, which is kept at 220v for domestic appliances, is rectified by using PN diodes. The following two rectifier circuit can be used.

    a. Half wave rectifier   b.   Full wave rectifier.

3.1 HALF WAVE RECTIFIER

In half wave rectification, the rectifier conducts current only during the positive half cycle of input a.c supply. The negative half cycle of a.c supply are suppressed i.e. during the negative half cycles, no current is conducted and hence no voltage appears across the load. Therefore current always flows in one direction (ie d.c) through the load though after every half cycle.

Fig. 3.1.1 shows the circuit where a single PN diode acts as a half wave rectifier. The a.c supply to be rectified is applied in series with the diode and load resistance R_L. Generally a.c supply is given through a transformer. The use of transformer permits two advantages. Firstly, it allows us to step up or step down the a.c input voltage as the situation

PN DIODE RECTIFIERS

demands. Secondly, the transformer isolates the rectifier circuit from power line and thus reduces the risk of electric shock.

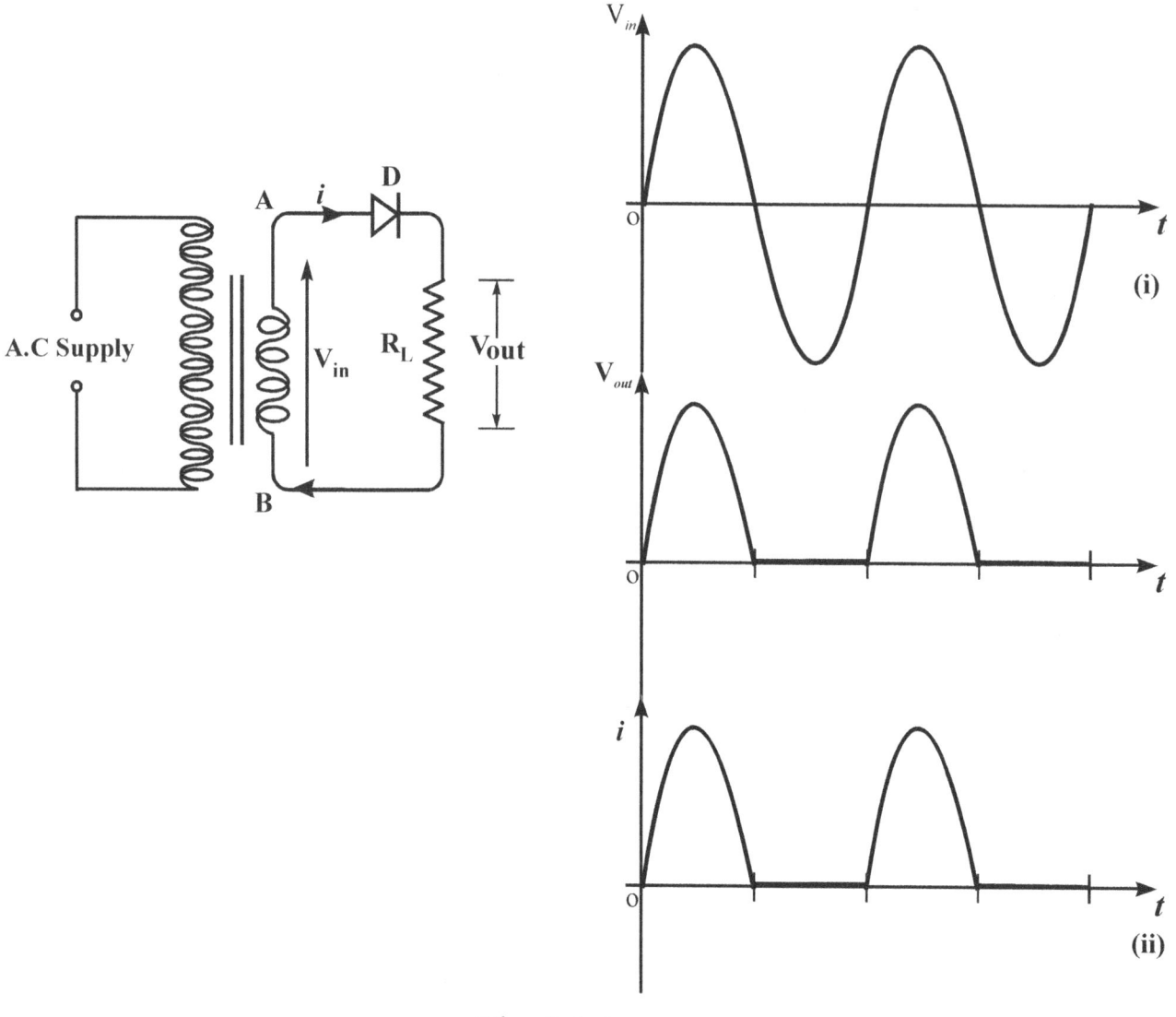

Fig. 3.1.1

The a.c voltage across the secondary winding AB is sinusoidal i.e. it changes polarities after every half cycle. During the positive half cycle of input a.c voltage, end A becomes positive with respect to end B. This makes the diode forward biased and hence it conducts current. But during the negative half cycle, end A is negative with respect to B.
Under this condition, the diode is reverse biased and it conducts no current. Therefore current flows through the diode during the positive

PN DIODE RECTIFIERS

half cycle of input a.c voltage only, it is blocked during the negative half cycle (see fig. 3.1.1 (ii).

In this way, current flows through the load R_L always in the same direction. Hence d.c output is obtained across, R_L. The output across the load is pulsating d.c and these pulsations are further smoothened with the help of filter circuits which will be discussed later.

The disadvantages of half wave rectifiers are;

1. The pulsating current in the load contains alternating component whose basic frequency is equal to the supply frequency. Therefore an elaborate filtering is required to produce steady direct current.

2. The a.c supply delivers power only half the time resulting in a low output power.

3.2 EFFICIENCY OF HALF WAVE RECTIFIER

The ratio of d.c power output to the applied input a.c power is known as rectifier efficiency i.e.

$$\text{Rectifier efficiency, } \partial = \frac{d.c \text{ power output}}{Input \text{ } a.c \text{ power}}$$

Consider a half wave rectifier shown in fig. 3.2.1

Let $V = V_m \sin\theta$ be the alternating voltage that appears across the secondary winding. Let r_f and R_L be the anode resistance and load resistance respectively. The diode conducts during the positive half cycle of a.c supply while no current conduction takes place during the negative half cycles. The output current is pulsating and in order to find d.c power, average current has to be found out.

PN DIODE RECTIFIERS

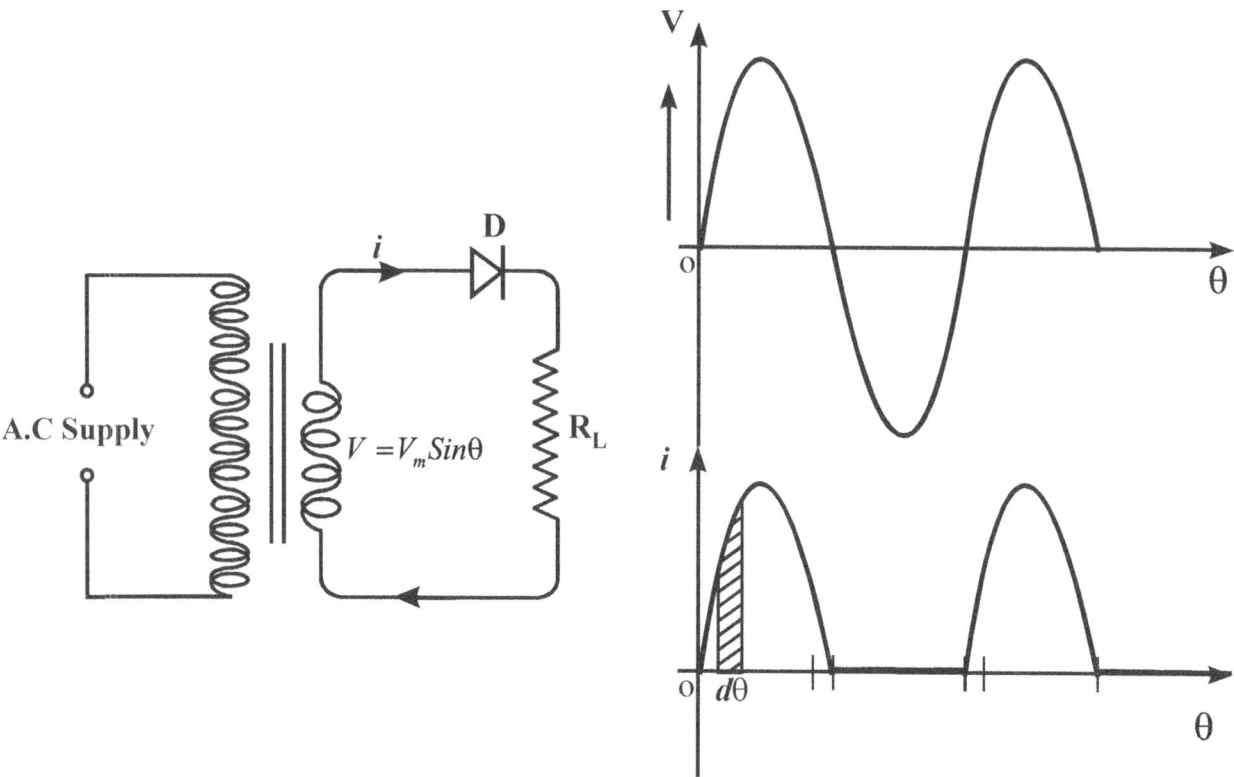

Fig. 3.2.1

$$Iav = I_{d.c} = \frac{1}{2\pi}\int_0^\pi i\,d\theta = \frac{1}{2\pi}\int_0^\pi \frac{V_m Sin\theta}{(r_f + R_L)}d\theta$$

$$= \frac{V_m}{2\pi(r_f + R_L)}\int_0^\pi Sin\theta\,d\theta$$

$$= \frac{V_m}{2\pi(r_f + R_L)}[-Cos\theta]_0^\pi$$

$$= \frac{V_m}{\pi(r_f + R_L)}$$

$$= \frac{I_m}{\pi} \quad \text{where } I_m = \frac{V_m}{r_f + R_L}$$

∴ d.c power, $P_{d.c} = I_{d.c}^2 \times R_L = \left(\frac{I_m}{\pi}\right)^2 \times R_L$

a.c power input $= P_{a.c} = I_{rms}^2 (r_f + R_L)$

for half-wave rectified wave, $I_{rms} = I_m/2$

∴ $P_{a.c} = \left(\frac{I_m}{2}\right)^2 \times (r_f + R_L)$

PN DIODE RECTIFIERS

$$\text{Rectifier efficiency} = \frac{d.c \text{ Output Power}}{a.c \text{ Input Power}}$$

$$= \frac{\left(I_m/\pi\right)^2 \cdot R_L}{\left(I_m/2\right)^2 \cdot (r_f + R_L)}$$

$$= \frac{0.406 R_L}{r_f + R_L} = \frac{0.406}{1 + \frac{r_f}{R_L}}$$

If r_f is negligible as compared to R_L (ie. $R_L \ll r_f$), maximum efficiency will be 40.6%.

Example 3.2.1: The applied input a.c power to a half wave rectifier is 100watts. The d.c output power obtained is 40 watts.

1. What is the rectification efficiency?
2. What happens to the remaining 60 watts?

Solution:

1. Rectification efficiency $= \dfrac{d.c \text{ Output power}}{a.c \text{ Input power}}$

    $= \dfrac{40}{100} = 0.4 = 40\%$

2. The 40% efficiency of rectification does not mean that 60% of power is lost in the rectifier circuit. Infact a crystal diode consumes little power due to its small internal resistance. The 100W a.c power is contained as 50watts in positive half cycles and 50 watts in negative half cycles. The 50 watts in the negative half cycle are not supplied at all. It is only the 50watts in the positive half cycle are converted into 40 watts.

    $\therefore$ Power Efficiency $= \dfrac{40}{50} \times 100 = \underline{80\%}$

PN DIODE RECTIFIERS

It can be seen that 100watts of a.c power was supplied; the half wave rectifier accepted only 50watts and converted it into 40 watts d.c power. It is therefore appropriate to say that the efficiency of rectification is 40% and not 80% which is power efficiency.

Example 3.2.2

An a.c supply of 220V is applied to a half wave rectifier circuit through a transformer of turn ratio 10:1

Find (1) The output d.c voltage and (ii) the peak inverse voltage. Assume the diode is ideal.

Solution.

Turn ratio = $\dfrac{N_1}{N_2} = 10$

RMS primary voltage = 220V

∴ Maximum primary voltage is $V_{pm} = \sqrt{2} \times rms\ primary\ voltage$.
$= \sqrt{2} \times 220 = 311.1V$

Maximum secondary voltage is,

$$V_{sm} = V_{pm} \times \dfrac{N_2}{N_1} = 311.1 \times \dfrac{1}{10}$$
$$= \underline{31.11V}$$

i) $I_{d.c} = \dfrac{I_m}{\pi}$

∴ $V_{d.c} = \dfrac{I_m}{\pi} \times R_L = \dfrac{V_{sm}}{\pi} = \dfrac{31.11}{\pi} = \underline{9.91V}$

ii) During the negative half-cycle of a.c supply the diode is reverse biased and hence conducts no current. Therefore, the maximum secondary voltage appears across the diode.

∴ Peak inverse voltage = $\underline{31.11V}$

Example 3.2.3

A PN diode having internal resistance $r_f = 20\Omega$ is used for half wave rectification. If the applied voltage $V = 50 \sin \omega t$ and load resistance $R_L = 800\Omega$, find

    (i) I_m, I_{dc}, I_{rms}
    (ii)   a.c power input and d.c power output.
    (iii) d.c output voltage.
    (iv) Efficiency of rectification.

Solution:

$V = 50 \sin \omega t$

∴ Maximum voltage, $V_m = 50V$

i) $I_m = \dfrac{V_m}{r_f + R_L} = \dfrac{50}{20 + 800} = 0.061 A = \underline{61 mA}$

$I_{dc} = \dfrac{I_m}{\pi} = \dfrac{61}{\pi} = \underline{19.4 mA}$

$I_{rms} = \dfrac{I_m}{2} = \dfrac{61}{2} = \underline{30.5 mA}$

ii) a.c power input $= I_{rms}^2 \times (r_f + R_L)$

$= (30.5 \times 10^{-3})^2 \times (20 + 800)$

$= \underline{0.763 \, Watt}$

d.c output power $= I_{dc}^2 R_L = (19.4 \times 10^{-3})^2 \times 800$

$= \underline{0.301 \, Watt}$

iii) d.c output voltage $= I_{dc} R_L = (19.4 mA \times 800)$

$= \underline{15.52 \, Volts}$

iv) Efficiency of rectification $= \dfrac{0.301}{0.763} \times 100$

$= \underline{39.5\%}$

Example 3.2.4:

A half-wave rectifiers is used to supply 45V dc to a resistive load of 700Ω. The diode has a resistance of 25Ω. Calculate the a.c voltage required.

Solution:

Output d.c voltage, $V_{dc} = 45V$

Diode resistance, $r_f = 25\Omega$

Load resistance, $R_L = 700\Omega$

Let V_m be the maximum value of a.c voltage required.

$$\therefore \quad V_{dc} = I_{dc} \times R_L$$

$$= \frac{I_m}{\pi} \times R_L = \frac{V_m \, R_L}{\pi(r_f + R_L)} \quad \left(note, \, I_m = \frac{V_m}{(r_f + R_L)}\right)$$

or $\quad 45 = \dfrac{V_m \times 700}{\pi(25 + 700)}$

$$\therefore \quad V_m = \frac{\pi(725) \times 45}{700} = \underline{146V}$$

Hence a.c voltage of maximum value 146V is required.

3.3 FULL-WAVE RECTIFIERS:

In full-wave rectification, current flows through the load in the same direction for both half-cycle of input a.c voltage. This can be achieved with two diodes working alternatively.

For the positive half-cycle of input voltage, one diode supplies current to the load and for the negative half-cycle, the other diode does so; current being always in the same direction through the load. Therefore, a full-wave rectifier utilises both half-cycles of input voltage to produce the d.c output.

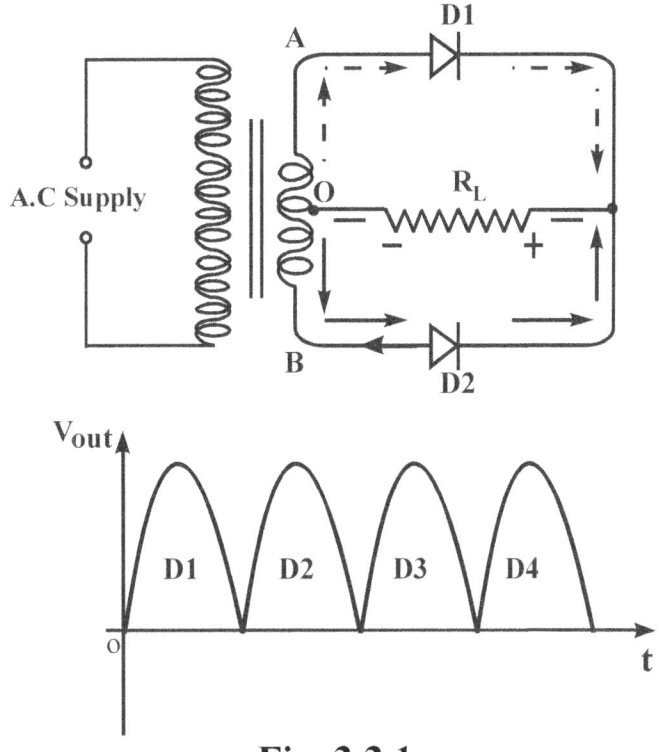

Fig. 3.3.1

The following two circuits are commonly used fro full-wave rectification.

(a) Circuit with two PN diodes

(b) Full-wave rectifier.

(a) **Circuit with two PN diodes:** The circuit employs two PN diodes connected so that each uses one half-cycle of input a.c voltage (see fig. 3.3.1).

In other words, diode D1 utilises the a.c voltage appearing across the upper half OA of the secondary winding for rectification while diode D2 uses the lower half winding OB.

During the positive half-cycle of secondary voltage the end A (fig. 3.3.1) of the secondary winding becomes positive and end B negative. This makes the diode D1 forward biased and diode D2 reverse biased. Therefore, diode D1 conducts while D2 does not.

The conventional current flow is through diode D1, load resistance, and the upper half arrows. During the negative half-cycle, end A of the secondary winding becomes negative and end B positive. Therefore, diode D2 conducts while diode D1 does not. The conventional current flow is through diode D2 load R_L and load half winding as shown by solid arrows. Referring to the diagram, it may be seen that current in the load R_L is in the same direction for both half-cycle of input a.c voltage. Therefore a d.c voltage is obtained across the load R_L. Also the polarities of the d.c output across the load should be noted. Experiment will prove that the peak inverse voltage (PIV) is twice the maximum voltage across the half secondary winding; PIV = 2Vm.

The disadvantages are (i) It is difficult to locate the centre tap on the transformers secondary winding. (2) The d.c output is small as each diode utilises only one-half of the transformer secondary voltage. (3) The diodes used must have high peak inverse voltage.

(b) **Full-Wave Bridge Rectifier:** Here, the need for a centre tapped power transformer is eliminated in the bridge rectifier. It contains four PN diodes, D1, D2, D3 and D4 connected to form a bridge (fig. 3.3.2). The a.c supply to be rectified is applied to the diagonally opposite ends of the bridge through the transformer. Between other two ends of the bridge, the load resistor R_L is connected.

During the positive half-cycle of secondary voltage, the end P of the secondary winding becomes positive and end Q negative. This makes the diodes D1 and D3 forward biased, while D2 and D4 are reverse biased. Therefore only diode D1 and D3 conduct. These two diodes will be in series through the load R_L as shown fig. 3.3.2a. The conventional current flow is shown by dotted arrows. It may be seen that current flows from A to B through the load, R_L. During the negative half-cycle of secondary voltage, end P becomes negative and end Q positive. This makes diodes D2 and D4 forward biased whereas diode D1 and D3 are reverse biased. Therefore, only D2 and D4 conduct.

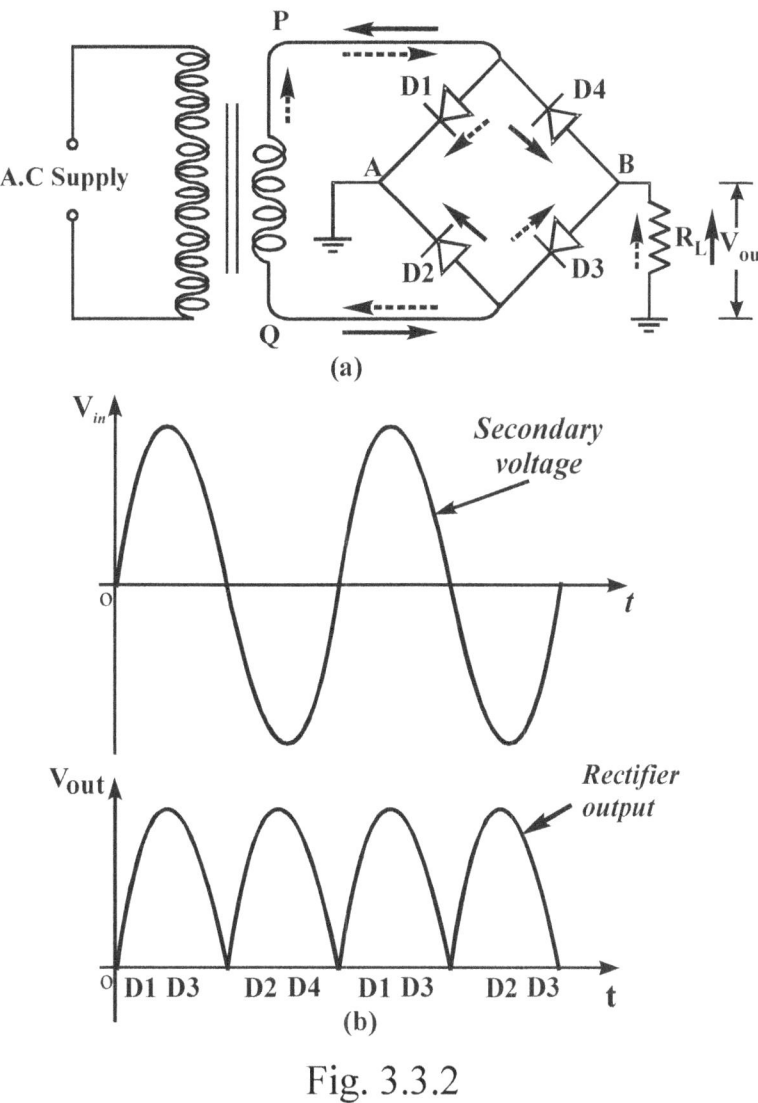

Fig. 3.3.2

These two diodes will be in series through the load, R_L as shown in fig. 3.3.2. The direction of the current flow is shown by the solid arrows. It may be seen that again current flows from A to B though the load i.e. in the same direction as for the positive half-cycle. Therefore, d.c output is obtained across the load R_L. It can be shown that the peak inverse voltage is equal to the maximum voltage (V_m) across the secondary winding.

The advantages of full-wave bridge rectifier are (i) the need for centre tapped transformer is eliminated (ii) the output is twice that of the centre-tapped circuit for the same secondary voltage. (iii) The PIV is one-half that of the centre-tapped circuit (for same d.c output).

The disadvantages are (i) it required fore diodes (ii) As during each cycle of a.c input two diodes that conduct are in series; therefore, voltage drop in the internal resistance of the rectifying unit will be as great as in the centre-tap circuit. This is objectionable when secondary voltage is small.

3.4 EFFICIENCY OF FULL-WAVE RECTIFIER: Fig. 3.4.1

Fig. 3.4.1 shows the output of full-wave rectification.

Let $V = V_m Sin\theta$ be the a.c voltage to be rectified.

Let r_f and R_L be the diode resistance and the load resistance respectively. The rectifier will conduct current through the load in the same direction for both half-cycles of input a.c voltage. The instantaneous current i is given by;

$$i = \frac{V}{R_f + R_L} = \frac{V_m Sin\theta}{r_f + R_L}$$

PN DIODE RECTIFIERS

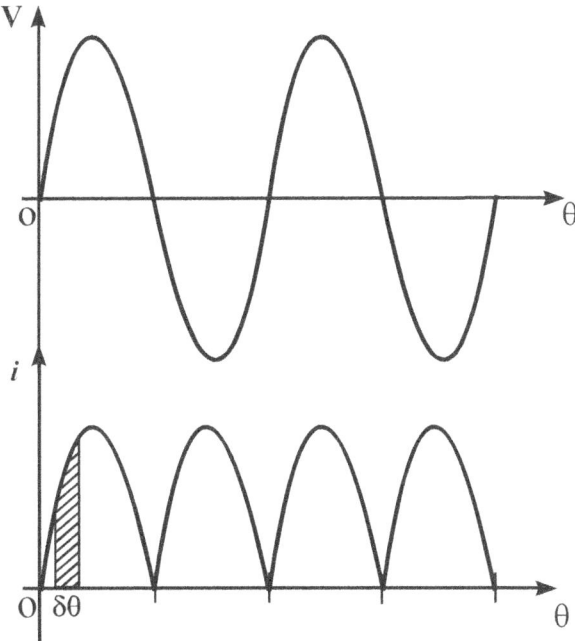

Fig. 3.4.1

The output current is pulsating direct current. Therefore in order to find the d.c power, average current has to be found. The average value of current is given by;

$$I_{dc} = \frac{2I_m}{\pi}$$

d.c power output, $P_{dc} = P_{dc}^2 \times R_L = \left(\frac{2I_m}{\pi}\right)^2 \bullet R_L$

$$P_{ac} = I_{rms}^2 (r_f + R_L)$$

$$Irms = \frac{I_m}{\sqrt{2}}$$

$$P_{ac} = \left[\frac{I_{rms}}{\sqrt{2}}\right]^2 (r_f + R_L)$$

Therefore, full-wave rectification efficiency is

$$\eta = \frac{P_{dc}}{P_{ac}} = \frac{\left(\frac{2I_m}{\pi}\right)^2 R_L}{\left(\frac{I_m}{\sqrt{2}}\right)^2 (r_f + R_L)}$$

$$= \frac{8}{\pi^2} \times \frac{R_L}{(r_f + R_L)} = \frac{0.812 R_L}{r_f + R_L} = \frac{0.812}{1 + \frac{r_f}{R_L}}$$

The efficiency will be maximum if r_f is negligible as compared to R_L. Therefore maximum efficiency is 81.2%. This is double the efficiency due to half-wave rectifier. Therefore a full-wave rectifier is twice as effective as a half-wave rectifier.

Example 3.4.1: A full-wave rectifier uses two diodes, the internal resistance of each diode may be assumed constant at 20Ω. The transformer r.m.s secondary voltage from centre tap to each end of secondary is 5V and load resistance is 980Ω. Find (i) the main load current (ii) the rms value of load current.

Solution:
$r_f = 20\Omega$, $R_L = 980\Omega$
Max a.c voltage $= V_m = \sqrt{2} \times 50 = 70.7V$
Max load current $= I_m = \dfrac{V_m}{r_f + R_L} = \dfrac{70.7V}{(20 + 980)\Omega}$
$= 70.7 mA$

(i) Mean load current, $I_{d.c} = \dfrac{2 \times I_m}{\pi} = \dfrac{2 \times 70.7}{\pi}$
$= 45 mA$

(ii) R.M.S value of load current;
$$I_{rms} = \frac{I_m}{\sqrt{2}} = \frac{70.7}{\sqrt{2}} = 50 mA$$

Example 3.4.2 In the centre-tap circuit shown in fig. 3.4.2, the diodes are assumed to be ideal i.e., having zero internal resistance. Find (i) d.c output voltage (ii) peak inverse voltage (iii) rectification efficiency.

PN DIODE RECTIFIERS

Solution:

Primary to secondary turn ratio, $\dfrac{N_1}{N_2} = 5$;

R.M.S primary voltage = 220V

∴ R.M.S secondary voltage $= 220 \times \left(\dfrac{1}{5}\right) = 44V$

Maximum voltage across secondary $= 44 \times \sqrt{2}$

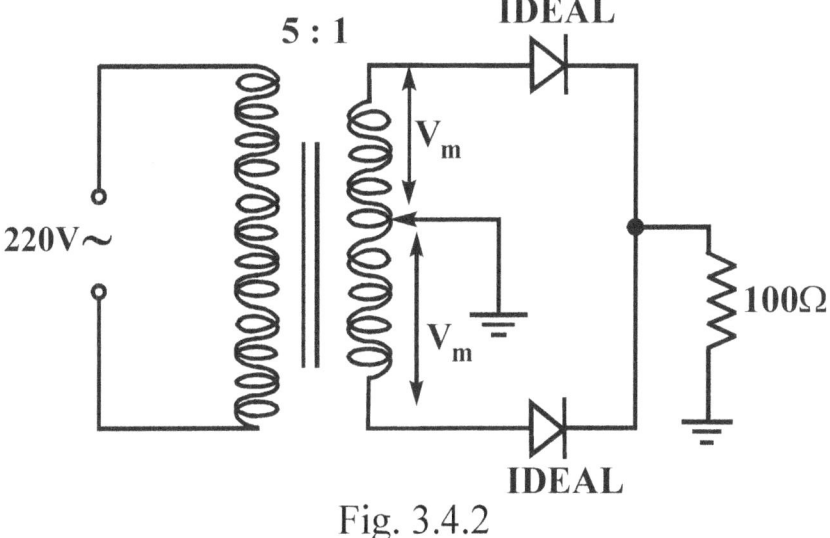

Fig. 3.4.2

Maximum voltage across half secondary winding;

$$V_m = \dfrac{62.2}{2} = 31.1V$$

(i) Average current, $I_{d.c} = \dfrac{2V_m}{\pi R_L} = \dfrac{2 \times 31.1}{\pi \times 100}$

$$= \underline{0.198A}$$

∴ d.c output voltage, $V_{dc} = I_{d.c} \times R_L$

$$= 0.198 \times 100$$

$$= \underline{19.8V}$$

(ii) The peak inverse voltage is equal o the maximum secondary voltage i.e.,

$$PIV = \underline{62.2V}$$

PN DIODE RECTIFIERS

(iii) Rectification efficiency = $\dfrac{0.812}{1 + \dfrac{r_f}{R_L}}$ $\quad (r_f = 0)$

∴ Rectification efficiency = <u>81.2%</u>

Example 3.4.3:

In the bridge type circuit shown in fig. 3.4.3, the diodes are assumed to be ideal

Find (i) d.c output voltage (ii) peak inverse voltage (iii) Output frequency. Assumed primary to secondary turn ratio to be 4.

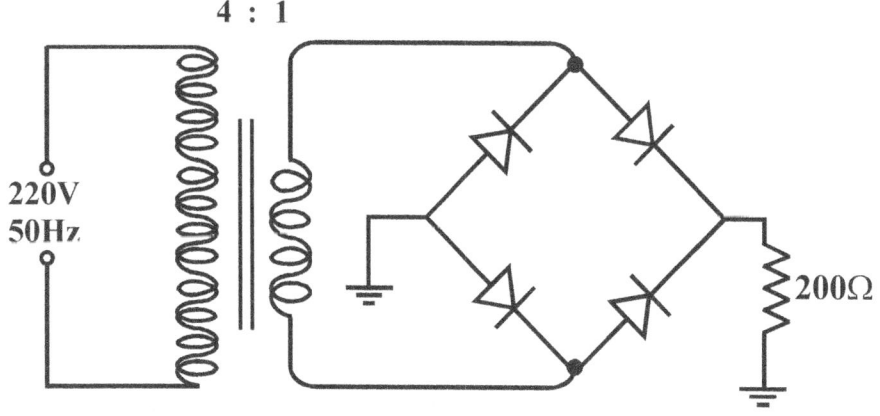

Fig. 1.4.5

Solution:

Turn ratio, $\dfrac{N_1}{N_2} = 4$

R.M.S primary voltage = 220V

R.M.S secondary voltage = $220 \left(\dfrac{N_2}{N_1} \right) = 220 \times \dfrac{1}{4}$

= 55V

Maximum voltage across secondary = V_m

(i) Average current, $I_{d.c} = \dfrac{2V_m}{\pi R_L} = \dfrac{2 \times 77.8}{\pi \times 200}$

= <u>0.25A</u>

∴ d.c output voltage, $V_{dc} = I_{d.c} \times R_L$

= 0.25 × 200

PN DIODE RECTIFIERS

$$= \underline{49.6V}$$

(ii) The peak inverse voltage is equal o the maximum secondary voltage i.e.,

$$PIV = \underline{77.8V}$$

(iii) In full-wave rectification, there are two output pulses for each complete cycle of the input voltage. Therefore, the output frequency is twice that of the a.c supply frequency i.e.,

$$f_{out} = 2f_{in} = 2 \times 50 = \underline{100Hz}$$

Example 3.4.4

The four diodes used in a bridge rectifier circuit have forward resistance which may be considered constant at 1Ω and infinite reverse resistance. The alternating supply voltage is 220Vrms and load resistance is 480Ω. Calculate (i) mean load current and (ii) power dissipated in each diode.

Solution:

Maximum a.c voltage = V_m = $220 \times \sqrt{2}V$

(i) At any instant in the bridge rectifier, two diodes in series are conducting. Therefore, total circuit resistance = $2r_f + R_L$

Maximum load current, $I_m = \dfrac{V_m}{2r_f + R_L}$

$$= \frac{220 \times \sqrt{2}}{2 \times 1 + 480}$$

$$= 0.65 A$$

∴ Mean load current, $I_{d.c} = \frac{2I_m}{\pi} = \frac{2 \times 0.65}{\pi}$

$$= \underline{0.41 A}$$

(ii) Since each diode conducts only half a cycle, diode rms current is; $\mathbf{I_{rms}} = I_m/2 = 0.65/2 = 0.33 A$

Power dissipated in each diode,

$$= I_{rms}^2 \times r_f = (0.33)^2 \times 1$$

$$= \underline{0.109 W}$$

3.5 RIPPLE FACTOR:

The output of a rectifier consists both d.c component and a.c component (also known as ripple). The ripple is undesirable and accounts for the pulsations in the rectifier output. The effectiveness of a rectifier depends upon the magnitude of the ripple in the output; the smaller the ripple, the more effective is the rectifier.

The ratio of rms value of the a.c component (ripple) to the d.c component in the rectifier output is known as ripple factor, i.e.,

$$\text{Ripple factor} = \frac{r.m.s.\ value\ of\ a.c\ component}{value\ of\ d.c\ component}$$

$$= \frac{I_{a.c}}{I_{d.c}}$$

Ripple factor is very important in deciding the effectiveness of a rectifier.

The smaller the ripple factor, the lesser the effective a.c component and hence more effective is the rectifier.

PN DIODE RECTIFIERS

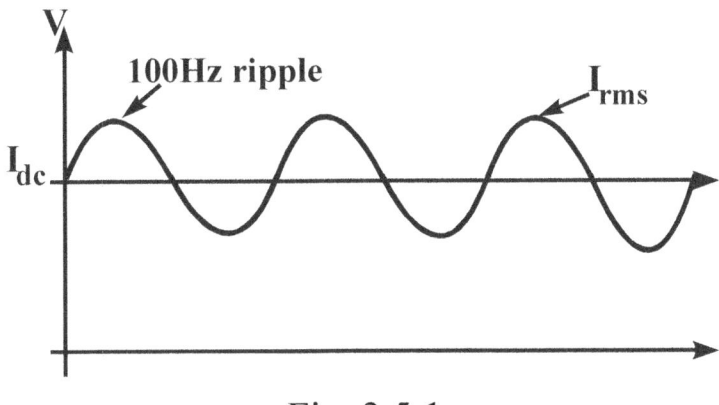

Fig. 3.5.1

The output current of a rectifier contains d.c as well as the ripple component. The undesired ripple or a.c component has a frequency of 100Hz (i.e., double the supply frequency – see fig. 3.5.1). This was explained in example 3.4.3. The ripple is a fluctuation superimposed on the d.c component.

By definition, the effective (i.e., rms) value of total load current is given by;

$$\text{Irms} = \sqrt{I_{dc}^2 + I_{ac}^2} \quad or \quad I_{ac} = \sqrt{I_{rms}^2 - I_{dc}^2}$$

Diving through by I_{dc} $= \dfrac{I_{ac}}{I_{dc}} = \dfrac{1}{I_{dc}}\sqrt{I_{rms}^2 - I_{dc}^2}$

But $\dfrac{I_{ac}}{I_{dc}}$ is the ripple factor,

$$\therefore \text{Ripple Factor} = \dfrac{1}{I_{dc}}\sqrt{I_{rms}^2 - I_{dc}^2}$$

$$= \sqrt{\left(\dfrac{I_{rms}}{I_{dc}}\right)^2 - 1}$$

(i) For half-wave rectification, $I_{rms} = \dfrac{I_m}{2}$ and $I_{dc} = \dfrac{I_m}{\pi}$

$$\therefore \text{Ripple Factor} \sqrt{\left(\dfrac{I_m/2}{I_m/\pi}\right)^2 - 1} = \underline{1.21}$$

(ii) For full-wave rectification, $I_{rms} = \dfrac{I_m}{\sqrt{2}}$ and $I_{dc} = \dfrac{2I_m}{\pi}$

$$\therefore \text{Ripple Factor } \sqrt{\left(\dfrac{I_m/\sqrt{2}}{2I_m/\pi}\right)^2 - 1} = \underline{0.48}$$

i.e., $\dfrac{\text{Effective a.c component}}{\text{Effective d.c component}} = 0.48$

In conclusion, it is clear that a.c component exceeds the d.c component in the output of a half-wave rectifier. This results in greater pulsations in the output. Therefore, half-wave rectifier is ineffective for conversion of a.c to d.c

In the output of a full-wave rectifier, the d.c component is more than the a.c component (ripple). Consequently, the pulsations in the output will be less than in half-wave rectifier. For this reason, full-wave rectification is invariably used for conversion of a.c to d.c

Example 3.4.5

A power supply A delivers $10V_{dc}$ with a ripple of 0.5V rms while the power supply B delivers $25V_{dc}$ with a ripple of $1mV_{rms}$. Compare the two power supplies and comment.

For power supply A,

$$\text{Ripple Factor} = \dfrac{V_{ac\ rms}}{V_{dc}} = \dfrac{0.5}{10} \times 100 = 5\%$$

Fro power supply B,

$$\text{Ripple Factor} = \dfrac{V_{a.c\ rms}}{V_{d.c}} = \dfrac{0.001}{25} \times 100 = 0.004\%$$

It is clear that power supply B with a lower ripple factor (0.004%) is better.

3.6 FILTER CIRCUITS:

In general, a rectifier is required to produce pure d.c supply required by electronic equipment. However, the output of a rectifier is pulsating i.e., it contains both a.c and d.c components. The a.c component is undesirable and must be kept away from the load. To do this requires the use of a filter which removes (or filter out) the a.c component and allow the d.c component to reach the load.

A filter circuit is normally installed between the rectifier and the load as shown in fig. 3.4.6.

A filter circuit is generally a combination of inductors (L) and capacitor (C). The filtering action of L and C depends upon the basic electrical principles. A capacitor passes a.c readily but does not allow d.c at all. On the other hand, an inductor opposes a.c but allow d.c to pass through it. It is then clear that suitable network of L and C can effectively remove the a.c component, allowing the d.c component to reach the load.

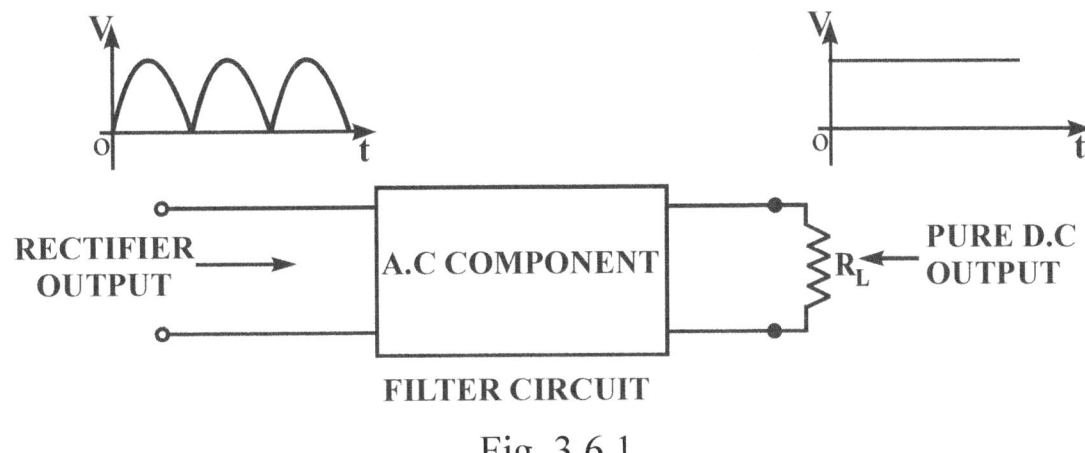

Fig. 3.6.1

The most commonly used filter circuits are capacitor filter, choke input filter and capacitor input filter or π-filter.

3.7 CAPACITOR FILTER CIRCUIT.

It consists of a capacitor C connected across the rectifier output in parallel with load R_L. The pulsating direct voltage of the rectifier is applied across the capacitor

as the rectifier voltage increases, it charges the capacitor and also ripples current to the load.

At the end of quarter-cycle (point A in fig. 3.7.1 (iii), the capacitor is charges to the peak value V_m of the rectifier voltage. Now the rectifier voltage starts to decrease. A this occurs, the capacitor discharges through the load and voltage across it (i.e., across parallel combination of R-C) decreases as shown by the line AB in fig. 3.7.1 (iii).

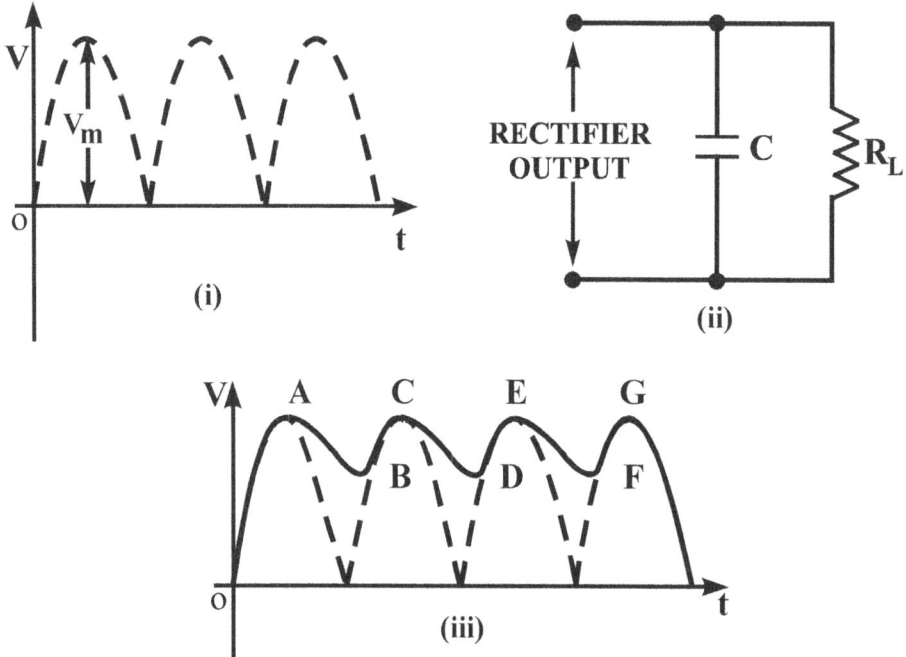

Fig. 3.7.1

The voltage across the load will decrease only slightly because immediately the next voltage peak is repeated again and again and the output voltage waveform becomes ABCDEFG. It may be seen that very little ripple is left in the output.

The capacitor filter circuit is extremely popular because of its low cost, small size, light weight and good characteristics. It is commonly used in transistor radio battery eliminators.

3.8 CHOKE INPUT FILTER:

Fig. 3.8.1 shows a typical choke input filter circuit. It consists of a choke L connected in series with the rectifier output and a filter capacitor across the load. Several identical filter sections are often used to reduce the pulsations effectively, as possible. But for the purpose of our explanation, only a single filter section is shown.

The pulsating output of the rectifier is applied cross terminals 1 and 2 of the filter circuit. As discussed before, the pulsating output of rectifiers contains a.c and d.c components. The choke offers high opposition to the passage of a.c component. The result is that most of the a.c components appears across the choke while the whole of d.c component passes through the choke on its way to the load. This results in the reduced pulsations at terminal 3.

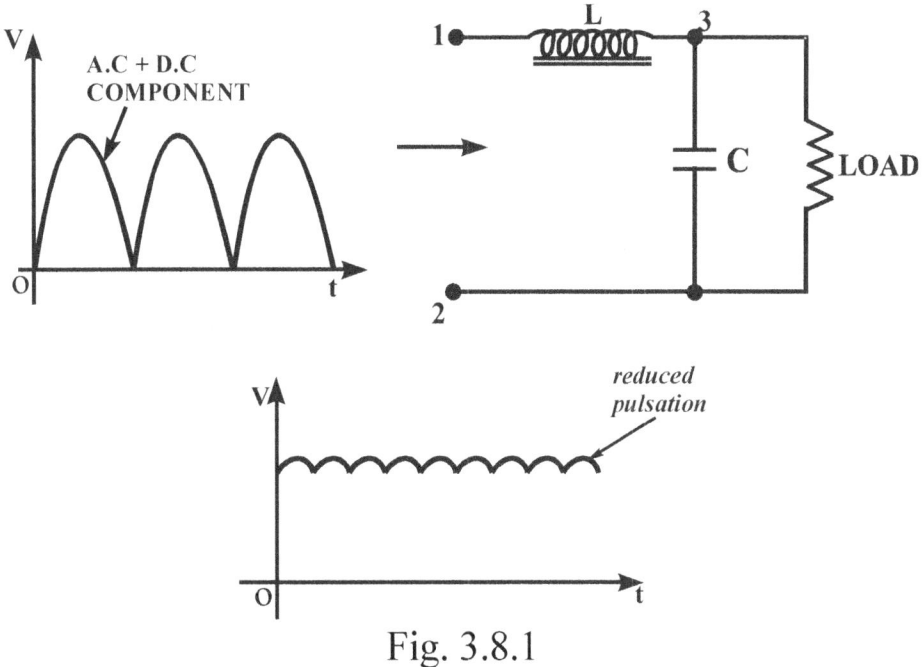

Fig. 3.8.1

At terminal 3, the rectifier output contains d.c components and the reaming part of a.c component which has managed to pass through the choke. Now the low reactance of the filter capacitor by passes a.c component by prevents the d.c component to flow though it. Therefore only d.c components reaches the load. In this way, the filter circuit filters out a.c component from the rectifier output allowing d.c component to reach the load.

3.9 CAPACITOR INPUT FILTER OR π-FILTER:

Fig. 3.9.1 shows a typical capacitor input filter or π-filter. It consists of filter capacitor C_1 connected across the rectifier output, a choke L in series and another capacitor C_2 connected across the load. Also here several identical sections are often used to improve the smoothness action but only one section is shown in the diagram.

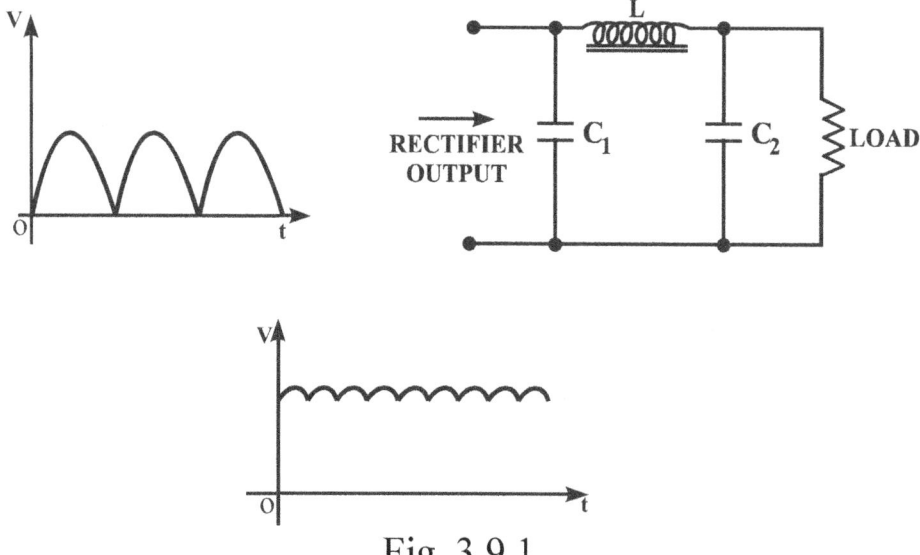

Fig. 3.9.1

The pulsating output from the rectifiers is applied across the input terminals of the filter. The filtering action of the three circuit components viz C_1, L and C_2 of this filter is described below:-

PN DIODE RECTIFIERS

a. The filter capacitor C_1 offers low reactance to a.c component of the rectifier output while it offers infinite reactance to the d.c component. Therefore, capacitor C_1 bypasses an appreciable amount of a.c component while the d.c component continues its journey to the choke L.

b. The choke L offers high reactance to the a.c component but it offers almost zero reactance to the d.c component. Therefore, it allows the d.c component to flow though it, while the bypassed a.c component is blocked.

c. The filter capacitor C_2 by passes the a.c component which the choke has failed to block. Therefore, only d.c component appears across the load and that is what we desire.

Example 3.9.1

The choke of fig. 3.9.1 has a d.c resistance of 25Ω. What is the d.c voltage of the full-wave signal into the choke has a peak value of 25.7V?

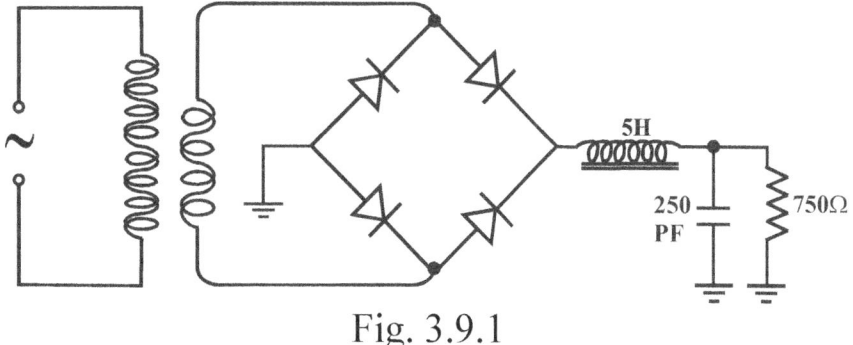

Fig. 3.9.1

Solution:

The output of a full-wave rectifier has a d.c component and an a.c component. Due to the presence of a.c component, the output of the rectifier has a pulsating characteristic as shown in fig. 3.9.2a. The maximum value of the pulsating output is V_m and d.c component is $V_{d.c} = \dfrac{2V_m}{\pi}$.

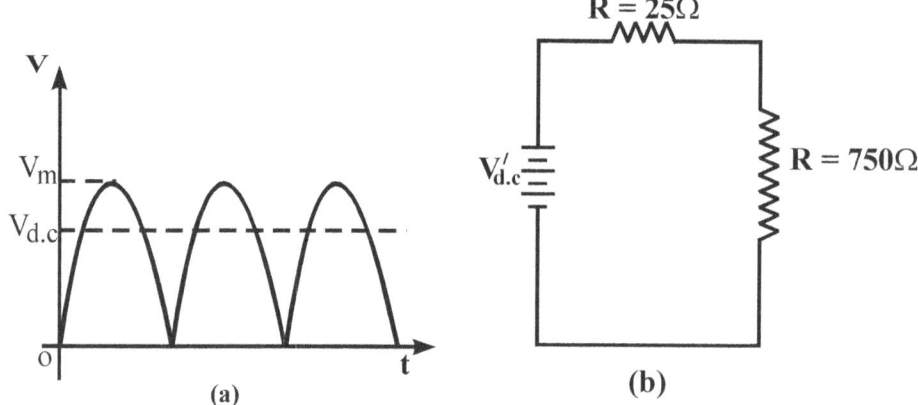

Fig. 3.9.2

For d.c component $V_{d.c}$, the choke resistance is in series with the load as shown in fig. 3.9.2b

$\therefore$ Voltage across load, $V_{d.c} = \dfrac{V'_{d.c} \cdot R_L}{R + R_L}$

In our example, $V'_{d.c} = \dfrac{2V_m}{\pi} = \dfrac{2 \times 25.7}{\pi} = 16.4V$

$\therefore$ Voltage across load $V_{d.c} = \dfrac{16.4}{25 + 750} \times 750 = \underline{\underline{15.9V}}$

The voltage across the load is 15.9V d.c pulse a small ripple.

CHAPTER FOUR

4.1 BIPOLAR TRANSISTORS

Suppose a piece of germanium crystal is doped as shown in fig. 4.1.1a, a PNP transistor will be realized. The crystal could also be doped in such a way that an NPN transistor is produced. The P region has a high conductivity and the N region low conductivity being thin with fewer electrons. It should be noted that two junctions will now be present in the crystal.

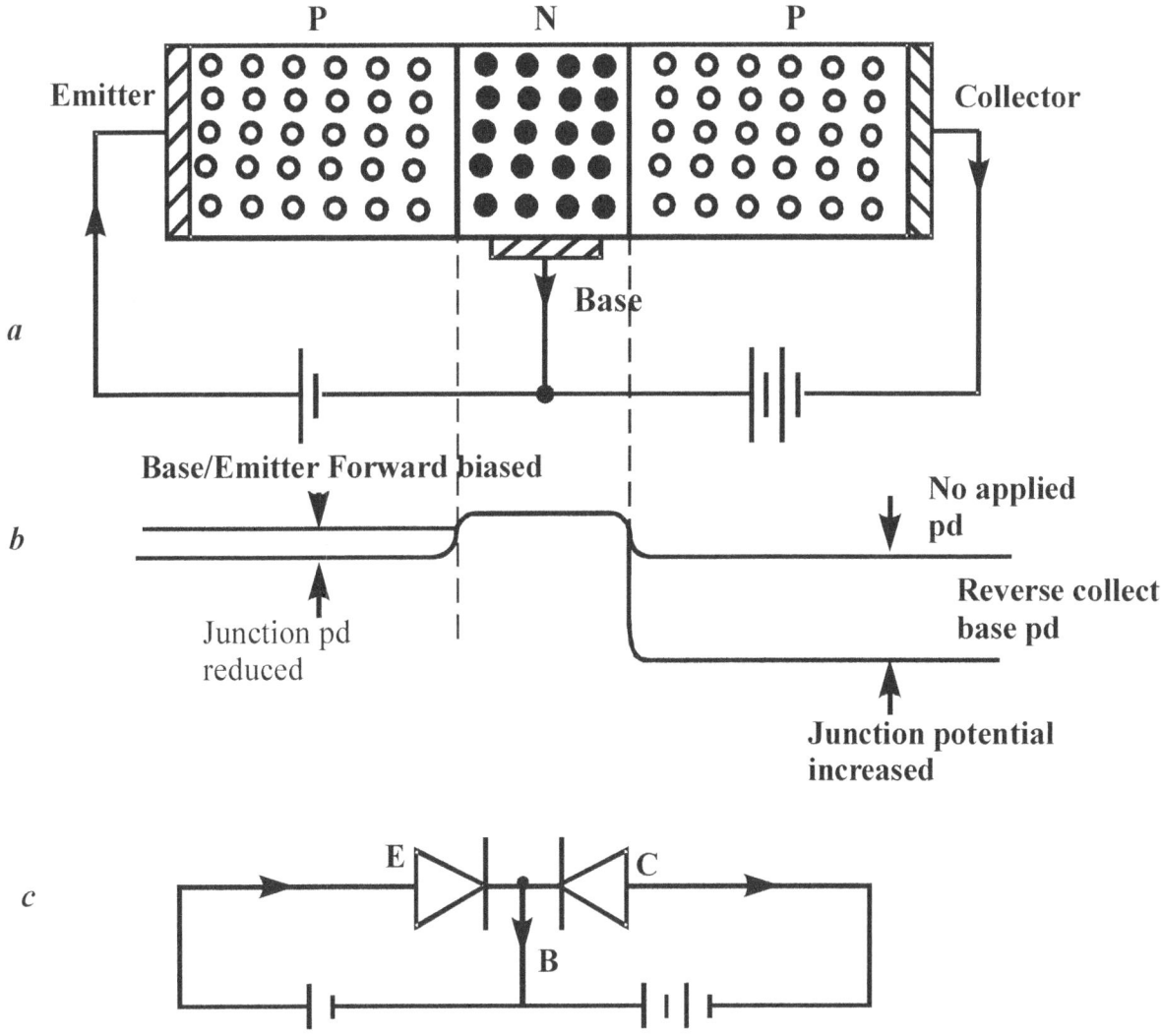

Fig. 4.1.1: PNP Transistor

BIPOLAR TRANSISTORS

The forward biased left hand PN junction results in the diffusion of holes from P to N region. But the corresponding injection of electrons from base to emitter is much smaller as there are fewer impurity centre in the base region. Once a hole arrives in the base region, they diffuse towards the collector and are swept out by the negative collector bias.

The diagram of fig. 4.1.1b shows the potential barriers at the junctions when they are biased as shown in fig 4.1.1a.

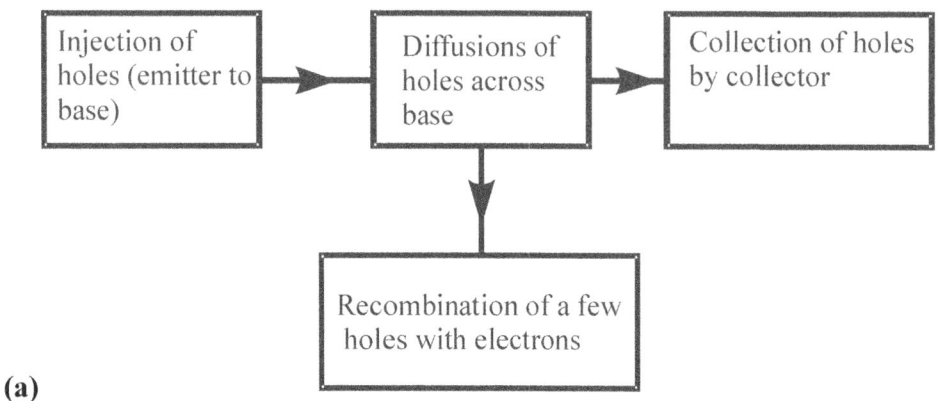

(a)

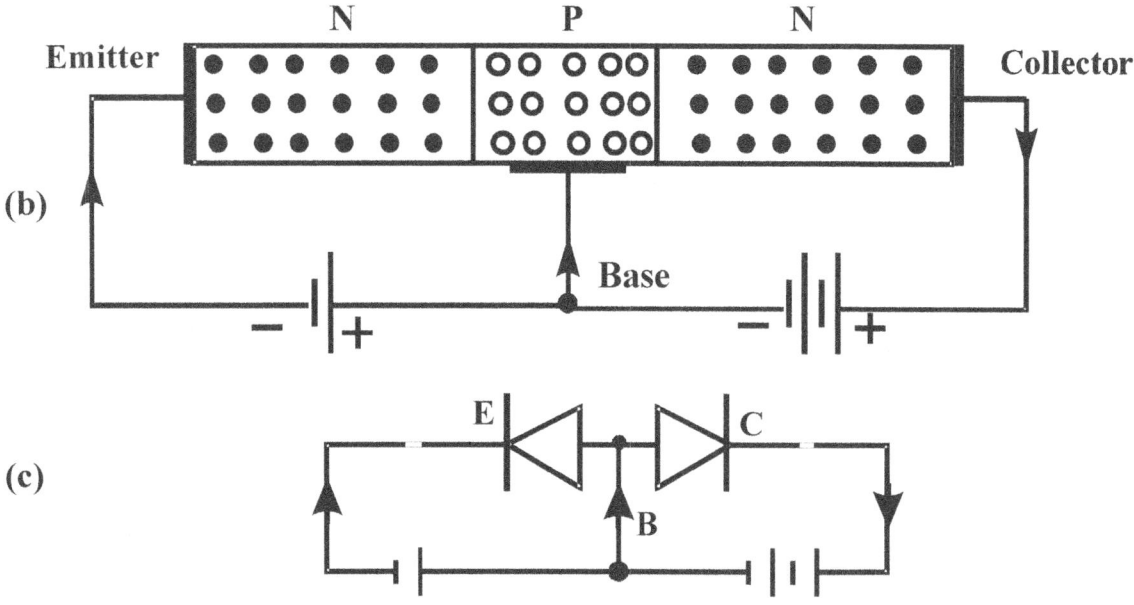

(b)

(c)

Fig. 4.1.2 NPN Transistor

Actually, some of the injected holes combine with the electrons in the base region and a small current result. The block diagram shown represents the behaviour of holes and electrons with the application of voltage. The PNP transistor can be represented as two PN junctions (PN diodes) to enable us to visualize the behaviour of PNP transistor when biased, see fig. 4.1.1C.

The intrinsic material can be doped in such a way as to give an NPN type of transistor as shown in fig 4.1.2b. Fig 4.1.2C shows the representation of back to back PN diodes representing NPN transistor when biased

4.2 CURRENTS IN EMITTER, COLLECTOR, BASE CIRCUIT OF PNP AND NPN TRANSISTORS

The symbol of a transistor is shown in fig. 4.2.1. Figure (a) is the representation of PNP, while b is NPN. The currents are as indicated in the diagram.

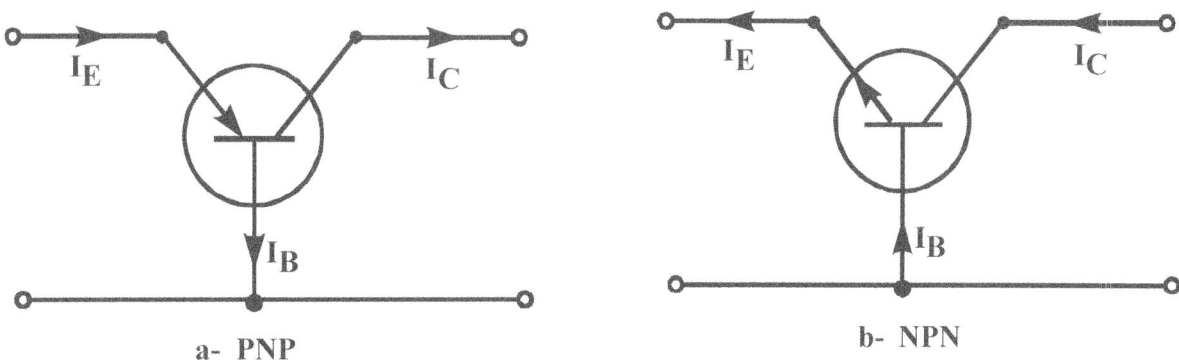

a- PNP   b- NPN

Fig. 4.2.1: Currents in emitter collector circuit of PNP and NPN transistors.

4.3 COMMON BASE CONFIGURATION

In fig. 4.3.1a, the NPN transistor is biased in common-base mode. Here the base is common to the emitter and collector. Fig 4.3.1b shows the currents in a common base NPN transistor while Fig. 4.3.1c shows the currents in a common base PNP transistor.

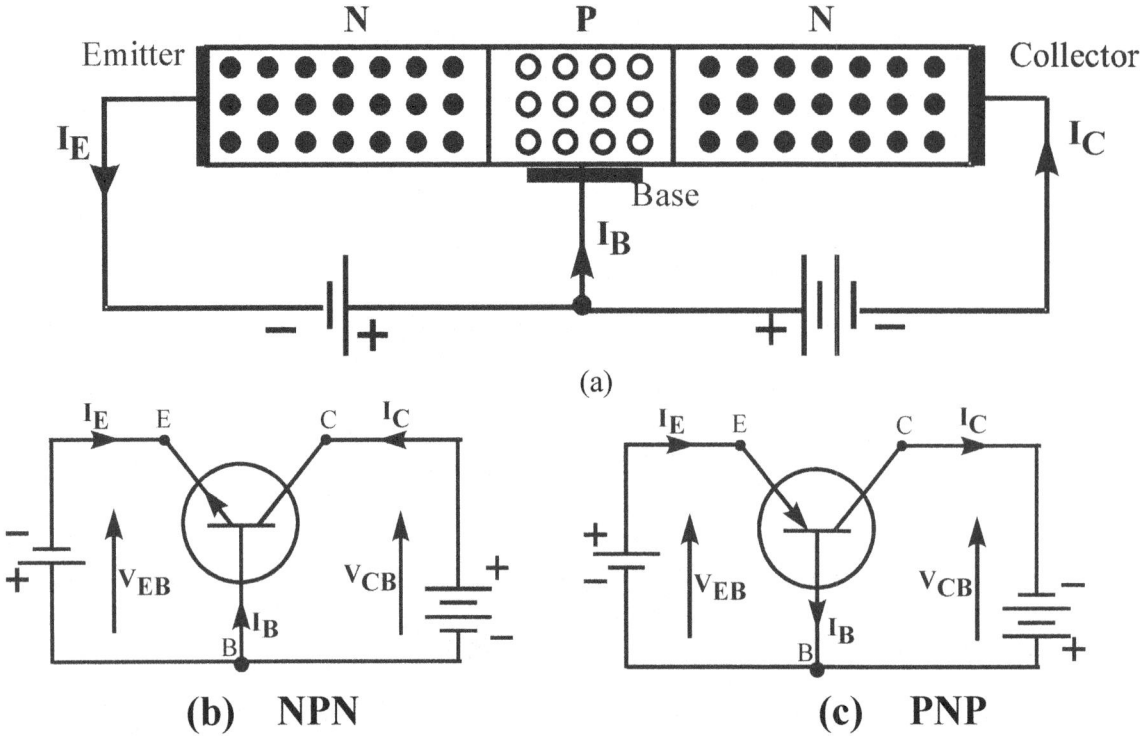

(b) NPN (c) PNP

Fig. 4.3.1 Common Base Circuit

4.4 COMMON EMITTER CONFIGURATION:

Here the emitter is common to the base and collector

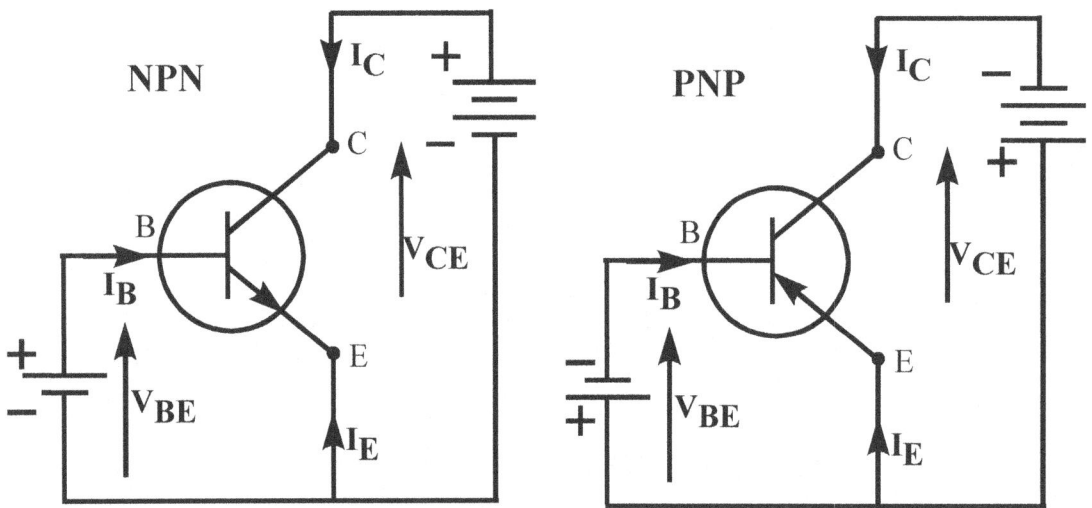

Fig. 4.4.1: Common emitter circuits for NPN and PNP

4.5 COMMON COLLECTOR CONFIGURATION

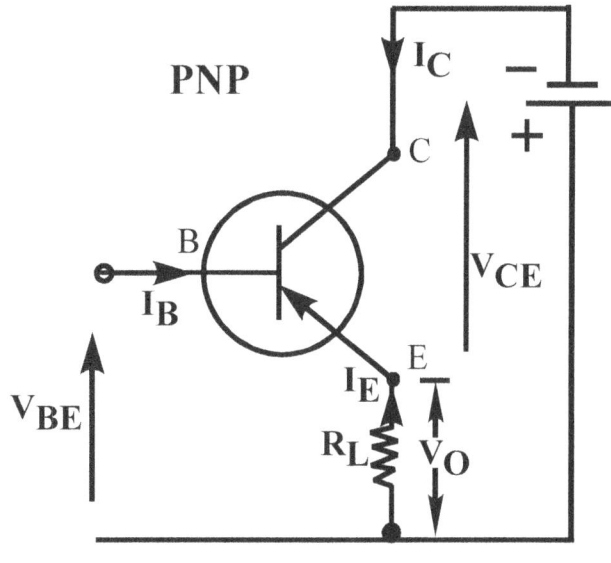

Fig. 4.5.1

In the common-emitter configuration, the emitter is connected directly to the base and collector circuits.

In the common base configuration, the base is connected directly to both the emitter and collector circuits.

Fig 4.3.1 and 4.4.1 show the biasing arrangements for both NPN and PNP in the common base and common emitter configuration.

In the case of common collector configuration of fig. 4.5.1, the load resistor is in the emitter lead rather than in the collector circuit. This circuit is sometimes called emitter follower.

4.6 STATIC CHARACTERISTICS FOR A COMMON BASE CIRCUIT

The circuit of fig. 4.6.1a shows the arrangement for the determination of the static characteristics of NPN transistor used in a common base configuration. The procedure is to maintain emitter current constant (A1), say 1mA by means of the slider R1. The readings on A2 are noted for various values of the collector-base voltages.

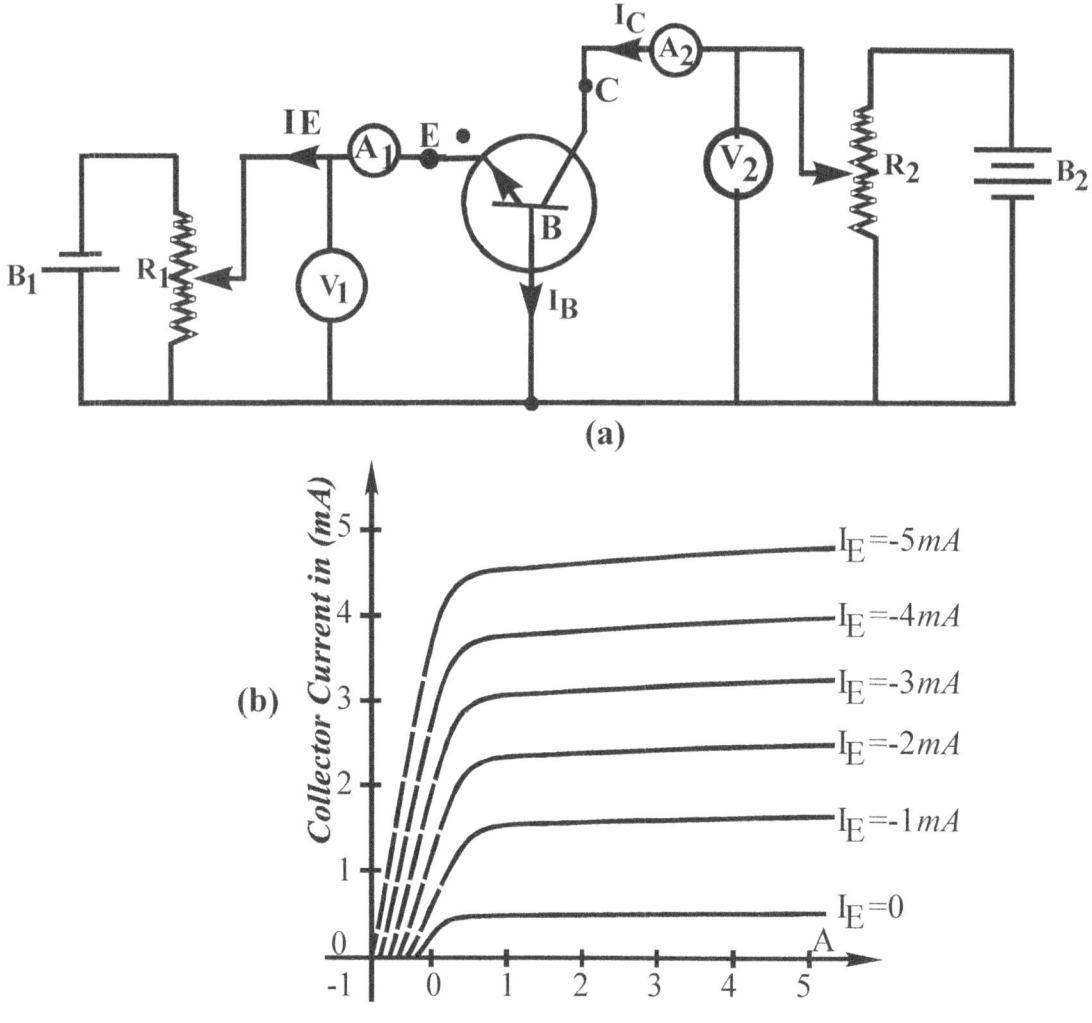

Fig. 4.6.2 (b) Static characteristic for a common base NPN transistor circuit
    (a)   Circuit for determination of static characteristics

The test is repeated for various values of the emitter current and the results are plotted as shown in fig. 4.6.1b.

It must be noted that the current is assumed to be positive when its direction is from the external circuit towards the transistor terminal, and the voltage V_{CB} is positive when C is positive relative to B. Hence for NPN transistor the collector current and collector base voltage are positive but the emitter current is negative. For PNP transistor, all the signs have to be reversed.

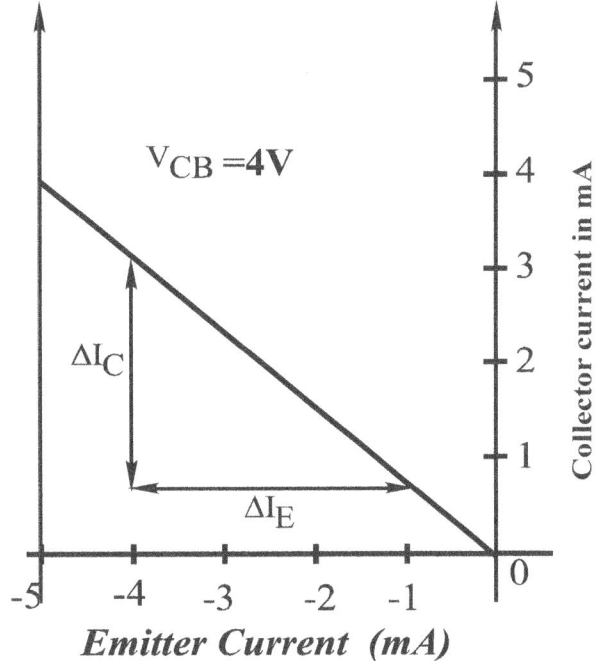

Fig. 4.6.2 Relationship between collector and emitter current for a given collector base voltage V_{CB}

For positive values of the collector base voltage the collector current remains almost constant i.e. nearly all the electrons entering the base of an NPN transistor are attracted to the collector. Also for a given collector base voltage, the collector current is practically proportional to the emitter current as shown in fig. 4.6.3.

The ratio of the change of collector current (ΔI_C) to the change of the emitter current ΔI_E, neglecting sign, for a given collector base voltage is termed the amplification factor for a common base circuit and it is represented by α i.e. $\alpha = \dfrac{\Delta I_C}{\Delta I_E}$ (for a given value of V_{CB})

$$= \text{Slope (neglecting sign) of } I_C/I_E \text{ graph}$$

4.7 STATIC CHARACTERISTICS FOR A COMMON EMITTER CIRCUIT

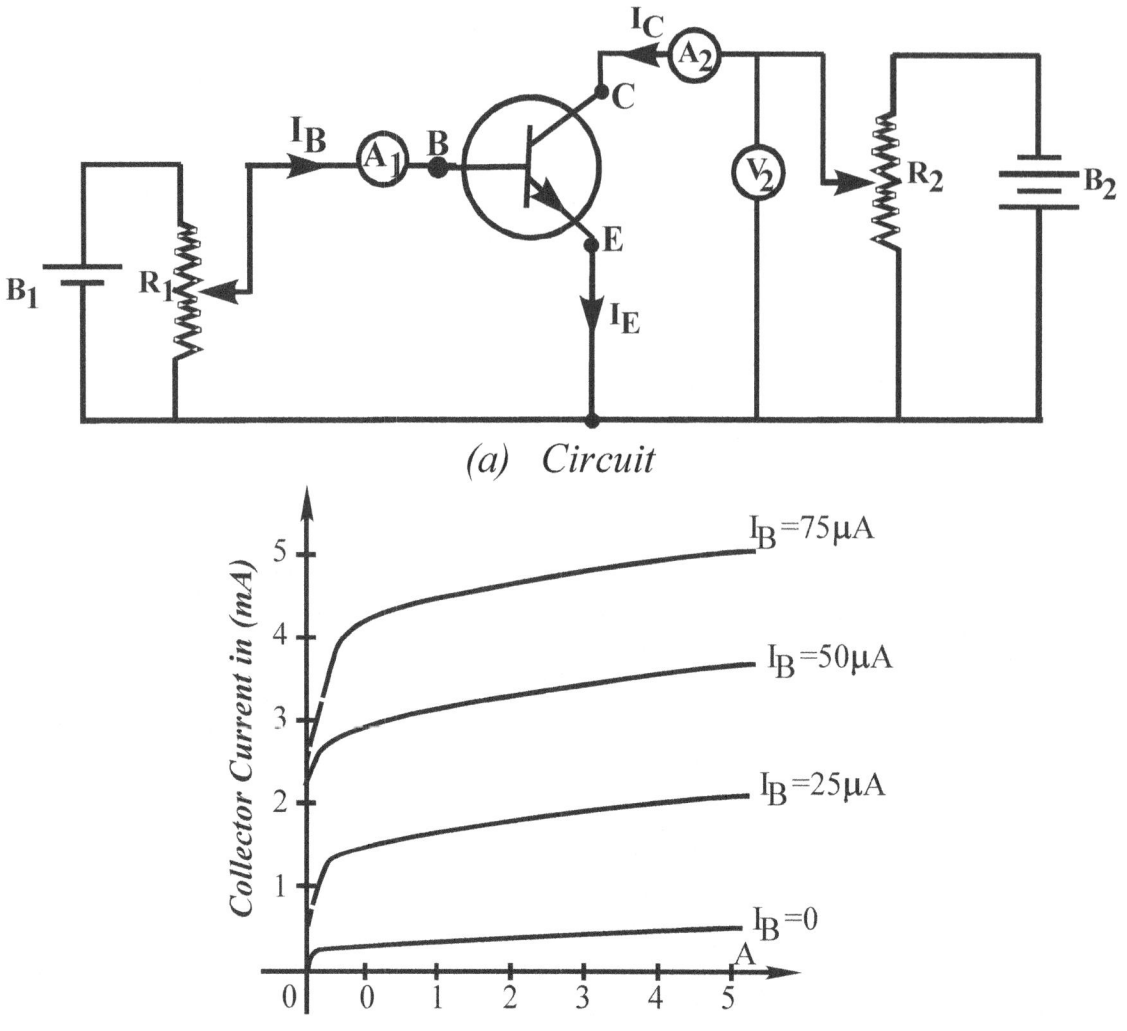

(a) Circuit

(b) Static characteristics for a common emitter NPN transistor

Fig. 4.7.1 (a) Circuit for the determination of the static characteristic for a common emitter configuration.
      (b) Static characteristics.

The circuit arrangement for the determination of the static characteristics of an NPN transistor used in a common emitter mode is given in fig. 4.7.1a.

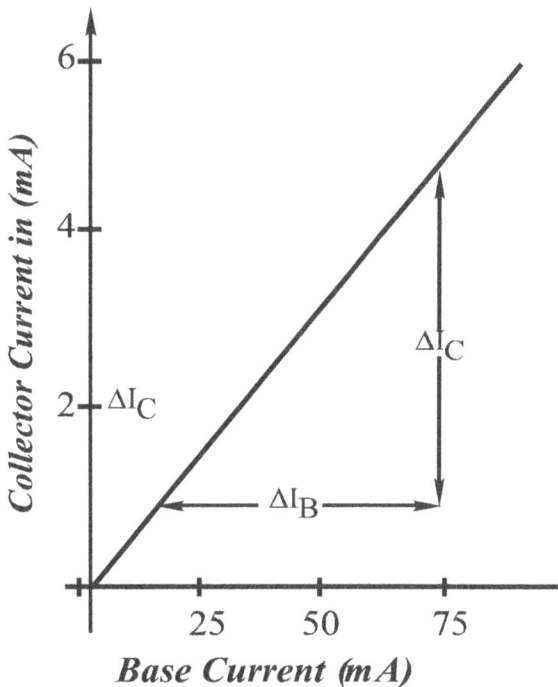

Fig. 4.7.2 relationship between collector and base currents for a given collector/emitter voltage

Here again the procedure is to maintain the base current constant and note the collector currents for various values of the collector emitter voltage (V_{CE}). The test is also repeated for various values of the base current and graph plotted as shown in fig. 4.7.1b.

For a given value of the voltage between collector and emitter (V_{CE} = 4v), the relationship between the collector and base current is practically linear as shown in fig. 4.7.2.

The ratio of the change of collector current ΔI_C to the change in base current, ΔI_B, for a given collector-emitter voltage is termed the current amplification factor for a common-emitter circuit and is represented by the symbol β i.e. $\beta = \dfrac{\Delta I_C}{\Delta I_B}$ (for a given value of V_{CE})

$$= \text{slope of } I_C/I_E \text{ graph (neglecting sign)}$$

4.8 RELATIONSHIP BETWEEN α AND β

From both common base and common emitter circuits above, it could be seen that the emitter current is equal to the sum of the collector and the base currents, i.e.

$$I_E = I_C + I_B$$
$$\therefore \Delta I_E = \Delta I_C + \Delta I_B$$

But $\alpha = \dfrac{\Delta I_C}{\Delta I_E}$ (common base configuration)

$$= \dfrac{\Delta I_C}{\Delta I_C + \Delta I_B}$$

Therefore $\dfrac{1}{\alpha} = 1 + \dfrac{\Delta I_B}{\Delta I_C} = 1 + \dfrac{1}{\beta} = \dfrac{(1+\beta)}{\beta}$

Hence $\alpha = \dfrac{\beta}{(1+\beta)}$

and $\beta = \dfrac{\alpha}{(1-\alpha)}$

A small variation of α corresponds to large variation in β.

Thus if α = 0.98, $\beta = \dfrac{0.98}{0.02} = 49$

And if β = 0.99, $\alpha = \dfrac{0.99}{0.01} = 99$

4.9 LOAD LINE FOR TRANSISTORS AND CURRENT AMPLIFICATION

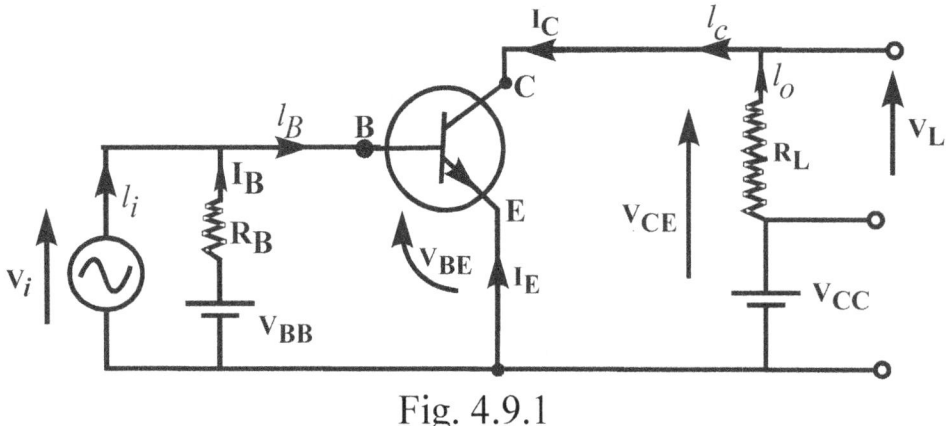

Fig. 4.9.1

BIPOLAR TRANSISTORS

Fig. 4.9.1 shows an NPN transistor biased in a common emitter mode. The voltage $Vi = V\, Sinwt$ is applied at the base as shown. Fig. 4.9.2 shows the characteristics and the representation of the input and out put signals.

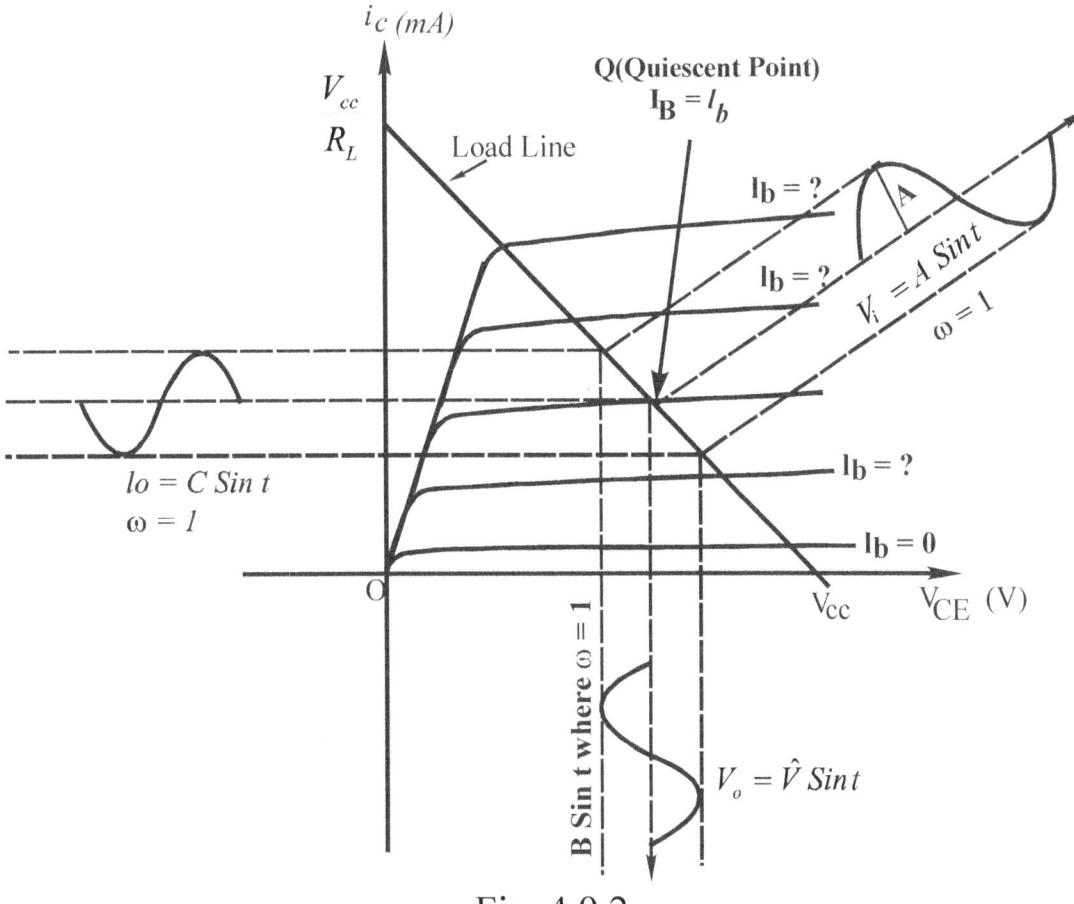

Fig. 4.9.2

The quiescent (operating point) point for $i_i = o$ from the equation of the base current.

i.e. $i_i + I_B = i_B$

For $i_i = 0$, $I_B = i_B$

Therefore for the base circuit

$$V_{BB} - I_B R_B - V_{BE} = 0$$

For a forward biased junction, V_{BE} is negligible.

$$\therefore I_B = \frac{V_{BB}}{R_B}$$

With the application of signal current

$$i_B = I_B + i_i$$

This is an a.c current superimposed on a d.c component.

The equation of the output circuit is

$$V_{CE} = V_{CC} - i_c R_L$$

This is also referred to as the equation of load line. The intersection of the load line and the characteristic curves of the non linear transistor represents the graphical solution of the equation of load line.

Thus when V_{CE} is zero, $i_C = \dfrac{V_{CC}}{R_L}$ and when i_c is zero, $V_{CE} = V_{CC}$

The signal output i_o is the sinusoidal component of i_c, the current in R.

The current amplification $= A_T = \dfrac{i_o}{i_i} = \dfrac{i_o \max}{i_i \max} = \dfrac{i_o\, rms}{i_i\, rms}$

The signal output voltage is the sinusoidal component of V_L i.e. V_o.

Voltage amplification $= \dfrac{V_o}{V_i} = \dfrac{V_o \max}{V_i \max} = \dfrac{V_o\, rms}{V_i\, rms}$

The output power P_0 is the product of the sinusoidal components of both current and voltage.

Therefore, Power Amplification $= \dfrac{V_o i_o}{V_i i_i} = \dfrac{V_o \max . i_o \max}{V_i \max . i_i \max}$

$$= \dfrac{V_o\, rms . i_o\, rms}{V_i\, rms . i_i\, rms}$$

$$= \dfrac{P_o}{P_i} = \dfrac{P_{O\max}}{P_i \max} = \dfrac{P_{o\, RMS}}{P_{i\, RMS}}$$

4.10 AMPLIFIERS

4.11 TRANSISTOR BIASING

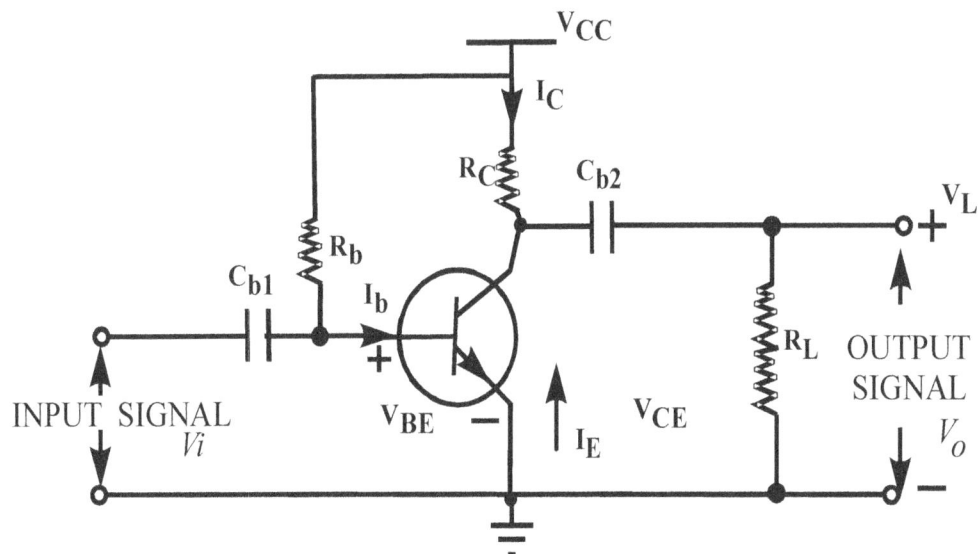

Fig. 4.11: Fixed bias circuit

Transistors function most linearly when it is constrained to operate in its active region. In order to establish an operating point in this region it is necessary to provide appropriate direct potentials and currents, using external sources.

Once an operating point Q is established such as shown in the figure 4.11.2, time varying excursions of the input signal (base current for example) should cause an output signal (collector voltage or collector current) of the same waveform. If the output signal is not a faithful reproduction of the input signal, for example if it is clipped on one side, the operating point is unsatisfactory and should be relocated in the collector characteristics.

Let us consider the common-emitter circuit in fig 4.11.1 and the output characteristics of fig. 4.11.2. It should be noted that our choice of R_C, R_L and R_b and V_{cc} may not enable us to operate the transistor everywhere in the active region because the various transistor ratings limit the range of useful operation.

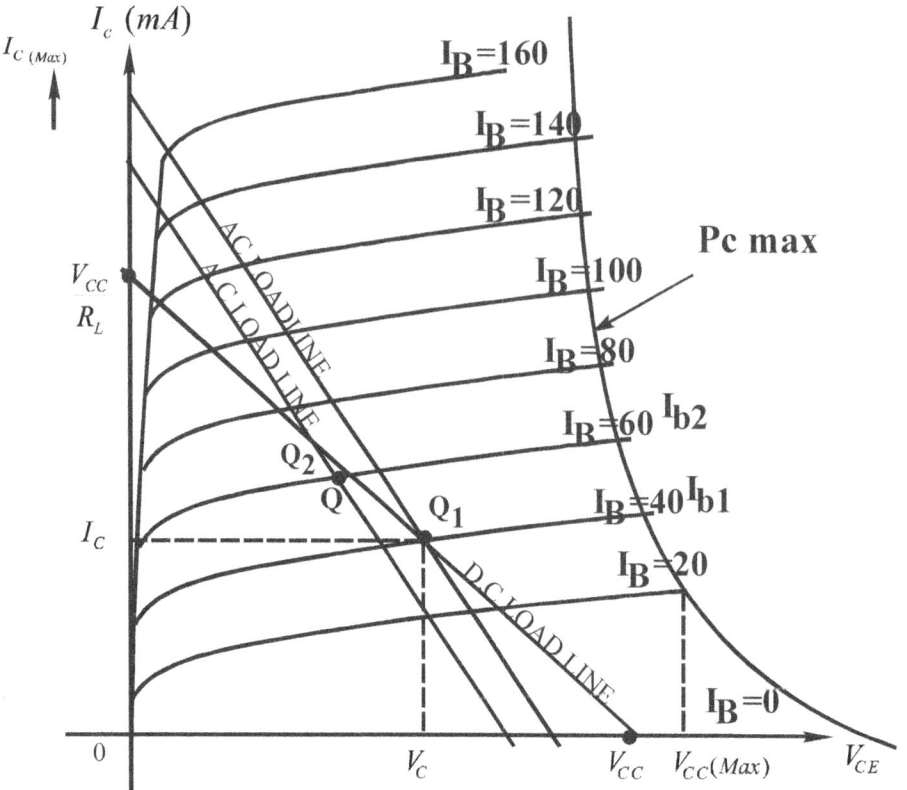

Fig. 4.11.2: Common emitter characteristics: dc and ac load lines

These ratings which are listed in the manufacturers specification are

1. Maximum Collector dissipation ($P_{c\,max}$)
2. Maximum Collector Voltage ($V_{c\,max}$)
3. Maximum Collector Current ($I_{c\,max}$)
4. Maximum Emitter –Base Voltage ($V_{EB\,max}$)

Figure 4.11.2 shows three of these bounds on a typical collector characteristic.

4.12 CAPACITIVE COUPLING

In the diagram of fig. 4.11.1 above, one end of Vi is at ground potential and the collector supply Vcc, also provides the biasing base current I_B. Under quiescent conditions (no input signal), C_{b1} acts as an open circuit i.e. blocking capacitor. This is because the reactance of a capacitor is infinite at zero frequency (dc). The capacitance C_{b1} and C_{b2} are chosen large enough, so that at the lowest frequency of excitation, their reactance are small enough so that they can be considered a short circuit. These coupling capacitors block d.c voltages and freely pass signal voltages (a.c). For example, the quiescent collector voltage does not appear at the output but Vo is an amplified replica of the input signal

4.13 STATIC AND DYNAMIC LOAD LINES

It is now known that under d.c condition C_{b2} acts as an open circuit. Hence the quiescent collector current and voltages are obtained by drawing a static (d.c) load line using the equation $V_{CE} = V_{CC} - i_C R_L$ as already discussed.

If $R_L = \infty$ and if the input signal (base current) is large and symmetrical, the operating point Q_1 must be located at the centre of the load line. In this way, the collector voltage and current may vary approximately and symmetrically around the quiescent values Vc and Ic respectively. If $R_L \neq \infty$, a dynamic load line must be drawn. It has already been noted that at the signal frequency, C_{b2} acts as a short circuit, therefore the effective load R_L at the collector is R_C in parallel with R_L. The dynamic load line must be drawn through the operating point Q_1 and must have a slope corresponding to $R'_L = R_c // R_L$. This a.c load line is indicated in fig. 4.11.2 where we observe that the input signal may swing approximately 40µA around Q_1 because if the base current decreases more than 40µA, the transistor is driven off. If a larger input swing is available, then in order to avoid cut off during a part of the cycle, the quiescent point must be relocated at a higher current on the dc load line. For example by simple trial and error, Q_2 is located on the d.c load line such that a line with a slope corresponding to the a.c resistance $R`_L$ drawn through Q_2 gives as large an output as possible without too much distortion. In our characteristics, the choice of Q_2 allows an input peak current swing of about 60μA.

4.14 THE FIXED BIAS CIRCUIT

The point Q_2 can be established by noting the required current I_{B2} in the characteristics of section 4.11 and choosing the resistance Rb in the circuit so that the base current is equal to I_{B2}

$$\therefore I_B = \frac{Vcc - V_{BE}}{Rb} = I_{B2}.$$

The voltage V_{BE} across a forward biased emitter junction is approximately 0.2v for germanium transistor and 0.7v for a silicon transistor in the active region. Since $Vcc \gg V_{BE}$

$$I_B = \frac{Vcc}{Rb}$$

In summary we see that the selection of the operating point Q depends on a number of factors. Among these factors are the a.c ad d.c load of the stage, the available power supply, the maximum transistor ratings, the peak signal excursion to be handled by the stage and the tolerable distortion.

4.15 TYPICAL TRANSISTOR JUNCTION VOLTAGE VALUES

The characteristics plotted in fig. 4.15.1 shows the relationship of the collector currents and the input voltages (V_{BE}) for NPN germanium and silicon transistors for common emitter circuits.

BIPOLAR TRANSISTORS

1. **CUT OFF REGION:** Cut off is defined to mean $I_E = 0$ and $I_c = I_{CO}$ and it is found that a reverse bias of $V_{EB\ (Cutoff)} = 0.1V(ov)$ will cut off a germanium (silicon) transistor.

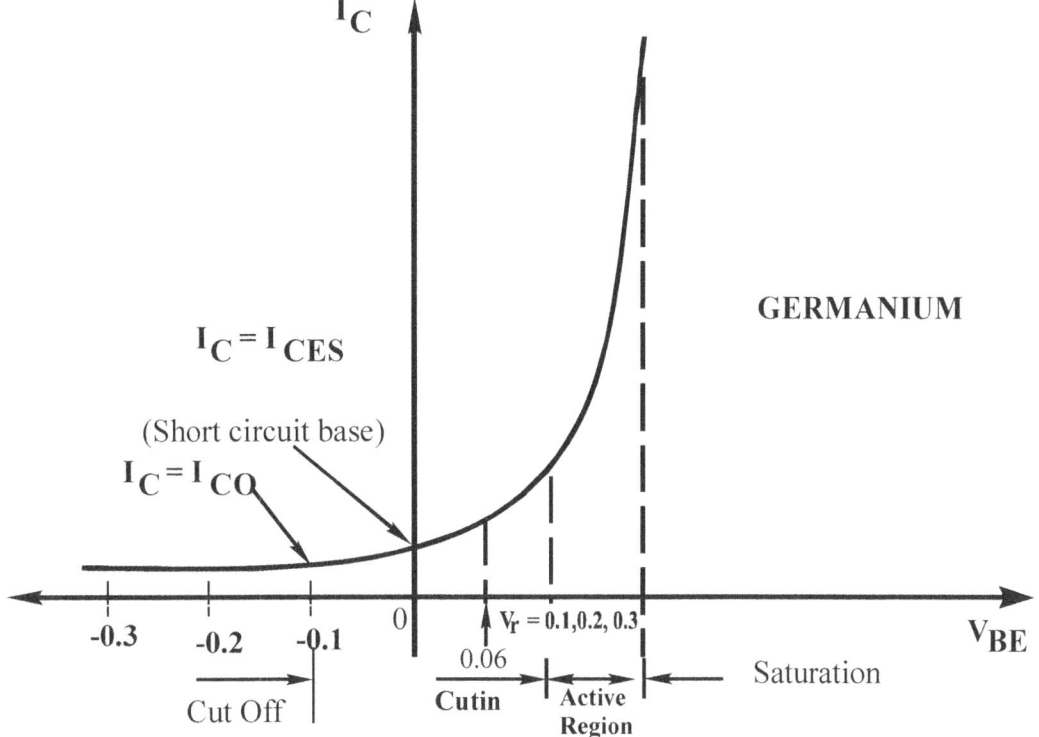

(a) PLOT OF COLLECTOR CURRENT AGAINST BASE TO EMITTER VOLTAGE FOR GERMANIUM

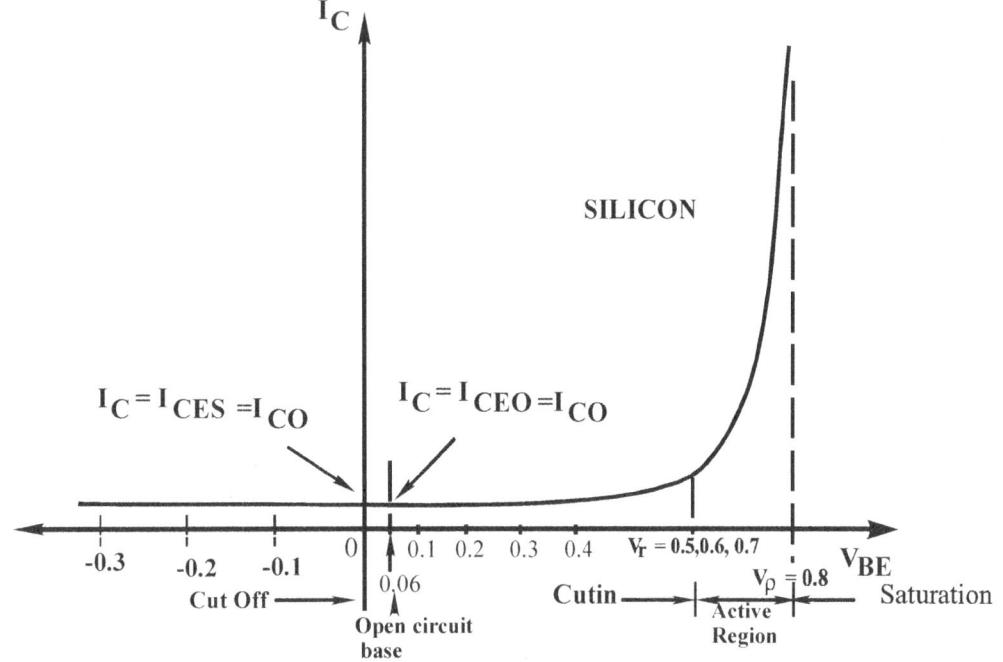

(b) PLOT OF COLLECTOR CURRENT AGAINST BASE TO EMITTER VOLTAGE FOR SILICON

Fig. 4.15.1

2. **SHORT CIRCUITED BASE:** Suppose that instead of reverse-biasing the emitter junction, we connect the base to the emitter so that $V_E = V_{BE} = 0$. As indicated $I_C \cong I_{CES}$ does not increase greatly over the cutoff value I_{CO}.

3. **OPEN CIRCUIT BASE:** Suppose we allow the base to float instead of short circuiting to the emitter ($I_B = 0$), $I_C = I_{CEO}$, At low current, $\alpha = 0.9$ for Ge and 0 for silicon.

4. **CUTIN VOLTAGE:** When the emitter junction is forward biased no appreciable base current flows until the emitter junction has been forward biased to the extent where $|V_{BE}| \geq |Vr|$ where Vr is called CUTIN voltage. Since the collector current is normally proportional to the base current, no appreciable collector current will flow until an appreciable base current flow. A plot of collector current against base to emitter voltage will normally exhibit a CUTIN voltage.

In principle, a transistor is in its active region whenever the base to emitter voltage is on the forward biasing side of the cut off voltage which occurs at a reverse bias voltage of 0.1v for germanium and 0v for silicon. In effect, the transistor enters its active region when $V_{BE} > Vr$. We may estimate the CUTIN voltage Vr by assuming that $V_{BE} = Vr$, when the collector current reaches 1% of the maximum (saturation) current.

Typical values of V_r are 0.1v for germanium and 0.5v for silicon.

5. **SATURATION VOLTAGES:** Manufacturers specify saturation values of input and output voltages in a number of various ways in addition to supplying characteristics curves. For example they may specify R_{CS} for several values of I_B or they may supply curves of $V_{CE, Sat}$ and $V_{BE, Sat}$ as a function of I_B and I_c. The saturation voltages depend not only on the operating point but also on the semiconductor material (germanium or silicon) and on the type of transistor construction.

4.16 SILICON CONTROLLED RECTIFIERS - THYRISTORS

The diode is an automatic unsophisticated switch. It is on whenever the applied voltage provides forward bias. The transistor also can be used as a switch, collector current flows whenever there is adequate base current to turn the transistor on.

Transistors are widely used as switches in digital circuits and other cases where moderate amount of power are to be controlled. Two transistors connected as shown in fig. 4.16.1 can be used for switching.

A major disadvantage; however is that transistors require high and continuous base current in the on state. In contrasts the multi layered semiconductor devices called thyristors have the ability to control large amounts of power with a minimum of control energy. As a result they are used in sophisticated applications to rectify ac to dc, inversion (dc to ac), relaying, timing, ignition and speed control at power levels ranging from a few milliwatts to hundreds of kilowatts.

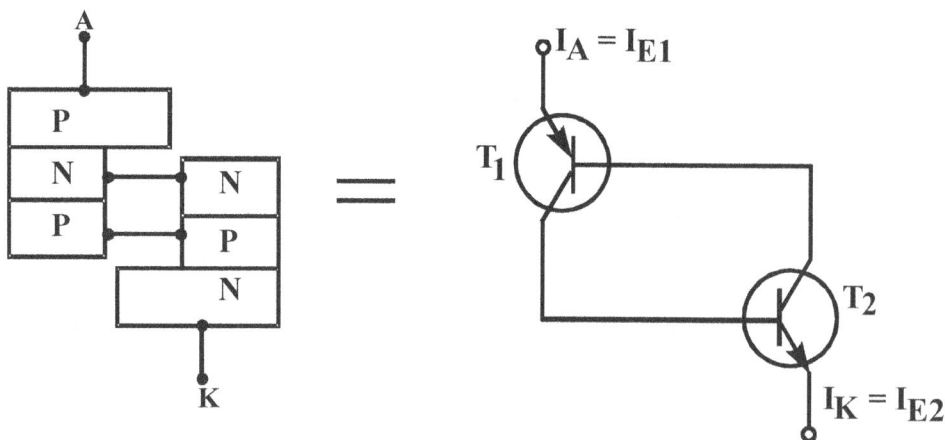

Fig. 4.16.1: Transistor Switch

4.17 THE MULTILAYER DIODE

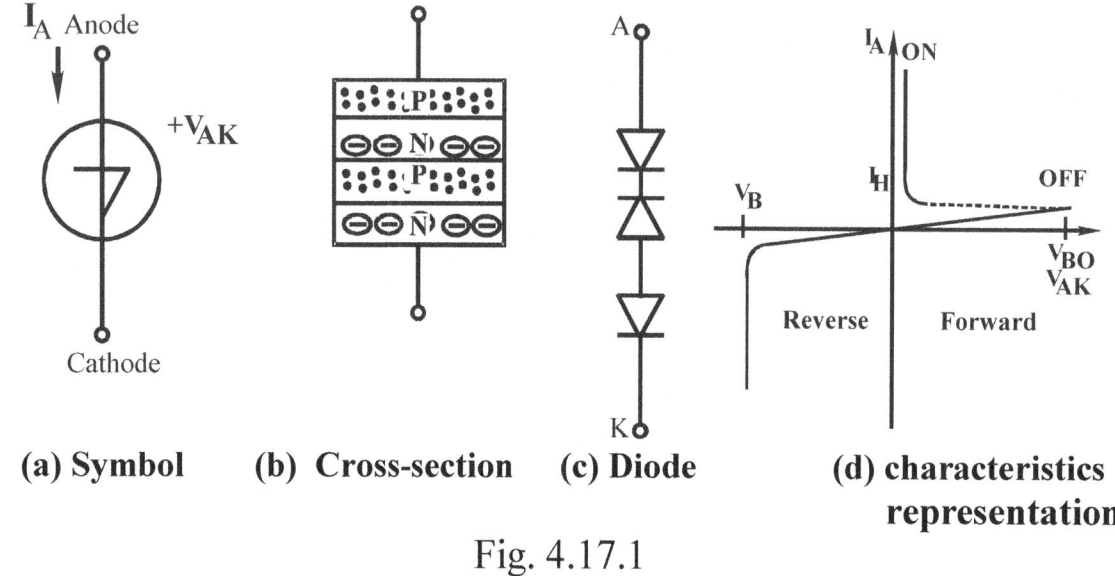

(a) Symbol   (b) Cross-section   (c) Diode   (d) characteristics representation

Fig. 4.17.1

Let us consider the diagrams of fig. 4.17.1.

For silicon at very low emitter current, α is small because of recombination of holes and electrons in the transition region. As I_A increases with increase in V_{AK}, α_1 and α_2 increase. As the quantity $(\alpha_1 + \alpha_2)$ approaches unity, the current tends to increase without limit and breakdown occurs. After switching on, the diode voltage is very small (~ 1v) and the current is limited by the resistance of the external circuit. If the applied voltage is reduced, the switch remains ON until the current has reached minimum holding current I_H. To switch the four layer diode OFF, therefore, the current is reduced below I_H by reducing the applied voltage or increasing the series resistance.

4.18 OPERATION OF SILICON CONTROLLED RECTIFIER

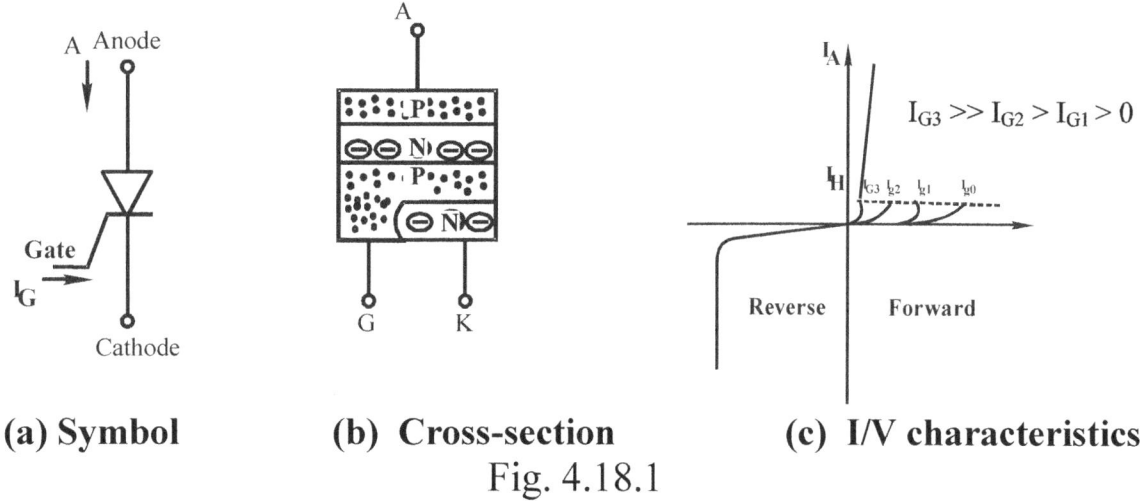

(a) Symbol   (b) Cross-section   (c) I/V characteristics

Fig. 4.18.1

4.19 THE SILICON CONTROLLED RECTIFIER (SCR)

In a PNPN diode represented in fig. 4.18.1 a gate connection to the inner P layer permits the introduction of a gate current that switches the Thyristor to the ON state at rated voltages. Such a silicon controlled rectifier SCR provides flexible switching and since the gate current is required only momentarily, very efficient control of larger amount of power is achieved.

CHAPTER FIVE

5.0 TRANSISTOR BIASING

5.1 INTRODUCTION

The basic function of transistor is to amplify week signals applied to the base of the transistor. The amplified output is obtained in the collector circuit. One important requirement during amplification is that only the magnitude of the signal should increase and there should be no change of signal shape. The increase in the signal magnitude without any change in shape is known as faithful amplification. In order to achieve this means are provided to ensure that input circuit (ie. base-emitter junction for common emitter circuit) of the transistor remain biased and output circuit (i.e. collector-base junction) always remains reverse biased during all parts of the signal. This is known as transistor biasing. We shall be discussing how transistor biasing helps in achieving faithful amplification.

Thus, the process of raising the strength of a weak signal without any change in its general shape is known as faithful amplification.

The theory of transistor reveal that it will function properly if its input circuit remains forward biased and output circuit remains reverse biased at all achieving faithful amplification. To ensure this the following basic conditions must be satisfied.

 i. Proper zero signal collector
 ii. Minimum proper base-emitter voltage V_{BE} at any instant
 iii. Minimum proper collector-emitter voltage V_{CE} at any instant.

TRANSISTOR BIASING

The conditions (i) and (ii) ensures that base-emitter junction shall remain properly forward biased during all parts of the signal. On the other hand condition (iii) ensures that base collector junction shall remain properly reverse-biased at all times. In order words, the fulfilment of these conditions will ensure that transistor works over the archive region of the output characteristics i.e. between saturation to cutoff.

5.2 PROPER ZERO SIGNAL COLLECTOR CURRENT

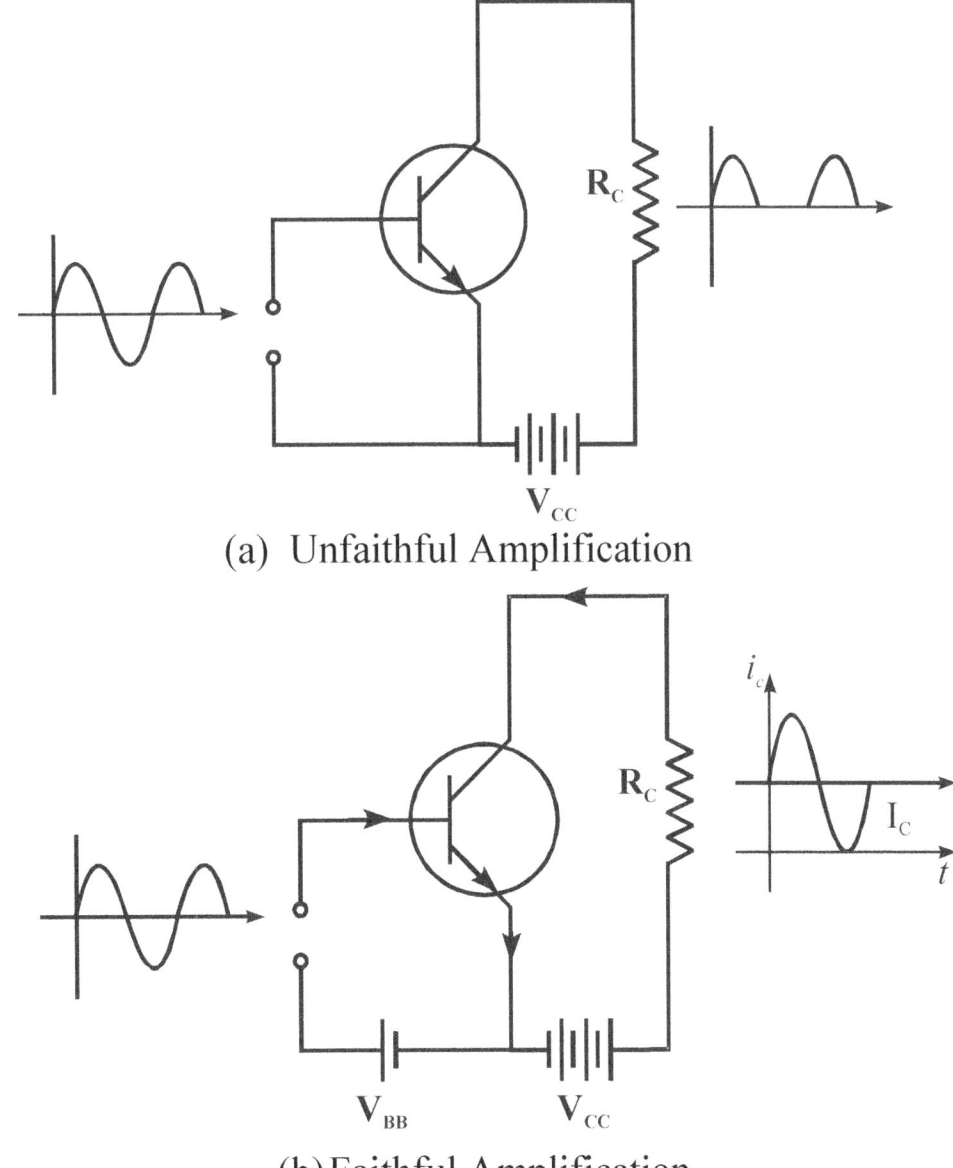

(a) Unfaithful Amplification

(b) Faithful Amplification

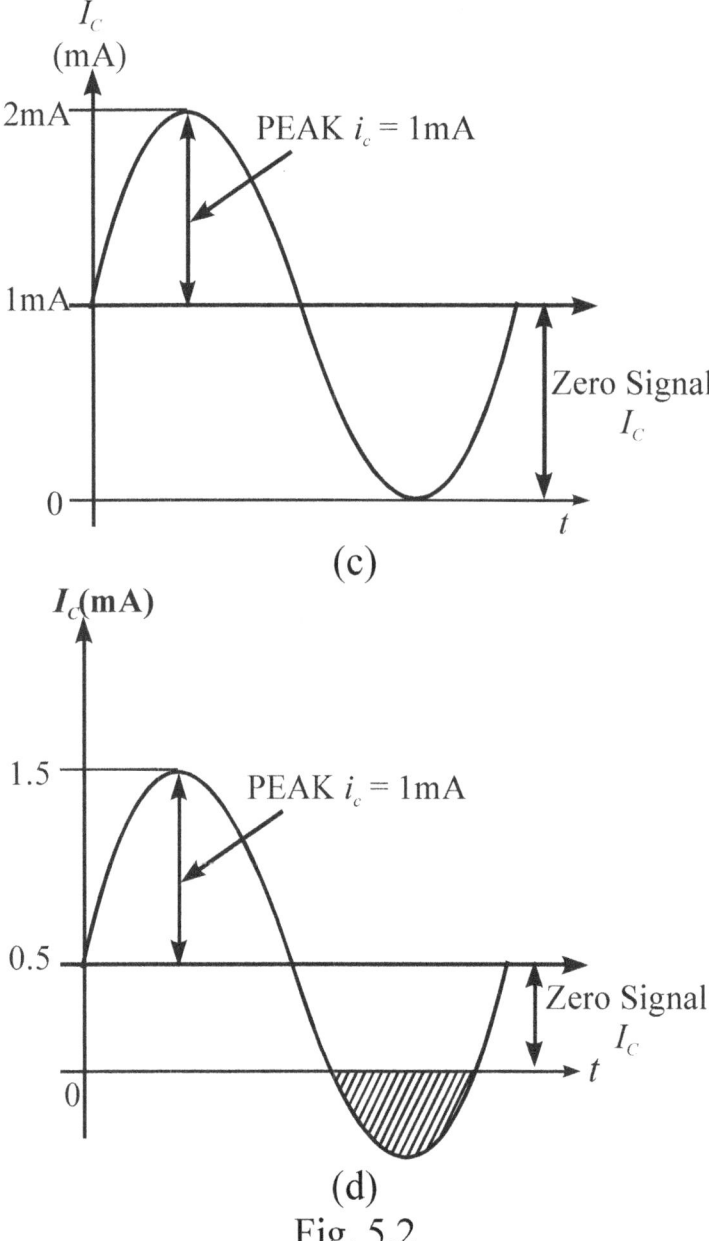

Fig. 5.2

Consider an NPN transistor circuit shown in fig 5.2.1a. During the positive half-cycle of the signal, base is positive.[1] w.r.t emitter and hence base emitter junction is forward biased. This will cause a base current and much larger collector current to flow in the circuit.

The result is that the positive half cycle of the signal is amplified in the collector as shown.

---

[1] w.r.t - with respect to

However, during the negative half-cycle of the signal, base-emitter junction in reverse biased and hence no current flows in the circuit. The result is that this is no output due to the negative half cycle of the signal. This we shall get an amplified output of the signal with its negative half cycles completely cut-off which is unfaithful amplification.

Now if we introduce a battery source V_{BB} in the base circuit as shown in fig. 5.2.1b, the magnitude of this voltage would be such that it keeps the input circuit forward biased even during the peak of the negative half-cycle of the signal. When no voltage. Is applied, a d.c current I_C will flow in the collector circuit due to V_{BB} as shown. This is known as zero signal collector current I_C. During the positive half cycle of the signal input circuit is more forward biased and hence collector current increases. However, during the negative half cycle of the signal, the input circuit is less forward biased and collector current decreases. In this way, negative half cycle of the signal also appears in the output and hence faithful amplification results. It follows therefore that for faithful amplification proper zero signal collector current must flow. The value of zero signal collector current should be at least equal to the maximum collector current due to signal alone ie.

*** Zero signal collector current $\geq$ maximum collectors current due to signal alone. ***

Consider a signal applied to the base of a transistor giving a peak collector current of 1mA. This gives zero signal collector current of 1mA. This zero signal collector current must be at least equal to 1mA so that even during the peak of negative half-cycle of the signal there is no cut off as shown in fig. 5.2.1c. If zero signal collector current is less, say

0.5mA as shown in fig 5.2.1d then some part (shaded portion) of the negative half-cycle of the signal will be cut off in the output

5.3 PROPER MINIMUM BASE-EMITTER VOLTAGE

TRANSISTOR BIASING

In order to achieve faithful amplification the base emitter voltage V_{BE} should not fall below 0.5V for germanium transistors and 0.7v for silicon transistors at any instant. The base current is very small until the input voltage overcomes the potential barrier at the base-emitter junction. The value of these potential barriers is 0.5v for *Ge* and 0.7 for *Si* transistors as shown in fig. 5.3.1

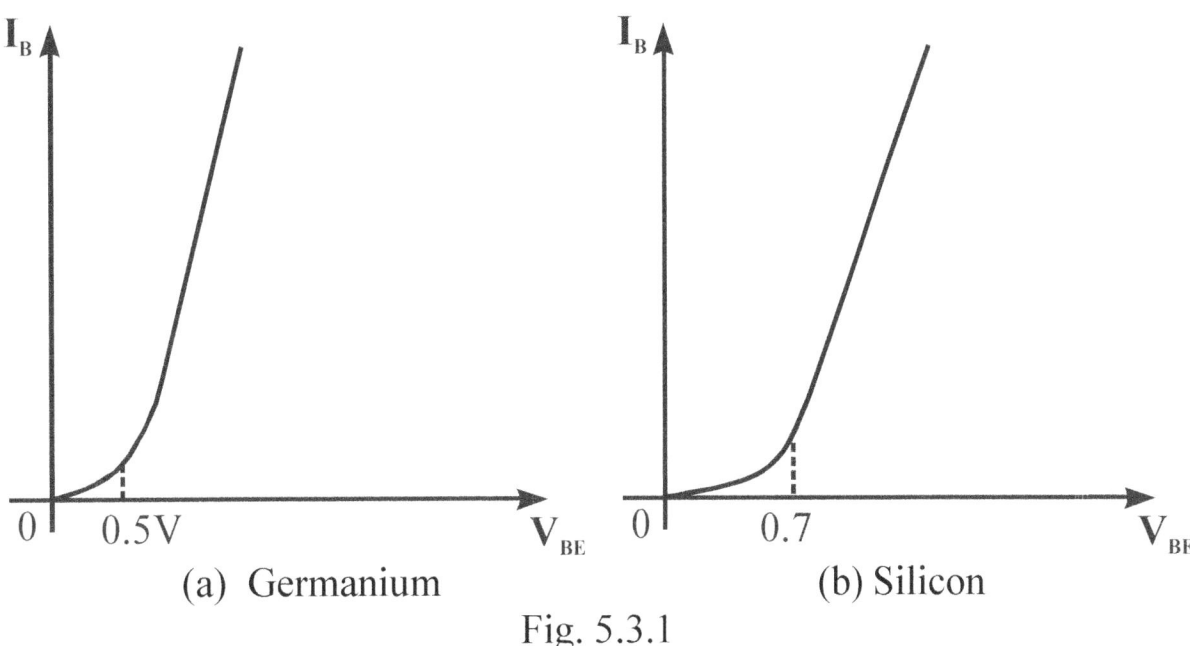

Fig. 5.3.1

Once the potential barrier is overcome, the base current and hence collector current increases sharply. Therefore if base emitter voltage V_{BE} falls below these values during any part of the signal, that part will be amplified to lesser extent due to small collector current. This will result in unfaithful amplification.

5.4 PROPER MINIMUM COLLECTOR-EMITTER VOLTAGE (V_{CE})

For faithful amplification, the collector-emitter voltage V_{CE} should not fall below 0/5v for germanium and 1v for silicon transistors. This is called **knee voltage** (see fig. 5.4.1). When V_{CE} is too low (less than 0.5v for **Ge** and 1v for **Si** transistors) the collector-base junction is not properly reverse biased. Therefore the collector cannot attract the charge carriers emitted by the emitter and hence a greater portion of them goes to the base

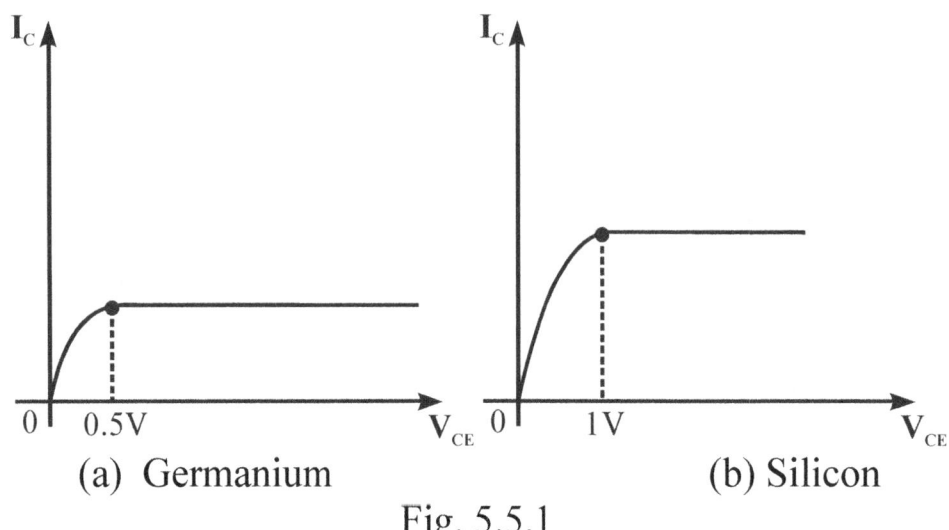

(a) Germanium (b) Silicon

Fig. 5.5.1

This decreases the collector current while the base current increases. Hence value of β falls. Therefore, if V_{CE} is allowed to fall below V_{BASE} during any part of the signal, that past will be less amplified due to reduced β. This will result in unfaithful amplification. However, when V_{CE} is greater than V_{Base}, the collector base junction is properly reserved biased and the value of β remains constant, resulting in faithful amplification.

5.5 TRANSISTOR BIASING

It has already been discussed that for faithful amplification, a transistor amplified must satisfy three basic conditions namely

    i)   Proper zero signal collector current

    ii)   Proper base-emitter voltage at any instant and

iii) Proper collector-emitter voltage at and instant

It is the fulfilment of these conditions which is known as transistor biasing. Thus, the proper flow of zero signal collector current and the maintenance of proper collector-emitter voltage during the passage of signal is known as transistor biasing. The basic purpose of transistor biasing is to keep the base-emitter junction properly forward biased and collector base junction properly reverse biased during the application of signal. This can be achieved with a bias battery or associating a circuit with a transistor. The circuit which provided transistor biasing is known as biasing circuit. It may be noted that transistor biasing is very essential for the proper operation of transistor in any circuit.

EXAMPLE 5.5.1:

An NPN silicon transistor has $V_{CC} = 6V$ and the collector load $R_C = 3k\Omega$. Find;

(i) The maximum collector current that can be allowed during the application of signal for faithful amplification.

(ii) The minimum zero signal collector current required.

Solution:

Collector supply voltage = 7v
Collector load = 3KΩ

(i) For faithful amplification, V_{CE} should not be less than 1V for *Si* transistor

$\therefore$ Max voltage allowed across $R_c = 7 - 1 = 6V$

$\therefore$ Max allowed collector current $= \dfrac{6V}{3K\Omega}$

$= \underline{2mA}$

If the collector current is allowed to rise above this value, V_{CE} will fall below 1V. Consequently value of β will fall, resulting in unfaithful amplification.

(ii) During the negative peak of the signal collector current can at the most be allowed to become zero. As the negative and positive half-cycles of the signal are equal, the change in collector current due to these will also be equal but in opposite direction. Therefore minimum zero signal collector current required $= \dfrac{2mA}{2} = 1mA$.

During the positive peak of the signal (point A in fig. 4.5b),

$i_C = 1 + 1 = 2mA$ and during the negative peak (point B),

$i_C = 1 - 1 = 0mA$

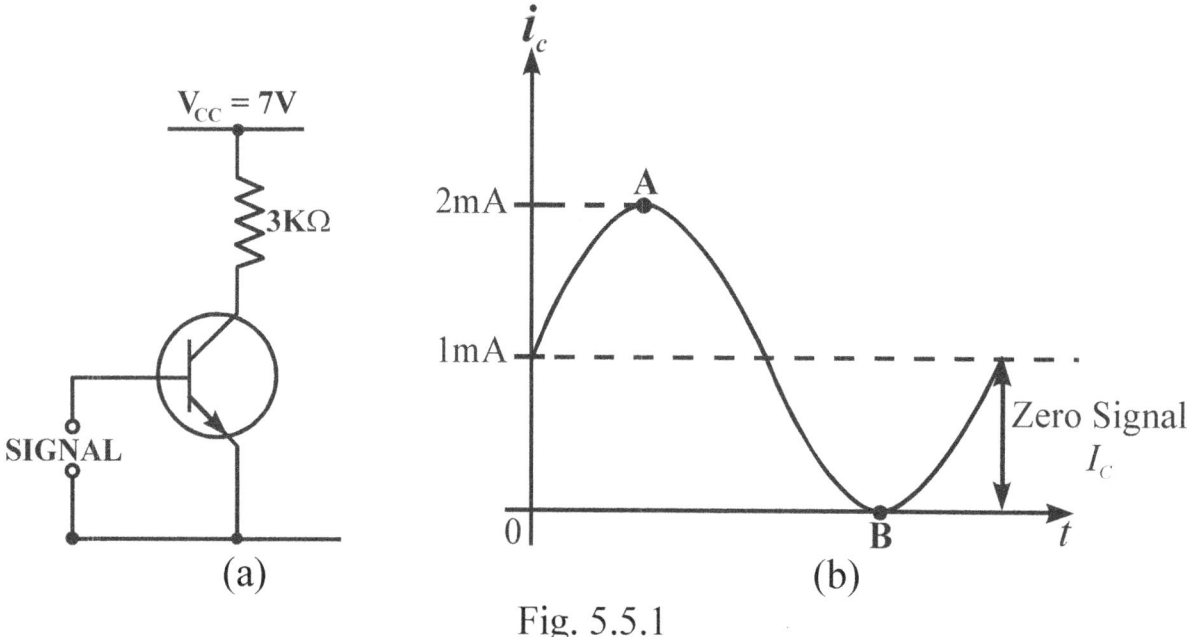

Fig. 5.5.1

Example 5.5.2:

A transistor employs a 4KΩ load and $V_{CC} = 13V$. What is the maximum input signal if β = 100? Given $V_{knee} = 1V$ and a change of 1V in V_{BE} causes a change of 5mA in collector current.

Solution:

Collector supply voltage, $V_{CC} = 13V$
Knee voltage, $V_{knee} = 1V$
Collector load $R_C = 4K\Omega$

$\therefore$ Maximum allowed voltage across $R_C = 13 - 1 = 12V$

$\therefore$ Maximum allowed collector current $= \dfrac{12V}{4K\Omega} = 3mA$

Maximum base current $i_B = \dfrac{i_C}{100} = \dfrac{3mA}{100} = 30\mu A$.

Now $\dfrac{Collector\ current}{Base\ Voltage\ (Signal\ Voltage)} = 5mA/V$

$\therefore$ Base voltage (signal voltage) $= \dfrac{Collector\ current}{5mA/V}$

$$= \dfrac{3mA}{5mA}$$
$$= \underline{600mV}$$

5.6 INHERENT VARIATION OF TRANSISTOR PARAMETERS

In practice, the transistor parameters such as β, V_{BE} are not the same for every transistor even if the same type. For example, BC147 is a silicon NPN transistor with β varying from 100 to 600 i.e. β for one transistor may be 100 and for the other it may be 600, although both of them are BC147. This large variation in parameters is a characteristic of transistor.

The major reason for these variations is that transistor is a new device and manufacturing techniques have not much advanced. For instance, it has not been possible to control the base width and it may vary, although slightly from one transistor to the other even of the same type. Such small variations results in large change in transistor parameters such as β, V_{BE} etc.

The inherent variations of transistor parameters may change the operating point, resulting in unfaithful amplification. It is, therefore very important that biasing network be so designed that it should be able to work with all transistors of one type whatever may be the spread in β or V_{BE}. In other words, the operating point should be independent of transistor parameters variation.

5.7 STABILISATION

The collector current in a transistor changes rapidly when

(i) The temperature changes

(ii) The transistor is replaced by another of the same type. This is due to the inherent variation of transistor parameter already discussed.

When the temperature changes or the transistor is replaced, the operating point (i.e. zero signals I_c and V_{CE}) also changes. However, for faithful amplification, it is essential that operating point remains fixed. This necessitates the making of the operating point independent of there variations. This is known as stabilisation. Therefore, the process of making operating point of transistor independent of temperature changes or variation in transistor parameter is known as stabilisation.

Stabilisation makes it possible for the zero signals I_C and V_{CE} to become independent of temperature variations or replacement of transistor i.e. the operating point is fixed.

A good biasing circuit always ensures the stabilisation of the operating point. Stabilisation of the operating point is necessary due to the following reasons:

TRANSISTOR BIASING

    (i)   Temperature dependence of I_C.

    (ii)   Individual variations

    (iii)   Thermal runaway

i. **Temperature Dependence of I_C:** The collector current I_C for CE circuit is given by;

$$I_C = \beta I_B + I_{CEO} = \beta I_B + (\beta + 1) I_{CBO}$$

The collector leakage current I_{CBO} is greatly influenced (especially in germanium transistor) by temperature changes. A rise of 100C doubles the collector leakage current, which may be as high as 0.2mA for low powered germanium transistors. As biasing conditions in such transistors are generally so set that zero signal I_C = 1mA, therefore, the change in I_C due to temperature variations cannot be tolerated. This necessitates the stabilisation of the operating point i.e. to hold I_C constant in spite of temperature changes.

ii. **Individual Variations:** The value of β and V_{BE} are not exactly the same for any two transistors even of the same type. Further, V_{BE} itself decreases when temperature increases. When a transistor is replaced by another of the same type, these variations change the operating point. This necessitates the stabilisation of the operating point i.e. to hold Ic constant irrespective of individual variations in transistor parameters.

iii. **Thermal Runway:** The collector current for a CE configuration is given by;

$$Ic = \beta I_B + (\beta + 1) I_{CBO} \quad \text{------------------------------(A)}$$

120

The collector leakage current I_{CBO} is strongly dependent on temperature. The flow of collector current produces heat within the transistor. This raises the transistor temperature and if no stabilization is done, the collector leakage current I_{CBO} also increases. It is clear from expression A that if I_{CBO} increases, the collector current Ic increases by $(\beta + 1) I_{CBO}$. The increased Ic will raise the temperature of the transistor, which in turn will cause I_{CBO} to increase. This effect is cumulative and in a matter of seconds, the collector current may become very large, causing the transistor to burn out. The self-distruction of an unstabilised transistor is known as **Thermal Runaway**.

In order to avoid thermal runaway and consequent destruction of transistor, it is essential that operating point is stabilised i.e. Ic is kept constant. In practice, this is done by causing I_B to decrease automatically with temperature increase by circuit modification. Then decrease in βI_B will compensate for the increase in $(\beta + 1) I_{CBO}$, keeping Ic nearly constant. In fact, this is what is always aimed at while building and designing a biasing circuit.

5.7 ESSENTIALS OF A TRANSISTOR BIASING CIRCUIT

It has been discussed that transistor biasing is required for faithful amplification. The biasing network associated with the transistor should meet the following requirements.

i. It should ensure proper zero signal collectors current.

ii. It should ensure that V_{CE} does not fall below 0.5V for *Ge* transistors and 1V for *Si* transistors at any instant.

iii. It should ensure the stabilization of operating point.

5.8 STABILITY FACTOR

It is desirable and necessary to keep collector current constant in the face of variations of I_{CBO} (sometimes represented as I_{CO}). The extent to which a biasing circuit is successful in achieving this goal is measured by stability factor *S*. It is defined as the rate of change of collector current Ic with respect to the collector leakage current I_{CO} at constant β and I_B.

i.e. Stability factor, $S = \dfrac{dIc}{dIco}$ at constant I_β and β.

The stability factor indicates the change in collector current Ic due to the change in collector leakage current I_{CO}. Thus a stability factor, 50 of a circuit means that Ic changes 50 times as much as any change in Ico.

The ideal value of *S* is 1, but is *S* never possible to achieve in practice. Experiments shows that values of *S* exceeding 25 results in unsatisfactory performance. The general expression of stability factor for a common emitter configuration can be obtained as follows;

$Ic = \beta I_B + (\beta + 1) I_{CO}$

Differentiating the above expression with respect to Ic, we have

$$1 = \beta \dfrac{dI_B}{dI_c} + (\beta+1)\dfrac{dI_{CO}}{dI_C}$$

$$= \beta \dfrac{dI_B}{dI_c} + \dfrac{(\beta+1)}{S} \quad \left[\text{Note } \dfrac{dI_{CO}}{dI_C} = \dfrac{1}{S}\right]$$

$$S = \frac{\beta + 1}{1 - \beta \frac{(dI_B)}{(dI_C)}}$$

5.9 METHODS OF TRANSISTOR BIASING

The most commonly used methods of obtaining transistor biasing are;

i. Base transistor method
ii. Biasing with feedback
iii. Voltage-divider bias

In all these methods, the same basic principle is employed i.e. required value of base current (and hence Ic) is obtained from Vcc in the zero signal conditions. The value of collector load Rc is selected to make V_{CE} not fall below 0.5V for germanium and 1V fro silicon transistors. For example if β = 100 and the zero signal collector current Ic is to be at 1mA, the I_B is made equal to $I_C/\beta = 1/1000 = 10\mu A$. Thus, the biasing network should be so designed that a base current of 10μA flows in the zero signal conditions.

5.10 BASE RESISTOR METHOD

Here a high resistance R_B (several hundred kilo ohm) is connected between the base and positive end of supply for NPN transistor (See fig 5.10.1) and between base and negative end of supply for PNP transistor. The required zero signal base current is provided by Vcc and it flows through R_B. The required value of zero signal base current I_B (and hence (Ic = βI_B) can be made to flow by selecting the proper value of base resistor R_B. It is necessary to find the value of R_B so that required collector current flows in the zero signal conditions. If Ic is the required zero signal collector current,

$$I_B = \frac{I_C}{\beta}$$

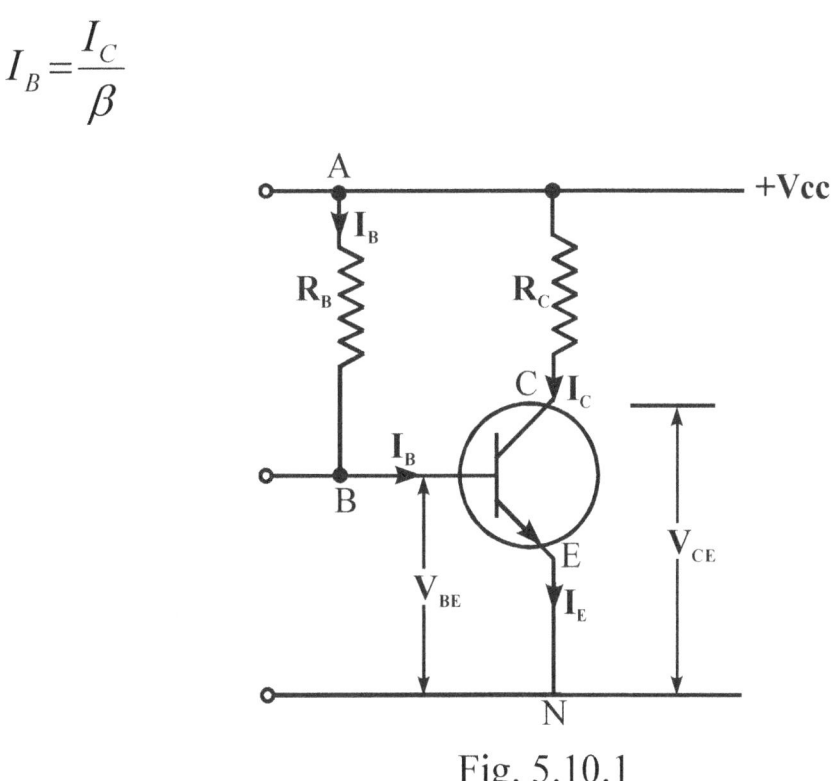

Fig. 5.10.1

Considering the closed circuit ABENA and applying Kirchoff's voltage law, we get

$V_{CC} = I_B R_B + V_{BE}$

$I_B R_B = V_{CC} - V_{BE}$

$$R_B = \frac{V_{CC} - V_{BE}}{I_B}$$

As Vcc and I_B are known and V_{BE} can be got from the transistor manual, the value of R_B can be readily from the above expression. Since V_{BE}, is generally very small compared to Vcc, the former can be neglected with little error, so that, $$R_B = \frac{V_{CC}}{I_B}$$

It may be noted that Vcc is a fixed known quantity and I_B is chosen at some suitable value. Hence R_B can always be found directly and for this reason, this method is sometimes called fixed biased method already discussed in section 3.11.

Stability Factor: In fixed-bias method of biasing, I_B is independent of Ic, so that $\frac{dI_B}{dI_c} = 0$, so that stability factor,

$$S = \frac{\beta + 1}{1 - \beta \frac{(dI_B)}{(dI_C)}} = \beta + 1$$

This means that Ic changed (β + 1) times as much as any change in I_{CO}. for instant, if β = 100, then S = 101 which means that Ic increases 101 times faster than I_{CO}. Due to the large value of S in a fixed bias, it has poor thermal stability.

Advantages;

i. It is very simple as only one resistor R_B is required.

ii. Biasing conditions can easily be set and the calculations are simple.

iii. There is no loading of the source by the biasing circuit since no resistor is employed across base-emitter junction.

Disadvantages:

i. It gives poor stability

ii. The stability factor is very high. Therefore, there are strong chances of thermal runaway.

Due to these disadvantaged, this method is rarely employed.

Example 5.10.1:

Fig. 5.10.1a shows biasing with base resistor method.

(i) Determine the collector current Ic and collector-emitter voltage V_{CB}. Neglect small base-emitter voltage. Given that $\beta = 50$.

(ii) If R_B in this circuit s changed to 50KΩ, find the new operating point.

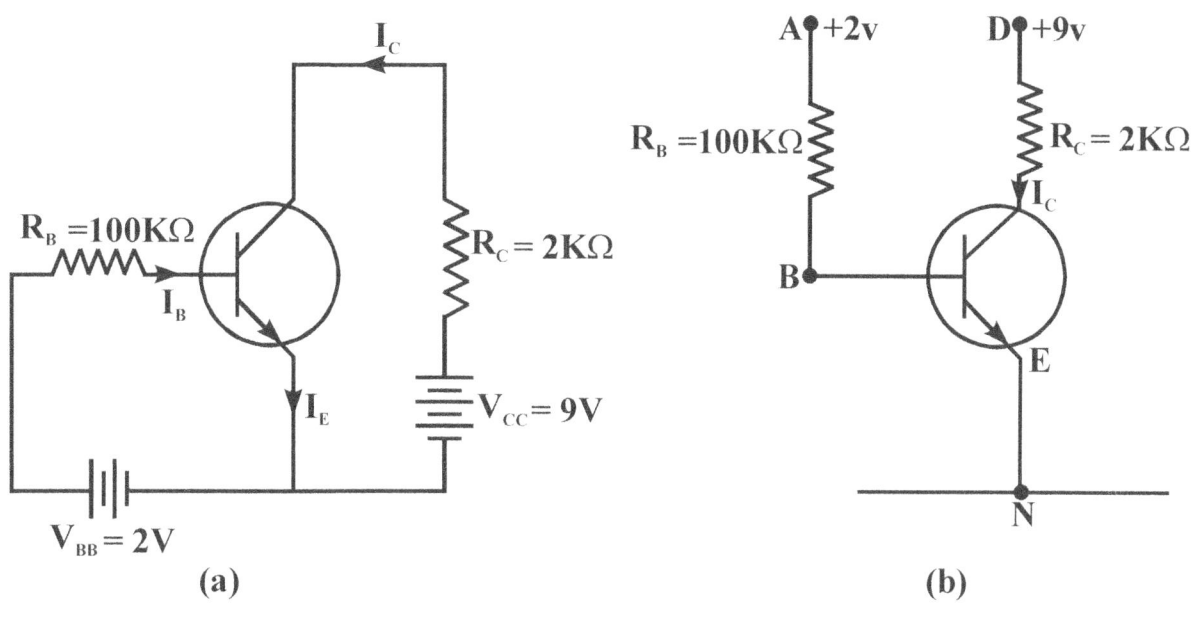

Fig. 5.10.1

Solution:

In fig. 5.10.1a biasing is provided by a battery $V_{BB} = 2V$ in the base circuit which is separated from the battery $V_{CC} = 9V$ used in the output circuit. The same circuit is shown in a simplified way in fig. 5.10.1b. here, we need to show only the supply voltages +2V and 9V. the negative terminals of the power supply are normally grounded to get a complete path of current.

i. Applying Kirchoff's law to circuit ABEN, we have,

$$I_B R_B + V_{BE} = 2V$$

As V_{BE} is negligible,

$$I_B = \frac{2V}{R_B} = \frac{2V}{100K\Omega} = 20\mu A$$

Collector current, Ic = βI_B = 50 × 20μA

$$= 1,000\mu A = 1mA.$$

Applying Kirchoff's voltage law to the circuit DEN, we have

$I_c R_c + V_{CE} = 9V$

$1mA \times 2k\Omega + V_{CE} = 9V$

∴ $\qquad V_{CE} = 9 - 2 = 7V$

(ii) When $R_B = 50k\Omega$, it is easy to see that the current is double

i.e. $I_B = 40\mu A$

Collector current = βI_B = 50 × 40
$$= 2,000\mu A = \underline{2mA}$$
Collector-emitter voltage, $V_{CE} = V_{cc} - I_c R_c$
$$= 9 - 2mA \times 2k\Omega$$
$$= \underline{5V}$$

∴ New operating point is 5V, 2mA.

Example 5.10.2

Fig 5.10.2a shows that a silicon transistor with β = 100 is biased by base resistor method. Draw the d.c load line and determine the operating point. What is the stability factor?

TRANSISTOR BIASING

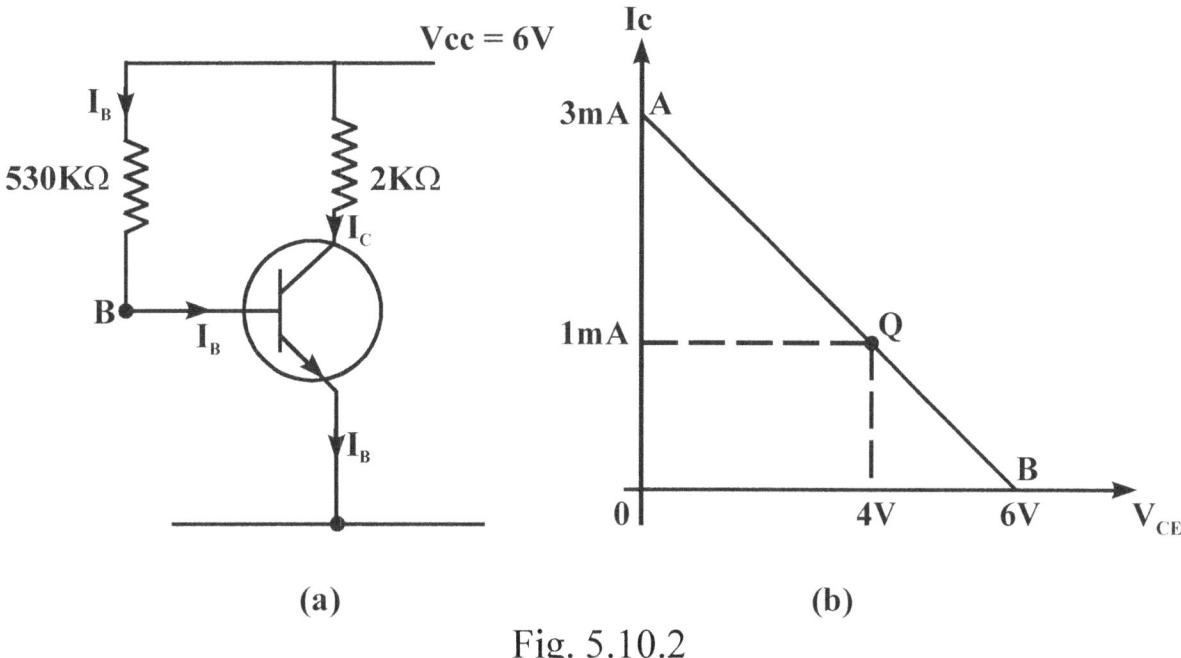

Fig. 5.10.2

Solution:

Vcc = 6V, R_B = 530KΩ, Rc = 2kΩ.

Referring to fig. 5.10.2a, the d.c load line is given by the expression,

$V_{CE} = Vcc - I_c R_c$

When Ic = 0, V_{CE} = Vcc = 6V. This locates the furst point OB = 6V of the loadline on the collector-emitter voltage axis as shown in fig 5.10.2b.

When V_{CE} = 0, Ic = Vcc/Rc = 6V/2KΩ = 3mA.

This located the second point A = 3mA of the load line on the collector current axis. By joining point A and B, d.c loadline AB is constructed (see fig. 5.10.2b).

Since it is silicon transistor, V_{BE} = 0.7V

Referring to fig 5.10.2a, it is clear that

$I_B R_B + V_{BE} = Vcc$

$$I_B = \frac{V_{CC} - V_{BE}}{R_B} = \frac{(6-0.7)V}{530 K\Omega} = \underline{10 \mu A.}$$

∴ Collector current Ic = βI_B = 100 × 10 = 1,000μA

Collector emitter voltage, $V_{CE} = V_{cc} - I_c R_c$

$$= 6 - 1mA \times 2k\Omega$$
$$= 6 - 2 = 4V$$

∴ Operating point is 4V, 1mA.

Fig. 5.10.2b shows the operating point Q on the d.c load line. Its coordinates are $I_c = 1mA$ and $V_{CE} = 4V$.

Stability factor $= \beta + 1 = 100 + 1 = 101$

Example 5.10.3:

i. A germanium transistor is to be operated at zero signal $I_c = 1mA$. If the collector supply $V_{cc} = 12V$, what is the value of R_B in the base resistor method? Take $\beta = 100$.

ii. If another transistor of the same batch with $\beta = 50$, is used, what will be the new value of zero signal I_c for the same R_B?

Solution:

$V_{cc} = 12V, \beta = 100$

Since it is *Ge*, $V_{BE} = 0.3V$

i. Zero signal $I_c = 1mA$

Zero signal $I_B = I_c / \beta = \dfrac{1mA}{100} = 0.01mA$

Using the relationship, $V_{cc} = I_B R_B + V_{BE}$

$$R_B = \dfrac{V_{CC} - V_{BE}}{I_B} = \dfrac{12 - 0.3}{0.01mA} = \dfrac{11.7V}{0.01mA} = 1170 K\Omega$$

iii. Now $\beta = 50$

$V_{cc} = I_B R_B + V_{BE}$

$$I_B = \frac{V_{CC} - V_{BE}}{R_B} = \frac{12 - 0.3}{1170 K\Omega} = 11.7V / 1170 K\Omega = \underline{0.01 mA.}$$

Zero signal $I_c = \beta I_B = 50 \times 0.01$
$$= \underline{0.5 mA}$$

It is clear from the above example that with the change in transistor parameter β, the zero signal collector current has changed from 1mA to 0.5mA. Therefore, base resistor method cannot provide stabilization;

Example 5.10.4

Calculate the values of three currents in the circuit shown in fig. 5.10.4.

Solution:

Applying Kirchoff's voltage law to the base side and taking resistance in KΩ and current in mA, we have,

$$V_{cc} = I_B R_B + V_{BE} + I_E \times 1$$

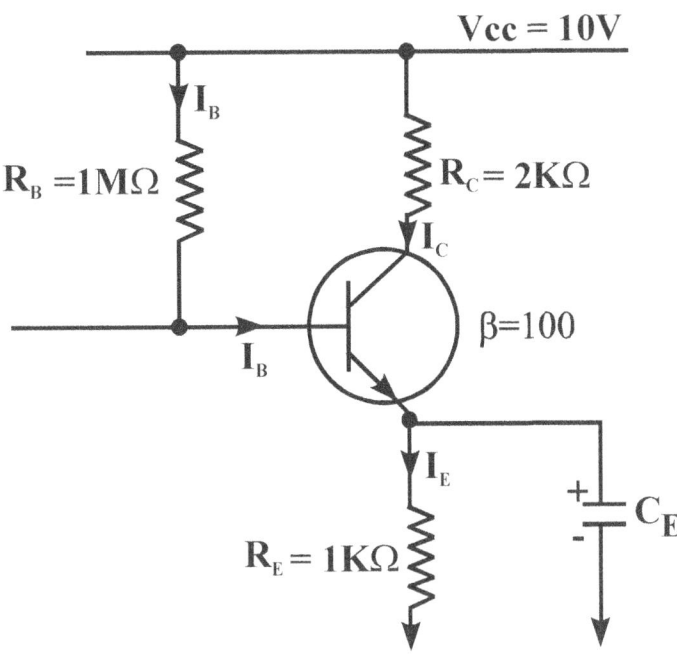

Fig. 5.10.4

TRANSISTOR BIASING

The value of V_{BE} is very small and can be neglected.

$$10 = 1000 I_B + I_E; \quad (I_E = I_C + I_B)$$
$$10 = 100 I_B + (\beta I_B + I_B)$$
$$10 = 100 I_B + (100 I_B + I_B)$$
$$10 = 1101 I_B$$

$$\therefore I_B = \frac{10}{1101} = \underline{0.0091 mA}$$

$$I_C = \beta I_B = 100 \times 0.0091 = \underline{0.91 mA}$$

$$I_E = I_C + I_B = 0.91 + 0.0091 = \underline{0.919 mA}$$

Example 5.10.5

Design base resistor bias circuit for a CE amplifier such that the operating point is $V_{CE} = 8V$ and $I_C = 2mA$. You are supplied with a fixed 15V d.c supply and silicon transistor with $\beta = 100$. Take base-emitter voltage $V_{BE} = 0.6$. Calculate also the value of load resistor that would be employed.

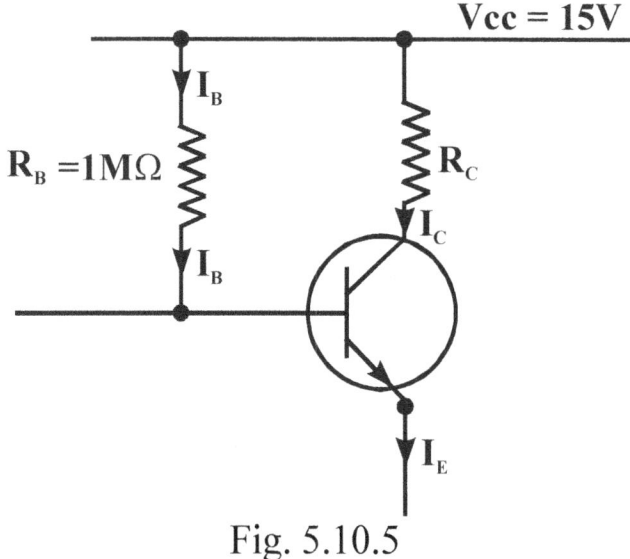

Fig. 5.10.5

Solution:

Fig. 5.10.5 shows CE amplifier using base resistor method of biasing.

$V_{CC} = 15V$; $\beta = 100$; $V_{BE} = 0.6$
$V_{CE} = 8V$; $I_c = 2mA$; $R_C = ?$; $R_B = ?$

But $V_{cc} = V_{CE} + I_C R_C$

$15V = 8V + 2mA \times R_C$

$$Rc = \frac{(15-8)V}{2mA} = \underline{3.5 K\Omega}$$

$V_{cc} = I_B R_B + V_{BE}$ and $I_B = \dfrac{I_C}{\beta} = \dfrac{2}{100} = 0.02 mA$

$$Rc = \frac{V_{CC} - V_{BE}}{I_B} = \frac{(15-0.6)V}{0.02 mA} = \underline{720 K\Omega}$$

5.11 BIASING WITH FEEDBACK RESISTOR:

In this method of biasing, one end of R_B is connected to the base and the other end to the collector shown in fig 5.11.1. Here, the required zero signal base current is determined not by Vcc but by the collector-base voltage, V_{CB}. It is clear that V_{CB} forward biases the base-emitter junction and hence base current I_B flows through R_B. This causes the zero signal collector current to flow in the circuit.

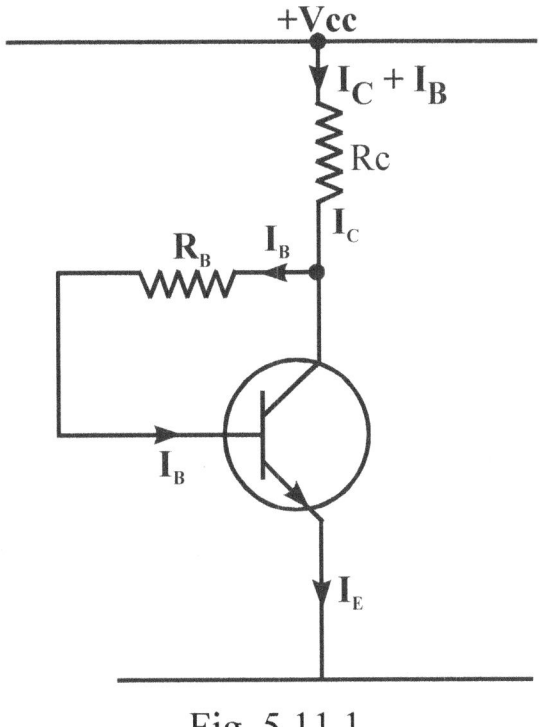

Fig. 5.11.1

TRANSISTOR BIASING

The required value of R_B needed to give the zero signal current I_C can be determined as follows;

$$V_{cc} = (I_c + I_B)R_C + I_B R_B + V_{BE} \quad (\text{But } I_E \ll I_C)$$

Therefore, $V_{cc} = I_C R_c + I_B R_B + V_{BE}$

$$R_B = \frac{V_{CC} - V_{BE} - I_C R_C}{I_B} = \frac{V_{CC} - V_{BE} - \beta I_B R_C}{I_B}$$

Note that $I_C = \beta I_B$

Alternately $\quad V_{CC} = V_{BE} + V_{CE}$

$$V_{CB} = V_{CE} - V_{BE}$$

$$R_B = \frac{V_{CE}}{I_B} = \frac{V_{CE} - V_{BE}}{I_B} \quad \text{where } I_B = \frac{I_C}{\beta}$$

It can be shown mathematically that stability factor **S** for this method of biasing is less than $(\beta + 1)$ ie.

Stability Factor $S < (\beta + 1)$

Thus, this method provide better thermal stability than the fixed bias.

Advantages:

ii. It is a simple method as it requires only one resistance, R_B.

iii. This circuit provided some stabilization of the operating point as shown below.

$$V_{CE} = V_{BE} + V_{CB}.$$

If the temperature increases, the collector leakage current will increase and hence the total collector current. But as soon as collector increases, V_{CE} decreases i.e. lesser voltage available across R_B. Hence the base current I_B decreases. The smaller I_B tends to decrease the collector current to original value.

TRANSISTOR BIASING

Disadvantages:

i. The circuit does not provide good stabilisation because stability factor fairly highs, though it is leaser than that of fixed bias. Therefore the operation point does change, although to lesser extent due to temperature variations and other effects.

ii. The circuit provides a negative feedback which reduces the gain of the amplifier. During the positive half-cycle of the signal, the collector current increases. The increased collector current would result in greater voltage drop across Rc. This will reduce the base current and hence collector current.

Example 5.11.1:

Fig. 5.11.2 shows a silicon transistor biased by feedback resistor method. Determine the operating point given that $\beta = 100$.

Solution:

Vcc = 20V, R_B = 100KΩ, Rc = 1KΩ

Since it is silicon, V_{BE} = 0.7V

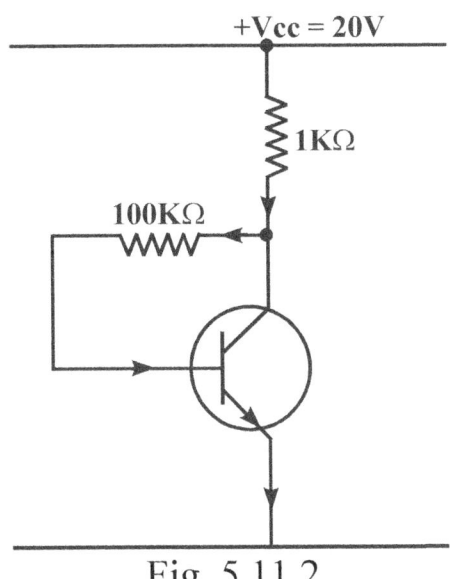

Fig. 5.11.2

Assuming I_B to be in mA and using the relationship,

TRANSISTOR BIASING

$$R_B = \frac{V_{CC} - V_{BE} - \beta I_B R_C}{I_B}$$

$$100 I_B = 20 - 0.7 - 100 \times I_B \times 1$$

$$200 I_B = 19.3$$

$$I_B = \frac{19.3}{200} = \underline{0.096 mA}$$

∴ Collector current, $I_C = \beta I_B = 100 \times 0.096 = 9.6 mA$.

Collector emitter voltage,

$$V_{CE} = V_{CC} - I_C R_C$$
$$= 20 - 9.6 mA \times 1 K\Omega.$$
$$= 10.4 V$$

∴ Operating point is 10.4V, 9.6mA

Example 5.11.2

i. It is required to set the point by biasing a transistor with feedback resistor at $I_C = 1mA$, $V_{CE} = 8V$. If $\beta = 100$, $V_{CC} = 12V$, $V_{BE} = 0.3V$, how will it be done?

ii. What will be the new operating point if $\beta = 50$, all other circuit values remaining the same.

Solution:

Vcc = 12V, $V_{CE} = 8V$, $I_C = 1mA$.

$\beta = 100$, $V_{BE} = 0.3V$

(i) To obtain the required operating point, we shall find the value of R_B (Please refer to fig 5.11.3).

TRANSISTOR BIASING

Now, collector load is

$$R_C = \frac{V_{CC} - V_{CE}}{I_C} = \frac{(12-8)V}{1mA} = 4K\Omega$$

Also $I_B = \frac{I_C}{\beta} = \frac{1mA}{100} = 0.01 mA$

Using the relationship, $R_B = \frac{V_{CC} - V_{BE} - \beta I_B R_C}{I_B}$

$$= \frac{12 - 0.3 - 100 \times 0.01 \times 4}{0.01}$$

$$= \underline{770 K\Omega}$$

(iii) If β = 50 and other circuit values remain the same,

$V_{cc} = V_{BE} + I_B R_B + \beta I_B R_C$

$12 = 0.3 + I_B (R_B + \beta R_C)$

$11.7 = I_B(770 + 50 \times 4)$

$$I_B = \frac{11.7V}{970K\Omega} = 0.012 mA$$

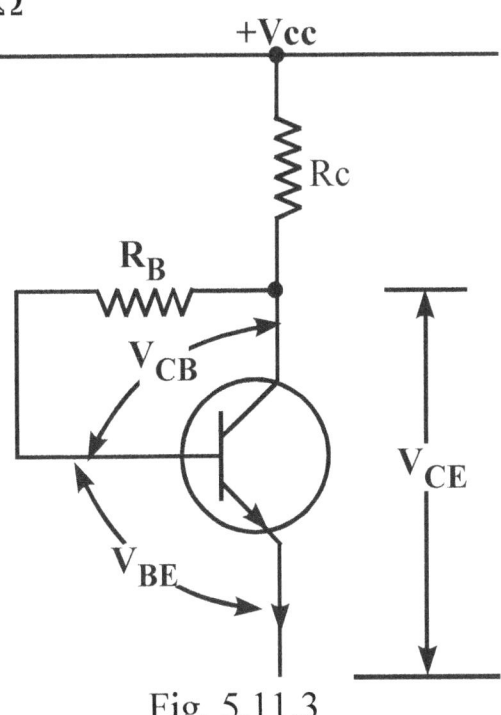

Fig. 5.11.3

TRANSISTOR BIASING

Therefore, collector current, $I_C = \beta I_B = 50 \times 0.012$

$= 0.6 \text{Ma}.$

Collector emitter voltage, $V_{CE} = V_{CC} - I_C R_C.$

$= 12 - 0.6 \times 4K\Omega$

$= 9.6V$

Now the new operating point is 9.6V, 0.6mA. It is necessary to note the change in operating point when a new resistor with lesser β is used. Therefore biasing with feedback resistor does not provide very good stabilisation.

Example 5.11.3

It is desired to set the operating point at 2V, 1mA by biasing a silicon transistor with feedback resistor R_B. If $\beta = 100$, find the value of R_B.

Solution:

For a silicon transistor, $V_{BE} = 0.7V$

$$I_B = \frac{I_C}{\beta} = \frac{1mA}{100} = 0.01mA$$

now, $V_{CE} = V_{BE} + V_{CB}$
$\quad\quad 2 \;\; = 0.7 + V_{CB}$
$\quad V_{CB} = 2 - 0.7 = 1.3V$

Therefore, $R_B = \dfrac{V_{CB}}{I_B} = \dfrac{1.3V}{0.01mA} = \underline{130K\Omega}$

5.12 VOLTAGE DIVIDER BIAS METHOD

This happens to be the most widely used method of providing biasing and stabilisation to a transistor. Here two resistors R_1 and R_2 are connected across the supply voltage, V_{CC} (see fig. 5.12.1) and provide biasing. The emitter resistance R_E provides stabilisation. The name *'voltage divider'* comes from the voltage divider formed by R_1 and R_2.

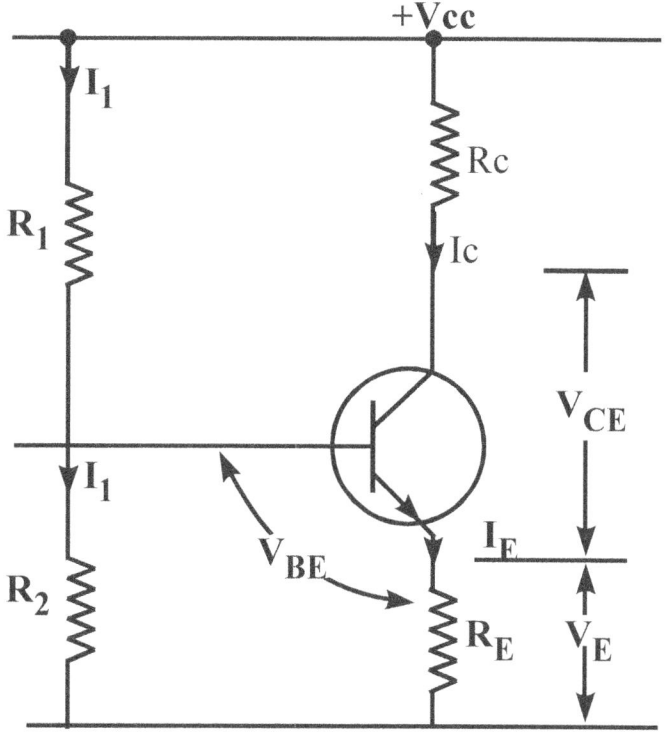

Fig. 5.12.1

The voltage drop across R_2 forward biases the base-emitter junction. This causes the base current and hence collector current flows in the zero signal conditions.

Assuming that the current flowing through resistor R_1 is I_1. As base current I_B is very small therefore it can be presumed that the current flowing through R_2 is also I_1

i. Collector current, I_C

   Collector current can be evaluated as follows:

$$I_1 = \frac{V_{CC}}{R_1 + R_2}$$

∴   Voltage across resistance R_2 is

TRANSISTOR BIASING

$$V_2 = \left(\frac{V_{CC}}{R_1 + R_2}\right) R_2$$

Applying Kirchoff's voltage law to the base circuit of fig 5.12.1,

$$V_2 = V_{BE} + V_E$$
$$V_2 = V_{BE} + I_E R_E$$

Therefore, $I_E = \dfrac{V_2 - V_{BE}}{R_E}$

Since $I_E = I_C$, $I_C = \dfrac{V_2 - V_{BE}}{R_E}$

It is clear from the last expression for I_C that it does not at all depend on β. Although I_c depends upon V_{BE} but in practice $V_2 \gg V_{BE}$, so that I_C is practically independent of V_{BE}. Thus I_C in this circuit is almost independent of transistor parameters and hence good stabilisation is ensured. It is due to this reason that potential divider bias has become universal method for providing transistor biasing.

(ii) Collector-emitter voltage, V_{CE}.

Applying Kirchoff's voltage law to the collector side,

$$V_{CC} = I_C R_C + V_{CE} + I_E R_E$$
$$= I_C R_C + V_{CE} + I_C R_E \qquad (I_E \cong I_C)$$
$$= I_C (R_C + R_E) + V_{CE}$$
$$V_{CE} = V_{CC} - I_C (R_C + R_E)$$

Stabilisation: In this circuit, excellent stabilisation is provided by R_E. It has been already shown that;

$$V_2 = V_{BE} + I_C R_E$$

Now suppose the collector current I_C increases due to use in temperature. This will cause the voltage drop across emitter resistance R_E to increase. As voltage drop across R_2 (i.e. V_2) is independent of I_C, therefore, V_{BE} decreases. This in turn causes I_B to decrease. The reduced value of I_B tends to restore I_C to the original value.

Stability Factor: It can be shown mathematically that stability factor of the circuit is given by;

$$S = \frac{(\beta+1)(R_T+R_E)}{R_T+R_E+\beta R_E}$$

$$= (\beta+1) \times \frac{1+\frac{R_T}{R_E}}{\beta+1+\frac{R_T}{R_E}} \quad \text{where} \quad R_T = \frac{R_1 R_2}{R_1+R_2}.$$

If the ratio $\frac{R_T}{R_E}$ is very small, it can be neglected and stability factor becomes,

$$S = (\beta+1) \times \frac{1}{\beta+1} = 1$$

This is the smallest possible values of **S** and leads to the maximum possible thermal stability. In actual practice, the circuit may have stability factor around 10.

Example 5.12.1:

Fig. 5.12.1a shows the voltage divider bias method. Draw the d.c loadline and determine the operating point. Assume the transistor to be of silicon.

Solution:

1. **D.C loadline**

$$V_{CE} = V_{CC} - I_C(R_C + R_E)$$

When $I_C = 0$, $V_{CE} = V_{CC} = 15V$

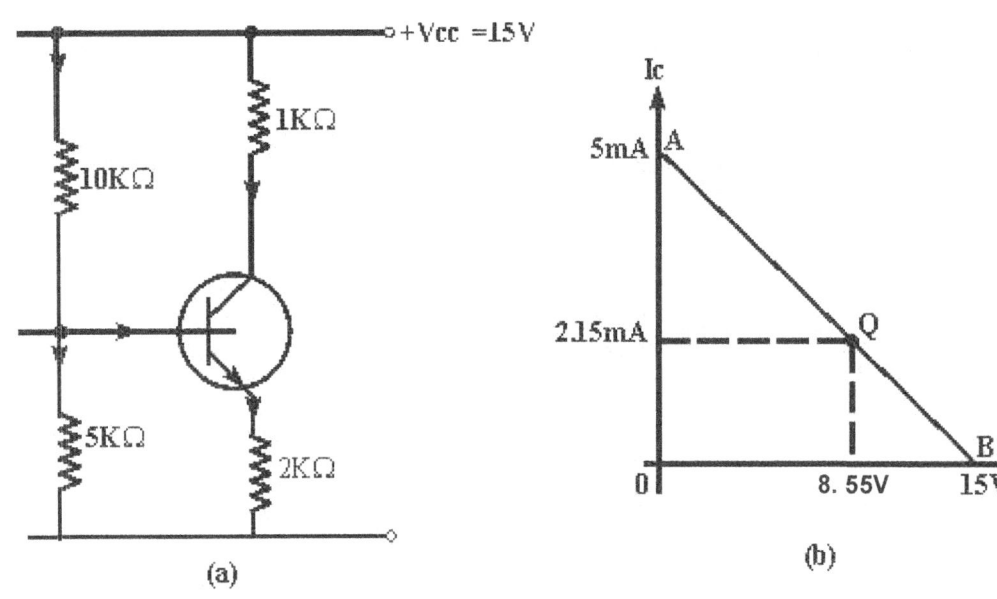

Fig. 5.12.1

This locates the first point B(OB = 15V) of the loadline on the collector-emitter voltage axis.

When $V_{CE} = 0$, $I_C = \dfrac{V_{CC}}{R_C + R_E} = \dfrac{15V}{(1+2)K\Omega} = 5mA$

This locates the second point A(OA = 5mA) of the loadline on the collector current axis. By joining point A and B, the d.c loadline AB is constructed as shown in Fig. 5.12.1b.

Operating point: For silicon, $V_{BE} = 0.7V$

Voltage across 5KΩ is,

$$V_2 = \dfrac{V_{CC}}{10+5} \times 5 = \dfrac{15 \times 5}{10+5} = 5V$$

Therefore, Emitter current $I_E = \dfrac{V_2 - V_{BE}}{R_E} = \dfrac{5 - 0.7}{2K\Omega}$

$= \dfrac{4.3V}{2K\Omega} = 2.15 \text{mA}$

Therefore, Collector current, $I_c = I_E = 2.15\text{mA}$

Collector emitter voltage, $V_{CE} = V_{CC} - I_C (R_C + R_E)$

$= 15 - 2.15\text{mA} \times 3K\Omega$

$= 15 - 6.45 = 8.55V$

Therefore, Operating point is 8.55V, 2.15mA

Fig. 5.12.1b shows the operating point Q on the loadline. Its coordinates are $I_c = 2.15\text{mA}$, $V_{CE} = 8.55V$.

CHAPTER SIX

6.0 ELECTRONIC ENGINEERING CIRCUITS

6.1 INTRODUCTION

The general operation and characteristics of electronic devices are considered from their terminal behaviour without a detailed reference to their internal or physical behaviour. Theses are Junction Diode, Bipolar Transistor, Field Effect Transistor, Zener Diode, Tuned Diode, Unijunction Transistor, Integrated Circuits. Since the operation of all of these devices is generally complex they are usually represented by an *Equivalent Circuit* which is a model usually derived mathematically or empirically from basic physical principles, or terminal characteristics. Most equivalent circuits conditions which must be borne in mind when used in circuit design or analysis. Electronic devices which are used in amplifiers are usually of the more complex type (transistor) having three terminals and may be considered as a two-port device.

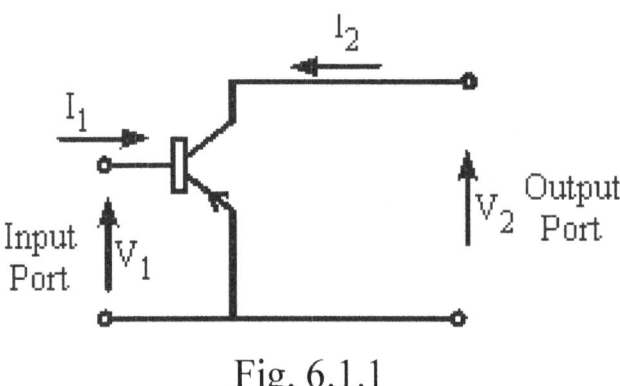

Fig. 6.1.1

In general these devices and many others may be represented by a 4 terminal network. (Usually two of the terminals are common see fig. 6.1.1)

6.2 ELECTRONIC DEVICES AS NETWORK OR CIRCUIT ELEMENTS

Fig. 6.2.1 shows the convention for linear network or "black box" representation of network or device.

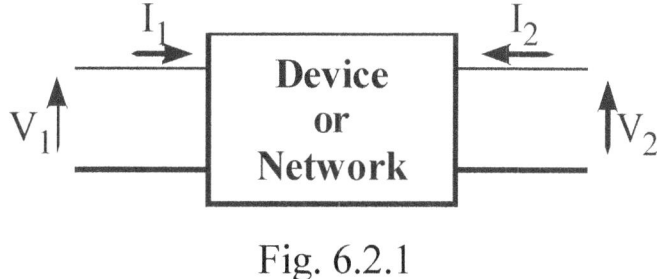

Fig. 6.2.1

It is not necessary to know the internal operation of the device since it is characterised by the four quantities V_1, I_1, V_2, I_2. Only two graphs are necessary to specify their complete behaviour since each graph can specify three quantities; two variables with third ad parameter.

There are many ways in which the four quantities may be related:-

1. $\begin{vmatrix} v_1 \\ v_2 \end{vmatrix} = \begin{bmatrix} z_{11} & z_{12} \\ z_{21} & z_{22} \end{bmatrix} \begin{bmatrix} i_1 \\ i_2 \end{bmatrix}$ O/C Impedance Parameters

2. $\begin{vmatrix} i_1 \\ i_2 \end{vmatrix} = \begin{bmatrix} y_{11} & y_{12} \\ y_{21} & y_{22} \end{bmatrix} \begin{bmatrix} v_1 \\ v_2 \end{bmatrix}$ S/C Admittance Parameters

3. $\begin{vmatrix} v_1 \\ i_2 \end{vmatrix} = \begin{bmatrix} h_{11} & h_{12} \\ h_{21} & h_{22} \end{bmatrix} \begin{bmatrix} i_1 \\ v_2 \end{bmatrix}$ Hybrid Parameters

4. $\begin{vmatrix} i_1 \\ v_2 \end{vmatrix} = \begin{bmatrix} g_{11} & g_{12} \\ g_{21} & g_{22} \end{bmatrix} \begin{bmatrix} v_1 \\ i_2 \end{bmatrix}$ Inverse Hybrid Parameters

5. $\begin{vmatrix} v_1 \\ i_1 \end{vmatrix} = \begin{bmatrix} a_{11} & a_{12} \\ a_{21} & a_{22} \end{bmatrix} \begin{bmatrix} v_2 \\ i_2 \end{bmatrix}$ Transmission (ABCD) Parameters

6. $\begin{vmatrix} v_2 \\ i_2 \end{vmatrix} = \begin{bmatrix} b_{11} & b_{12} \\ b_{21} & b_{22} \end{bmatrix} \begin{bmatrix} v_1 \\ i_1 \end{bmatrix}$ Inverse Transmission Parameters

These equations are linear descriptions and are only valid fro electronic circuit when the parameters are obtained from the linear regions of the device characteristics. This is termed "small signal" operation. Fro large changes and d.c conditions one must observe the characteristics, draw and loadline etc.

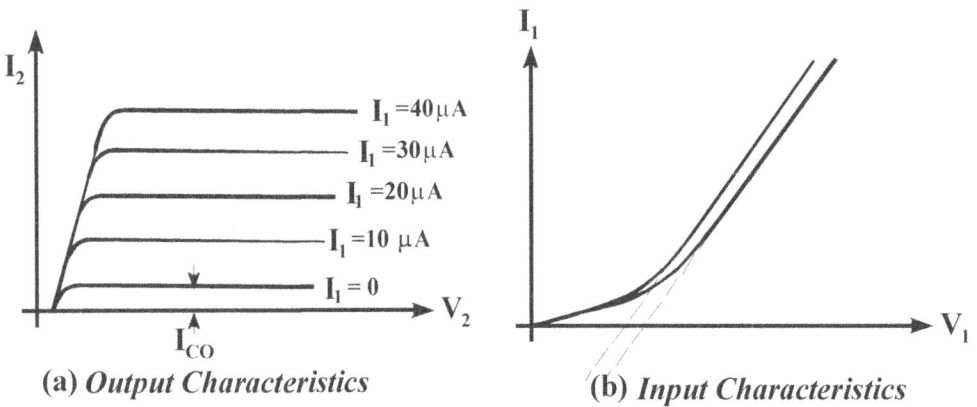

(a) Output Characteristics (b) Input Characteristics

(a) Output Characteristics (b) input Characteristics
Fig. 6.2.2 Shows typical characteristics of a transistor

For transistors (bipolar) the hybrid parameters are most widely used because of their ease of practical measurement and physical interpretation. When related to transistors the hybrid parameters ware often written as

$v_1 = h_i \, i_1 + h_r \, i_2$
$i_2 = h_f \, i_1 + h_o \, v_2$ **where**

$h_i = h_{11}$
$h_r = h_{12}$
$h_f = h_{21}$
$h_o = h_{22}$

6.2.1 DERIVATION OF CHARACTERISTICS OF STRAIGHT LINE APPROXIMATION

From graph (fig. 6.2.2a) in the linear region a straight line approximate.

$$I_2 = m_1 V_2 + c_1 \quad \text{-----------(A)}$$

Note ($y = mx + C$, *equation of a straight line*)

For equivalent spaced characteristics

$$C_1 = k_1 + k_1' I_1$$

Substituting (A), $\quad I_2 = m_1 V_2 + k_1^1 I_1 + k_1$

When $I_1 = 0$, $\quad I_2 = I_{CO}$ (leakage current)

For small changes

$$\delta I_2 = m_1 \delta V_2 + k_1^1 \delta I_1 \quad \text{-----------(B)}$$

For constant input current $\delta I_1 = 0$

$$m_1 = \left. \frac{\delta I_2}{\delta V_2} \right|_{\delta I_1 = 0} = h_o \begin{Bmatrix} \text{O/P Admittance with I/P} \\ \text{opened circuited a.c or} \\ \text{small changes} \end{Bmatrix}$$

For constant output voltage $\delta V_2 = 0$

$$k_1^1 = \left. \frac{\delta I_2}{\delta I} \right|_{\delta V_2 = 0} = h_f \begin{Bmatrix} \text{Forward current gain} \\ \text{with O/P short circuited} \\ \text{to a.c or small changes.} \end{Bmatrix}$$

Substituting in (B) gives $\delta I_2 = h_f \delta I_1 + h_o \delta V_2$

Writing i_2 for δI_2, i_1 for δI_1, v_2 for δv_2 etc. for small changes or a.c signals

$$\underline{\underline{i_2 = h_f i_1 + h_o v_2}}$$

From the characteristics this equation represents the small signal of a.c operation of a transistor and also described the behaviour of the equivalent circuit of fig 6.2.3.

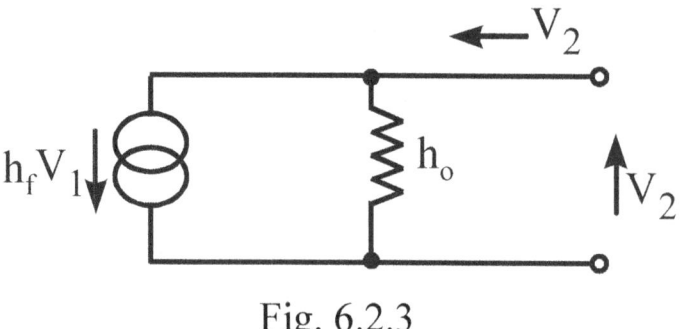

Fig. 6.2.3

From graph of fig. 6.2.2b $\quad I_1 = m_2 V_1 + C_2$ --------------------(C)

Again for eqvispaced characteristics,

$$C_2 = -\left(k_2^1 V_2 + k_2\right)$$; Note intercept on axis is –ve

Substituting in (C), $I_1 = m_2 V_1 - k_2^1 V_2 - k_2$

Again for small changes

$$\delta I_1 = m_2 \delta V_1 - k_2^1 \delta V_2$$ ---------------------------- (D)

For constant O/P voltage, $\delta V_2 = 0$

$$\frac{1}{m_2} = \left.\frac{\delta V_1}{\delta I_1}\right|_{\delta V_2 = 0} = h_i \begin{cases} \text{input impedance with} \\ \text{O/P short circuited to a.c} \\ \text{or small changes} \end{cases}$$

For constant I/P current $\delta I_1 = 0$

$$m_2 \delta V_1 = k_2^1 \delta V_2$$

$$\left.\frac{\delta V_1}{\delta V_2}\right|_{\delta V_2 = 0} = \frac{k_2^1}{m_2} = h_i \begin{cases} \text{reverse voltage} \\ \text{feedback ratio with} \\ \text{I/P opened circiuted to} \\ \text{a.c or small changes} \end{cases}$$

Substituting in (D) gives

$$\delta I_1 = \frac{1}{h_i}\delta V_1 - \frac{h_r}{h_i}\delta V_2$$

Again writing i_1 for δI_1, etc

$$i_1 = \frac{1}{h_i}V_1 - \frac{h_r}{h_i}V_2$$

$$\underline{\underline{v_1 = h_i i_1 + h_r v_2}}$$

This description of the input behaviour of a transistor to small signals may be represented by the circuit of fig 6.2.4. The direction of $h_i v_2$ indicates that when v_2 increases v_1 decreases.

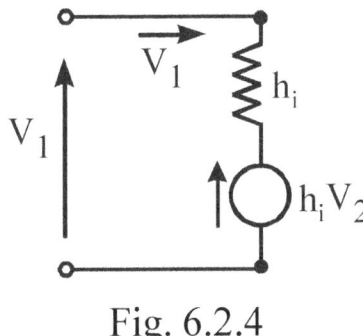

Fig. 6.2.4

The complete equivalent circuit of both the input and output behaviour of the transistor is shown in fig. 6.2.5 below. It is known as the hybrid equivalent circuit.

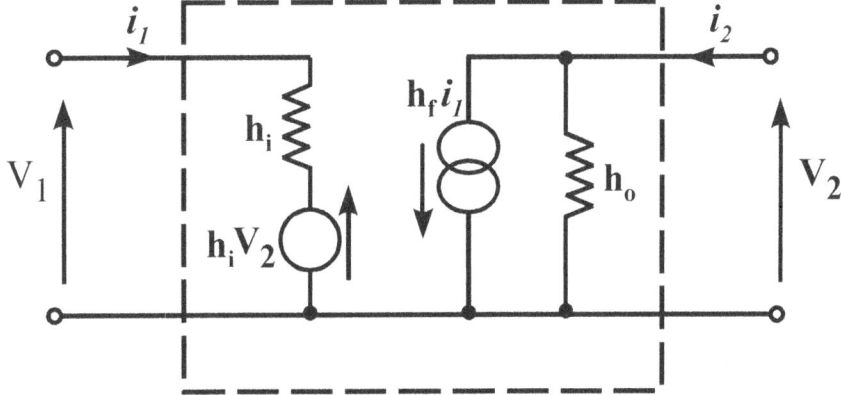

Fig. 6.2.5: Hybrid Equivalent Circuit

Equations are;

$$v_1 = h_i i_1 + h_r i_2$$

$$i_2 = h_f i_1 + h_o v_2$$

h parameters are not usually or measured for some special operating condition and as such may change if operating condition change.

Typical vales of **h** parameters from transistor characteristics are

$h_i = 1K\Omega$ $\qquad h_r = 10^{-3}$

$h_f = 100$ $\qquad h_o = 50 \times 10^{-6} S$ (or υ)

Alternative Mathematical Derivation

From graph of fig. 6.2.2a, $\quad I_2 = f(I_1, V_2)$

$$\delta I_2 = \left(\frac{\delta I_2}{\delta I_1}\right)\delta I_1 + \left(\frac{\delta I_2}{\delta V_2}\right)\delta V_2 \text{ ------- (total differential)}$$

For constant V_2, $\delta V_2 = 0$

$$\left.\frac{\delta I_2}{\delta I_1}\right|_{\delta V_2 = 0} = \frac{\delta I_2}{\delta I_1} = h_f$$

For constant I_1, $\delta i_1 = 0$

$$\left.\frac{\delta I_2}{\delta V_2}\right|_{\delta i_1 = 0} = \frac{\delta I_2}{\delta V_2} = h_O$$

For graph of fig. 6.2.2b $I_1 = f(V_1, V_2)$

$$\delta I_1 = \left(\frac{\delta I_1}{\delta V_1}\right)^{\delta V_1} + \left(\frac{\delta I_1}{\delta V_2}\right)^{\delta V_2}$$

For constant V_2, $\delta V_2 = 0$

$$\left.\frac{\delta V_1}{\delta I_1}\right|_{\delta V_2 = 0} = 1 \Big/ \frac{\delta I_1}{\delta V_1} = h_i$$

For constant I_1, $\delta I_1 = 0$

$$-\frac{\delta I_1}{\delta V_1}\delta V_1 = \frac{\delta I_1}{\delta V_2}\delta V_2$$

Assume mathematically possible; $\quad -\dfrac{\delta V_1}{\delta V_2} = + \left.\dfrac{\delta V_1}{\delta V_2}\right|_{\delta I_1 = 0} = h_r$

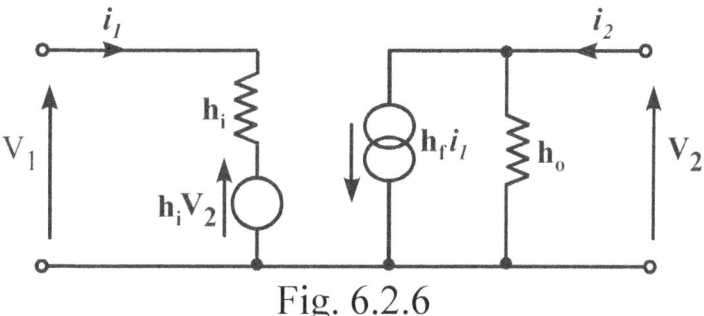

Fig. 6.2.6

Note +ve sign indicates direction.

$$\delta I_1 = \frac{1}{h_I}\delta V_1 + \left(\frac{\delta V_1}{\delta V_2} \cdot \frac{\delta I_1}{\delta V_1}\right)\delta V_2$$

$$= \frac{1}{h_I}\delta V_1 - h_r \cdot \frac{1}{h_i}\delta V_2$$

$$\delta V_1 = h_i\,\delta I_1 + h_r\,\delta V_2$$

$$\delta I_2 = h_f\,\delta I_1 + h_o\,\delta V_2$$

Again writing i_1 for δI_1 etc small changes

$$v_1 = h_i\,i_1 + h_r\,i_2$$

$$i_2 = h_f\,i_1 + h_o\,v_2$$

There are the same equations as before and hence the same equivalent circuit.

6.3 EQUIVALENT T CIRCUITS

Many equivalent circuits are derived from either device characteristics or a mathematical model of their behaviour. The equivalent T network has the advantage over many circuit models (which are more convenient for analysis, although not necessarily for design) in that the network quantities can be related to parameters which can be physically visualised in the transistor. The circuit given in fig. 6.3.1 may be derived from the general equations of diffusion taking into effect recombination, hole concentration gradient and diffusion constant. It is valid for low frequency operation only where the transistor parameters may be regarded as real.

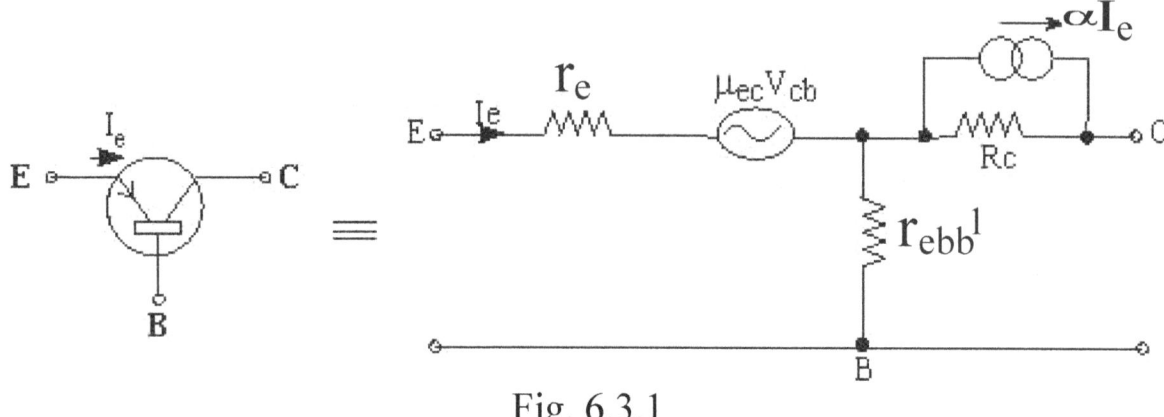

Fig. 6.3.1

Where;
i. r_e the emitter resistor is approximately forward biased diode and approximately equal to $\dfrac{KT}{ei_e} \propto \dfrac{1}{i_e}$

~25 - 300Ω

ii. r_C collector resistor, that of reverse biased diode; varying from 500KΩ ~ several MΩ.

iii. r_{bb}' is due to the presence of base bulk material and depends on the transistor geometry; 50 - 1000Ω.

The effect of the feedback voltage $\mu_{ec}V_{cb}$ generator is usually replaced by a feedback resistor $r_{bb''}$ as shown in fig. 6.3.2.

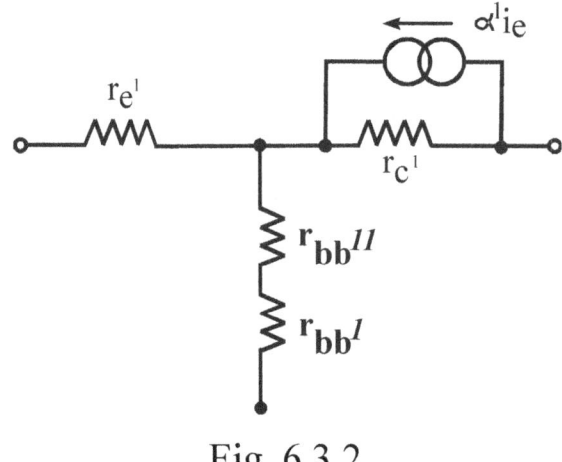

Fig. 6.3.2

The generator is inserted in the emitter lead because constant emitter current changes in V_{cb} affect V_{eb}. The voltage is in phase with V_{cb} by an amount μ_{ec} which is much smaller ($\sim 10^{-4}$). The effect of this is usually neglected. The slight modification of r_e, $\propto$, r_C to r_e', $\propto'$, r_C' is also ignored and $r_{bb''}$ and $r_{bb'}$ lumped together in a straight resistor, r_b.

iv. A current source $\propto i_e$ is placed in parallel with r_C to account for transistor action. Although other effects such as surface leakage have not been taken into account reasonably accurate results are obtained with the circuit given in fig. 6.3.3.

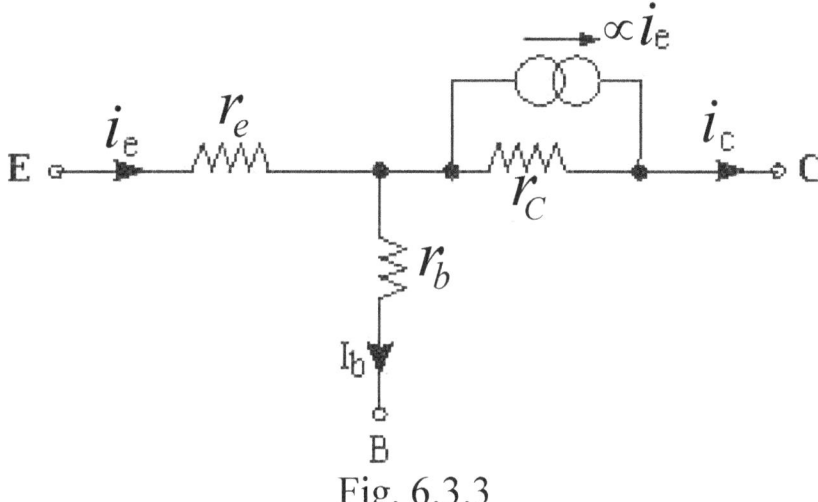

Fig. 6.3.3

It is of course valid for small signals only and cannot in general be used to calculate the d.c conditions.

The main difficulty lies in the measurement of the circuit parameters which in general are not readily available.

An alternative representation is to use a voltage source as shown in fig. 6.3.4

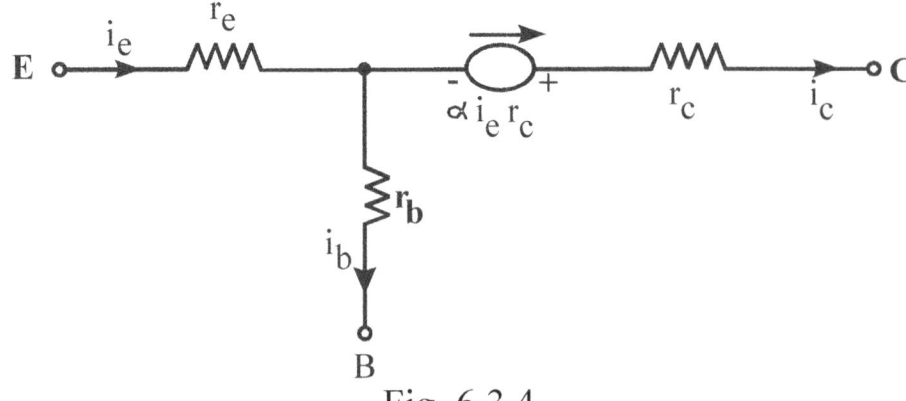

Fig. 6.3.4

Further, the model can be manipulated to form a model based on base current control. Consider the diagram shown in fig. 6.3.5

Fig. 6.3.5

But $i_e = i_b + i_c$

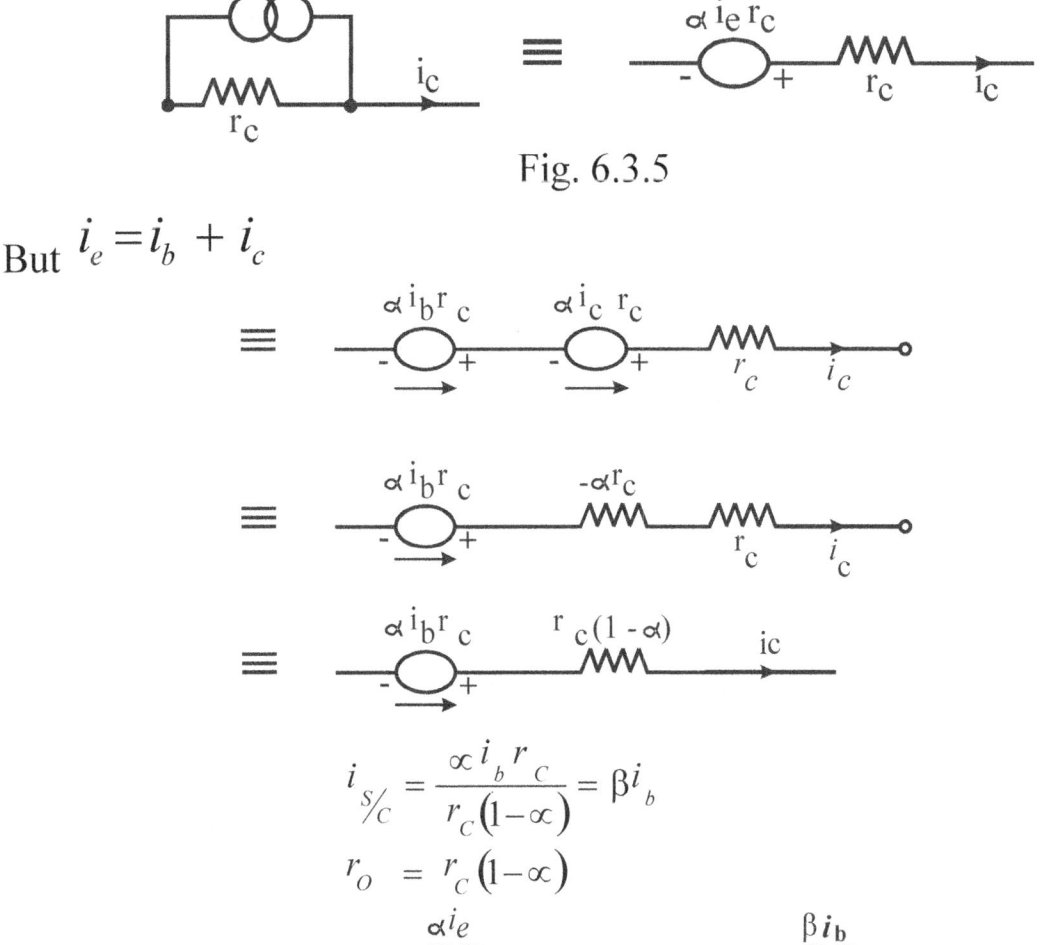

$$i_{s/c} = \frac{\alpha i_b r_c}{r_c(1-\alpha)} = \beta i_b$$

$$r_o = r_c(1-\alpha)$$

Therefore,

Fig. 6.3.6

ELECTRONIC ENGINEERING CIRCUITS

Thus analysis may be described in terms of either of the two commonly used current gains, α or β

Analysis Using the Simplified Equivalent T- Circuit
Consider a common emitter amplifier shown in fig. 6.3.7a

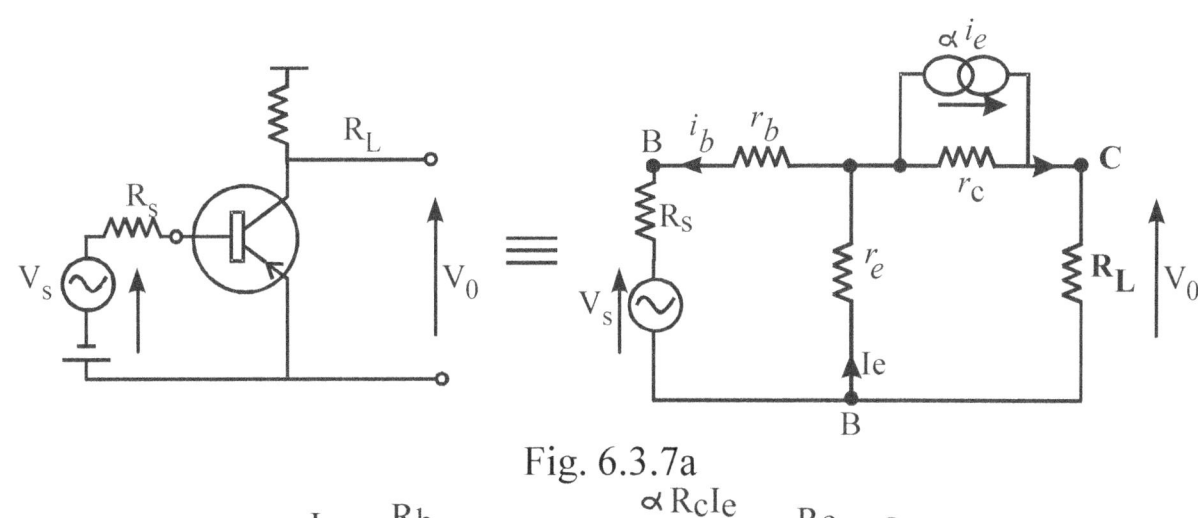

Fig. 6.3.7a

Fig. 6.3.7b

Note,
$$i_e = i_2 - i_1$$
$$I_b = -i_1$$
$$i_c = i_2$$

Applying kirchoff's voltage law to circuit of fig 6.3.7b,

$$V_S = i_1(r_S + r_b + r_e) - i_2 r_e \quad \text{-----------(i)}$$
$$O = -i_1 r_e - \alpha r_e(i_2 + i_1) + i_2(r_e + r_C + r_L) \text{--------(ii)}$$

From (ii)

$$i_2 = \frac{r_e - \alpha R_C}{R_e + R_L + R_C(1-\alpha)} \cdot i_1 \quad \text{----------------(iii)}$$

Current gain of amplifier usually defined as $\dfrac{i_C}{i_b} = -\dfrac{i_2}{i_1}$

Therefore, $A_i = -\dfrac{i_2}{i_1} = \dfrac{\alpha r_C - r_e}{r_e + r_L + r_C(1-\alpha)}$

Approximations: If $\alpha r_C \gg r_e$ and $r_C(1-\alpha) \gg r_L + r_e$ i.e. small collector loads.

$A_i = \dfrac{\alpha}{1-\alpha} = \beta$ (the current gain of transistor alone).

Substituting (iii) in (i) gives after rearrangement.

$$i_i = \dfrac{V_S}{R_S + r_b + r_e - \dfrac{r_e(r_e - \alpha R_C)}{R_L + r_e + r_C(1-\alpha)}}$$

Considering the input circuit shown in fig. 6.3.8,

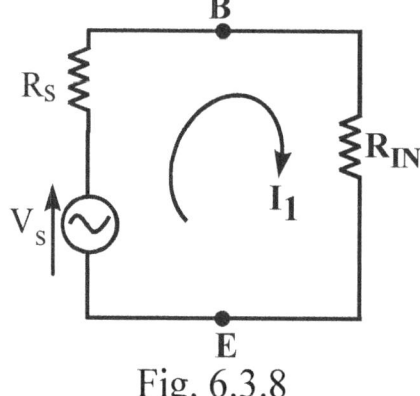

Fig. 6.3.8

If R_{IN} = input resistance of transistor amp.

$$i_1 = \dfrac{V_S}{R_S + R_{IN}}$$

Equating these two equations for i_i gives, Input Resistance,

$$R_{IN} = r_b + r_e - \dfrac{r_e(r_e - \alpha r_C)}{R_L + r_e + r_C(1-\alpha)}$$

If r_e is neglected as small, $R_{IN} = r_b + r_e \cdot \dfrac{\alpha r_C}{R_L + r_C(1-\alpha)}$

Further if $R_C(1-\alpha) \gg R_L$ $R_{IN} = r_b + \dfrac{\alpha}{(1-\alpha)} r_e = r_b + \beta r_e$

$r_b = 500\Omega$ $r_e = 25\Omega$ $\alpha = 0.98$ $r_c = 1M\Omega$ $R_{IN} = 1725\Omega$

Substituting (iii) in (i).

$$V_S = \frac{R_L + re + r_C(1-\alpha)}{r_e - \alpha r_C} \cdot (R_S + r_b + r_e) - i_2 r_e$$

Solving for i_2 gives

$$i_2 = \frac{\dfrac{r_e - \alpha r_C}{R_S + r_b + r_e} \cdot V_S}{R_L + r_e + r_C(1-\alpha) + \dfrac{r_e(\alpha r_C - r_e)}{R_S + r_b + r_e}} \quad\text{------------(IV)}$$

Consider again the output circuit shown in fig. 6.3.9;

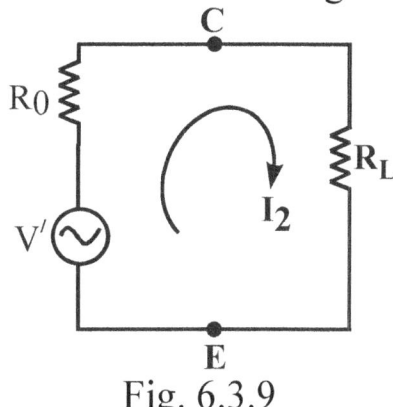

Fig. 6.3.9

$$i_2 = \frac{V'}{R_L + R_O}$$

Equating values of i_2 gives expression for output resistance

$$R_O = R_L + r_e + r_C(1-\alpha) + \frac{r_e(\alpha r_C - r_e)}{R_S + r_b + r_e}$$

$$= r_C(1-\alpha) + \frac{r_e(R_L + r_b + \alpha r_C)}{R_S + r_b + r_e}$$

For very high R_S,
$$R_O = R_L + r_e + r_C(1-\alpha) = r_C(1-\alpha) \approx 20K\Omega, \text{ approx.}$$

and for R_S V low $\approx$ zero, $R_O \to r_C(1-\alpha) + \dfrac{r_e(r_b + \alpha r_c)}{r_b + r_e} \approx 87K\Omega$ approx.

Output voltage $V_O = i_2 R_L$

Voltage gain, $A_V = \dfrac{V_O}{V_S} = \dfrac{i_2 R_L}{V_S}$ (from (iv) gives -)

$$A_V = \frac{-(\alpha r_C - r_e)R_L}{[R_L + r_e + r_C(1-\alpha)](R_S + r_b + r_e) + r_e(\alpha r_C - r_e)}$$

Approximation: $-\alpha r_C \gg r_e$

$$A_V = \frac{-\alpha r_C R_L}{[R_L + r_C(1-\alpha)](R_S + r_b + r_e) + \alpha r_C r_e}$$

For $R_L = 10K$, $R_S = 1K$, $A_V = -140$

If $R_L \ll r_C(1-\alpha)$

$$A_V = \frac{V_O}{V_S} = \frac{-\alpha R_L}{r_E + (R_S + r_b)(1-\alpha)}$$

Similarly analysis can be applied to other configurations. The expressions for the values of A_V, R_{IN}, R_O and A_V are given below.

(i) COMMON BASE

$$A_i = \frac{r_b + \alpha r_C}{r_b + r_C + R_L} \qquad \text{for } r_C \gg r_b + R_L/A_i = \alpha$$

$$R_{IN} = r_e + r_b\left[\frac{R_L + r_C(1-\alpha)}{r_b + r_C + R_L}\right] \qquad \begin{cases}(\sim r_e)\\ 30 \sim 300\Omega\end{cases}$$

$$R_O = r_C + r_b\left[\frac{R_S + r_e - \alpha r_C}{R_S + r_e + r_b}\right] \qquad (200K \sim 1M\Omega)$$

$$A_V = \frac{(r_b + \alpha r_C)R_L}{(R_L + r_b + r_C)(R_S + r_e + r_b) - r_b(r_b + \alpha r_C)}$$

(ii) COMMON COLLECTOR

$$A_i = \frac{r_C}{r_C(1-\alpha) + r_e + R_L} \qquad \begin{cases}\text{for } r_e + R_L \ll r_C(1-\alpha)\\ A_i \sim (1+\beta)\end{cases}$$

$$R_{IN} = r_b + r_C\left[\frac{r_e + R_L}{r_C(1-\alpha) + r_e + R_L}\right] \qquad \begin{cases}\text{for } R_L \gg r_e \quad r_C(1-\alpha) \gg R_L\\ R_{IN} = r_b + \frac{r_C}{1-\alpha} = (1+\beta)R_L\end{cases}$$

$$A_V = \frac{r_C R_L}{[R_L + r_e + r_C(1-\alpha)](R_S + r_b + r_C) - r_C^2(1-\alpha)}$$

for $r_b \ll r_C + R_S$ and $R_e \ll R_L + r_C(1-\alpha)$

$$A_V = \frac{r_C R_L}{R_L r_C + R_L R_S + r_C R_S(1-\alpha)}$$

$$= \frac{R_L}{R_L + R_S(1-\alpha) + \frac{R_L R_S}{r_C}} = 1 \text{ unity}$$

$$R_O = r_e + r_C(1-\alpha)\left[\frac{r_b + R_S}{r_b + r_C + R_S}\right]$$

For $r_C \gg R_S$, $r_C \gg r_b$

$$= r_e + (r_b + R_S)(1-\alpha) \sim 55\Omega$$

Relationship between h and T Parameters

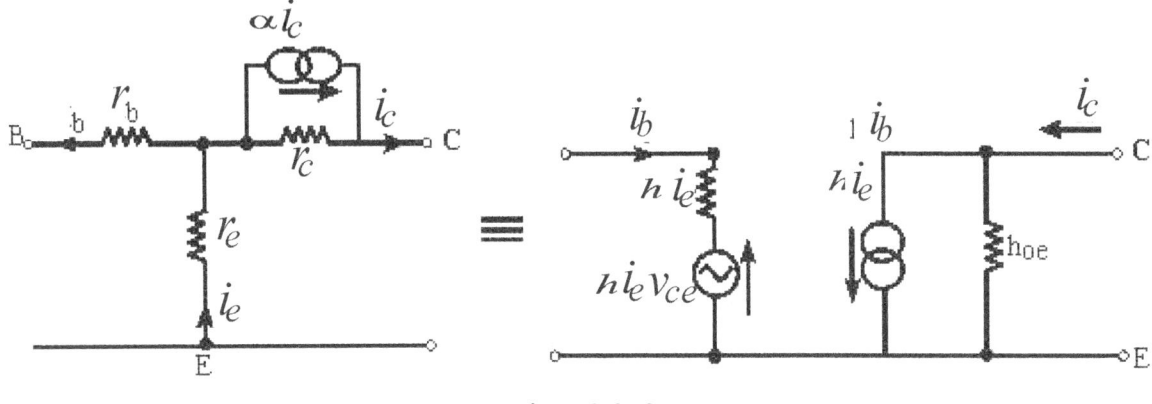

Fig. 6.3.9

hie = input resistance and output short circuited to a.c

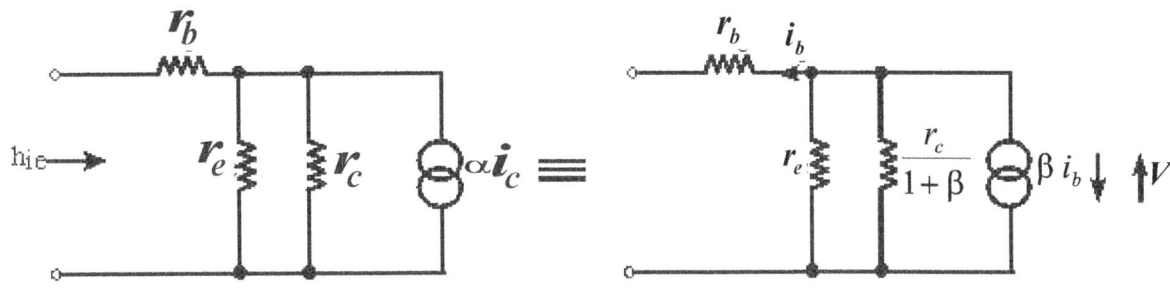

Fig. 6.3.10

$$V = (i_b + \beta i_b) \cdot \frac{r_e r_c / 1+\beta}{r_e + r_c/1+\beta} \quad, \quad \frac{V}{i_b} = \frac{r_e r_C}{r_e + \frac{r_C}{1+\beta}}$$

Therefore, $h_{ie} = r_b + \dfrac{r_e r_c}{r_e + \dfrac{r_c}{1+\beta}} = r_b + r_e(1+\beta)$

(ii) h_{ie} = reverse voltage. Forward bias ratio with input a.c open circuited

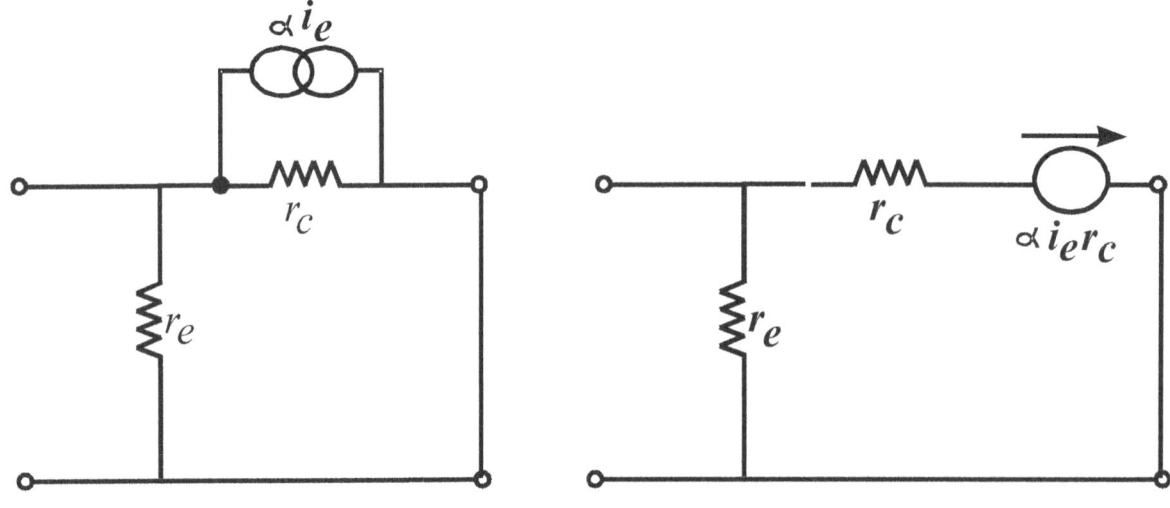

Fig. 6.3.11

$$h_{ie} = \dfrac{r_e}{r_e + r_C}$$

h_{fe} = forward current gain when output in a.c short circuited

$$h_{fe} = -\beta$$

h_{oe} = output admittance. When input is a.c open circuited.

$$h_{ie} = \dfrac{1}{r_e + r_C} = \dfrac{1}{r_C}$$

Obviously it is possible to convert from **T** to **h** parameters. From the foregoing

$$r_C = \dfrac{1}{h_{oe}}$$

$$\beta = -h_{fe}$$

Since $h_{re} = \dfrac{r_e}{r_e + r_C} = \dfrac{r_e}{r_e + \dfrac{1}{h_{oe}}}$

$$r_e(1-h_{re}) = \frac{h_{re}}{h_{oe}}$$

$$r_e = \frac{h_{re}}{h_{oe}(1-h_{re})} = \underline{\underline{\frac{h_{re}}{h_{oe}}}}$$

Since

$$h_{ie} = r_b + r_e(1+\beta)$$

$$r_b = h_{ie} - \frac{h_{re}}{h_{oe}}(1-h_{fe})$$

$$= h_{ie} + \frac{h_{re}h_{fe}}{h_{oe}} - \frac{h_{re}}{h_{oe}}$$

$$= \underline{\underline{h_{ie} + \frac{h_{re}h_{fe}}{h_{oe}}}}$$

The above show the approximate relationship for parameters in the common emitter configuration. Obviously many sets of interrelations can be derived to convert from a given set of parameters to any other irrespective of the circuit configurations.

Capacitance collector to base c_{cb} (miller capacitance)

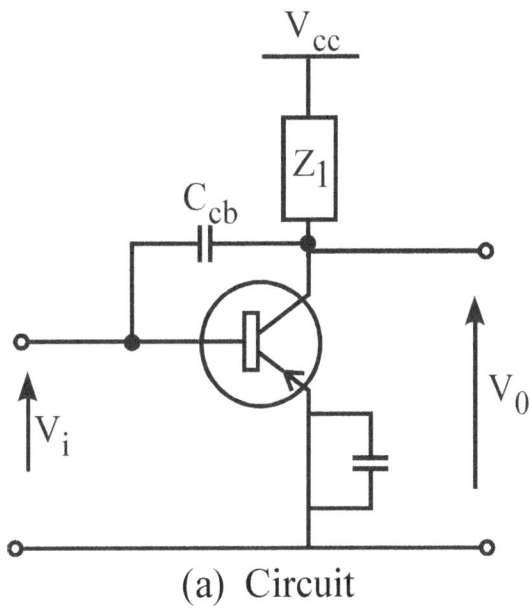

(a) Circuit

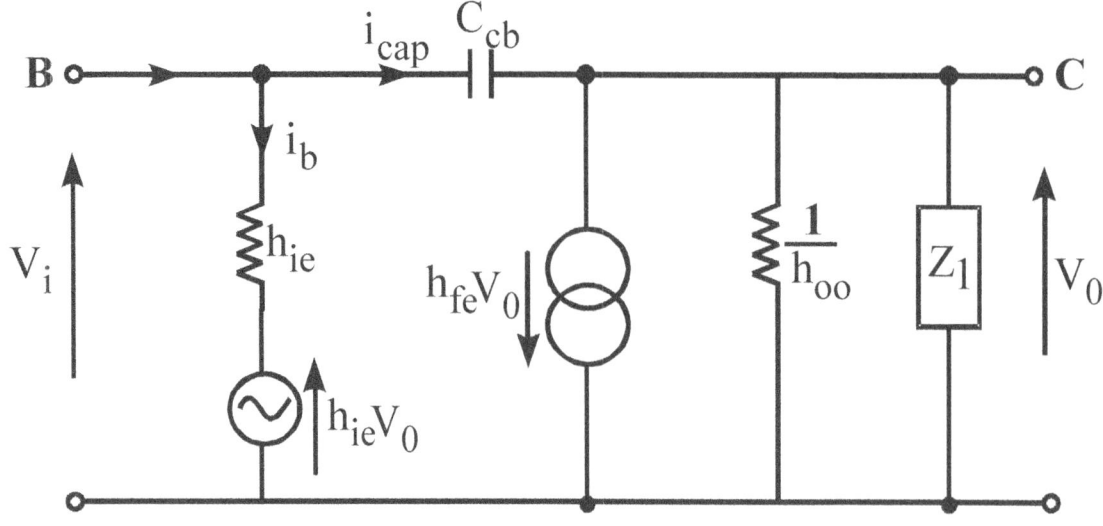

(b) High frequency equivalent circuit.

Fig. 6.3.12

In fig. 6.3.12, C_{Cb} couples the input and output circuits of the amplifier together allowing energy interchange to take place. According to the frequency, circuit component value etc two things may happen

i. The input impedance may change (lowered) and
ii. The amplifier conditions of gain and phase may cause it to oscillate.

From the above the voltage across the capacitor C_{Cb} is the sum of the voltage input and output voltage ($V_i + V_O$). Since these is nominally in antiphase current flowing in capacitor is given by;

$$i_{Cap} = C_{Cb}\frac{d}{dt}(v_i + v_O)$$

$$= C_{Cb}\frac{dv_i}{dt}\left(1 + \frac{v_O}{v_i}\right)$$

Since $\frac{v_O}{v_i} = A_V$ the voltage gain of the stage, $i_{Cap} = (1+A_V)C_{Cb}\frac{dv_i}{dt}$

Thus the effect for small signals is such that C_{Cb} appears as an effective (equivalent) input capacitance of $(1 + A_V) C_{Cb}$ as shown in fig. 6.3.13

161

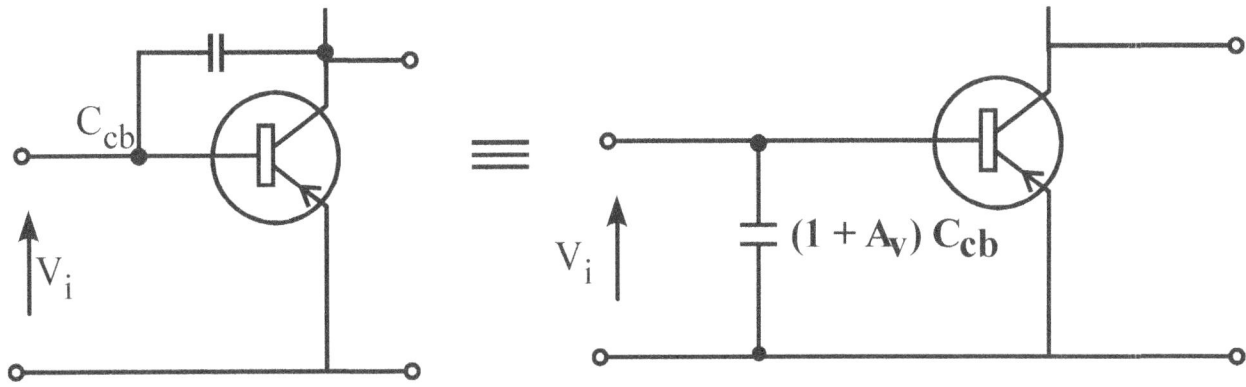

Fig. 6.3.14

The foregoing implies that for a multistage configuration shown in fig. 6.2.15a the preceding stage has $(1 + A_{v2})C_{Cb2}$ in parallel with its effective collector load thus reducing its high frequency performance substantially- see equivalent circuit fig. 6.3.15b.

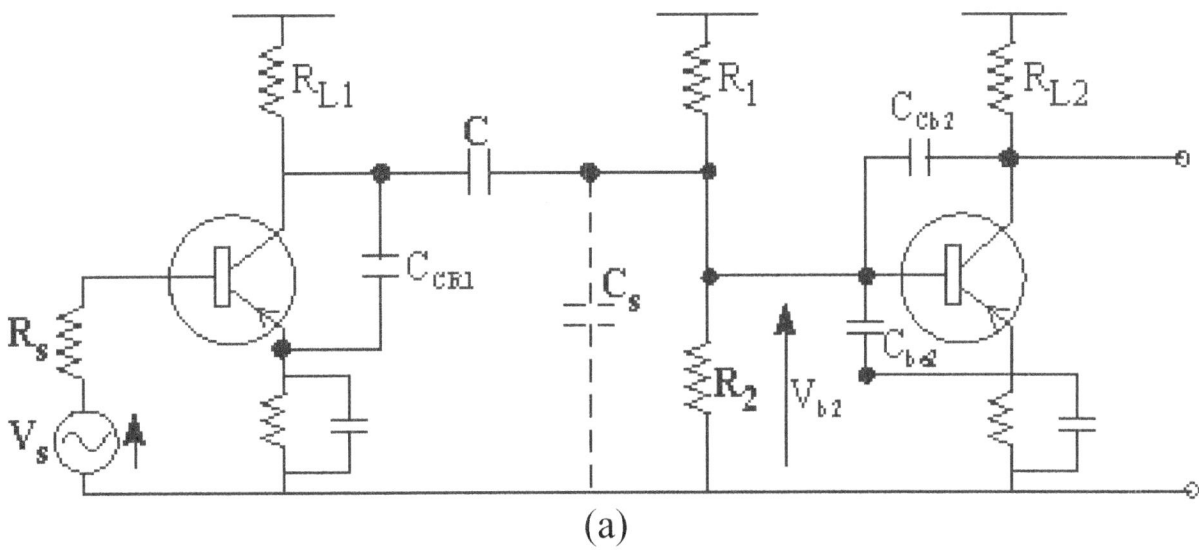

(a)

Assume emitters adequately decoupled C_s = stray capacitor

(b)

Fig. 6.3.15

At frequencies where reactance of coupling capacitor C is low the equivalent circuit reduces to that shown in fig. 6.2.16

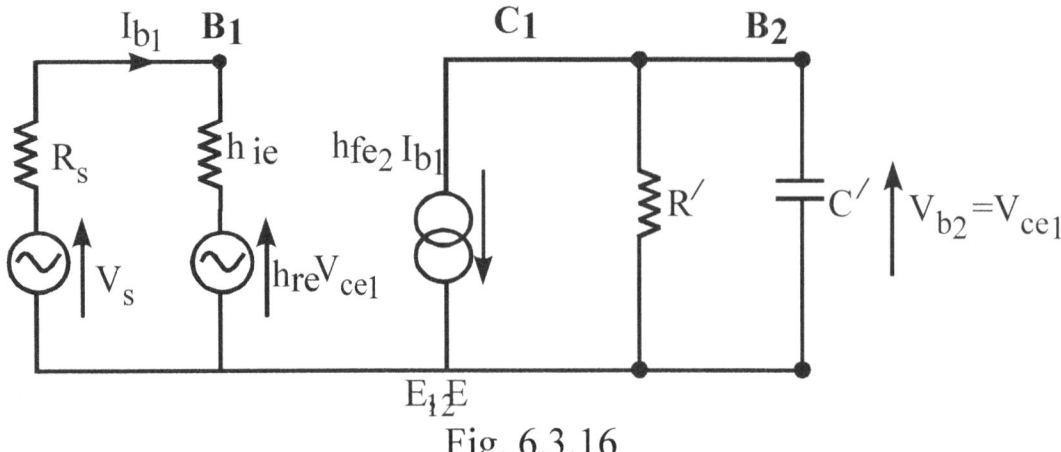

Fig. 6.3.16

Where $R' = \dfrac{1}{h_{oe}} \| R_{L1} \| R_1 \| R_{IN2} \|$

$C' = C_{Ce1} + C_{be2} + C_{Cb2}(1 + A_{V2}) + \boxed{C_S}$

$5pF + 5pF + 2pF(1+99) \Rightarrow 210\,pF$

$Z = R' // C' = \dfrac{R'}{1 + j\omega C' R'}$

Voltage at base B_2, $V_{b2} = -h_{fe}\, i_{bi} \cdot Z$

Base current $i_b = \dfrac{V_S - h_{re} V_{be2}}{R_S + h_{re}}$

Voltage at base 2, $V_{b2} = \dfrac{-h_{fe} R'}{(1 + j\omega C' R')} \cdot \left[\dfrac{V_S - h_{re} V_{be2}}{h_{ie} + R_S} \right]$

$V_{b2}\left[1 - \dfrac{h_{fe} h_{re} R'}{(h_{ie} + R_S)(1 + j\omega C' R')}\right] = \dfrac{-h_{fe} R' V_S}{(h_{ie} + R_S)(1 + j\omega C' R')}$

Overall gain $\dfrac{V_{b2}}{V_S} = \dfrac{-h_{fe} R'}{h_{ie} + R_S - h_{fe} h_{re} R' + j\omega C' R'(h_{ie} + R_S)}$

$\dfrac{V_{b2}}{V_S} = \dfrac{\dfrac{-h_{fe} R'}{h_{ie} + R_S - h_{fe} h_{re} R'}}{1 + \dfrac{j\omega C' R'(h_{ie} + R_S)}{h_{ie} + R_S - h_{fe} h_{re} R'}} = \dfrac{K}{1 + j\omega T}$

$h_{ie} = 1.5K$, $R_S = 1K$, $h_{fe} = 100$, $h_{re} = 10^{-4} \times 2$,

$R' = 5K \Rightarrow h_{fe}h_{re}R' = 100 \ll h_{ie} + R_S$

For $(h_{ie} + R_S) \gg h_{fe}h_{re}R'$ approximately turnover frequency given by

$$\omega = \frac{1}{C'R'}$$

Thus the bandwidth of the amplifier 1st stage is severely reduced because of the Miller effect cursed by the second stage. In extreme cases the reduction can be alleviated somewhat by using an emitter follower a buffer stage. (See fig. 6.2.17)

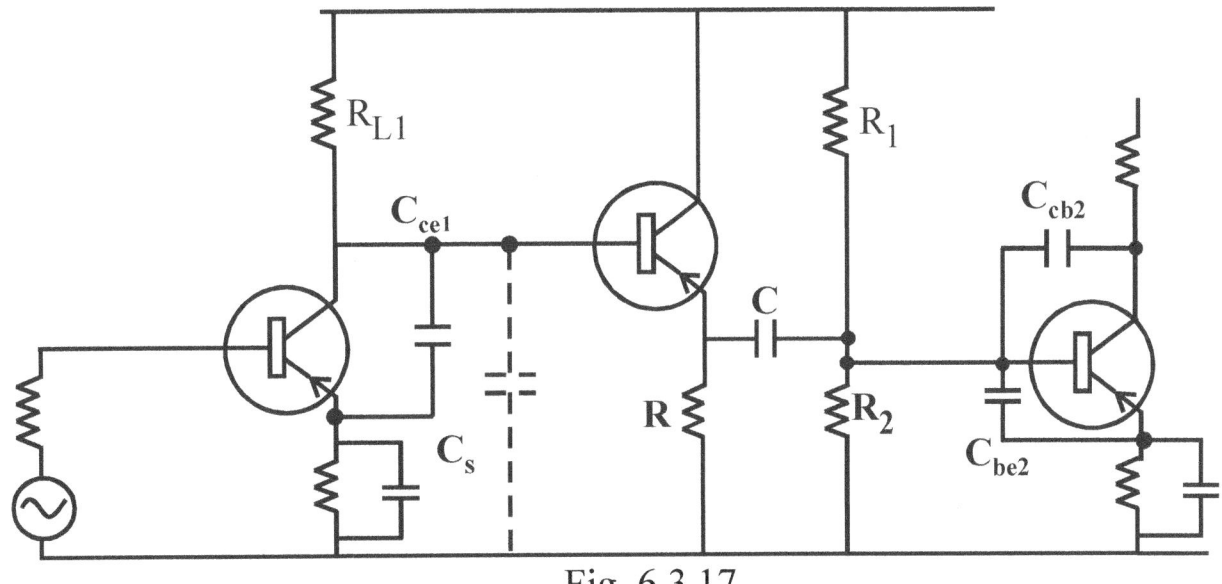

Fig. 6.3.17

C_{IN} of last stage is equal $C_{be2} + (1 + A_2) C_{Cb2}$. Which is effectively in parallel with the output impedance of the emitter follower which is low (<100Ω) and is thus approximately short circuited. Since volt gain of emitter follower is approximately unity and input impedance very high its effect on collector of 1st stage is relatively small.

6.4 THE HYBRID π EQUIVALENT CIRCUIT

The *h* parameter equivalent circuit is very useful at low frequency. At high frequency it suffers the disadvantage that the parameters change with frequency and not easily measured. A circuit which is useful at high frequency and includes the effects of diffusion and junction capacitance is the hybrid π shown in fig. 6.4.1.

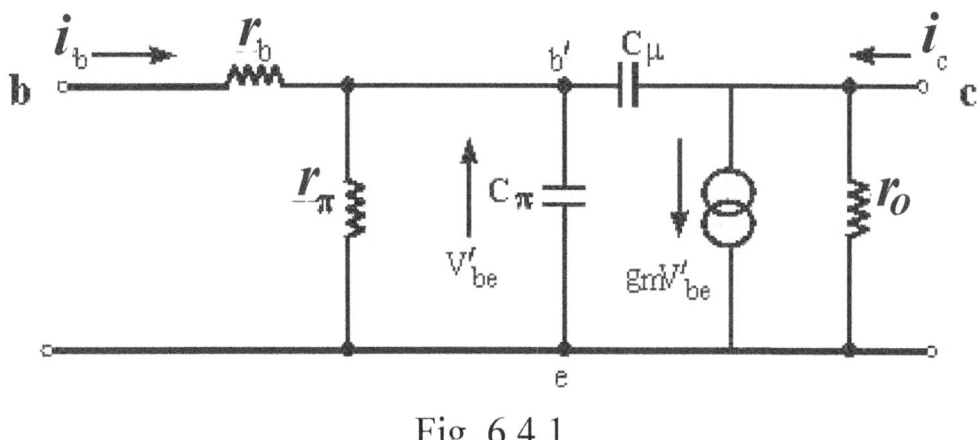

Fig. 6.4.1

Note that C_μ = the junction capacitance (C_{jc}).

The current generator is expressed in terms of voltage across emitter junction *gm* is the transconductance. The parameters above may be obtained from a few simple measurements of;

i. The *l*.F Common emitter short circuited current gain (h_{fe}, β_O)

ii. The Common emitter unity gain frequency w_T (the G.B product)

iii. The d.c Collector current I_{dc} at which the above were measured.

iv. The Common base output capacitance C_{ob}. From these it is possible to calculate, $gm = \dfrac{q}{kT} \cdot |I_C|$ valid if $I_C \gg I_{CO}$ but small enough for assumption of small signal
$\beta_O = h_{fe} = h_{fe}$

From short circuited test $r_\pi = \beta_O / gm$

$$C_\mu = C_{Ob}$$

$$r_b = h_{ie} - r_\pi$$

C_Π is calculated from $h_{fe} = gm / [g_\pi + j\omega(C_\Pi + C_\mu)]$

6.5 FREQUENCY RESPONSE PLOTTING WITH LOG CO-ORDINATE

So far analysis of amplifier has shown that the gain or transfer function has been of the form;

$$\left|\frac{V_O}{V_I}\right| = \left|\frac{+K}{1+jwT}\right| = \left|\frac{K}{\sqrt{1+(wT)^2}}\right|$$

The response is of the form shown in fig. 6.5.1

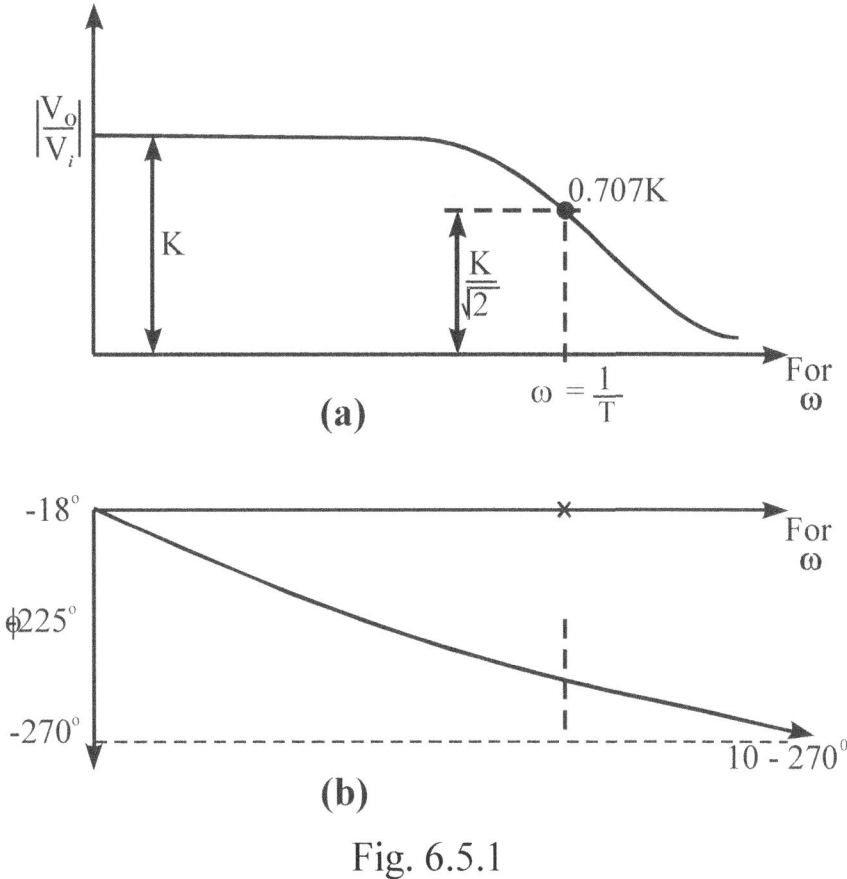

Fig. 6.5.1

ELECTRONIC ENGINEERING CIRCUITS

If we consider plotting $Log\left|\dfrac{V_O}{V_I}\right|$ in dB as they are in common usage against $\log_{10}\omega$ or $\log_{10} f$, we have,

$$\log_{10}\left|\dfrac{V_O}{V_i}\right| = \log_{10}\left|\dfrac{K}{\sqrt{1+(wT)^2}}\right|$$

$$= \log_{10} K - \log_{10}\left(\sqrt{1+(wT)^2}\right)$$

Multiplying through by 20

$$20\log_{10}\left|\dfrac{V_O}{V_i}\right| = \left|\dfrac{V_O}{V_i}\right|_{dB} = 20\log_{10} K - 20\log_{10}\left(\sqrt{1+(wT)^2}\right)$$

Considering the terms separately.

1. Graph of $20 \log_{10} K$. this is shown in fig. 6.5.2

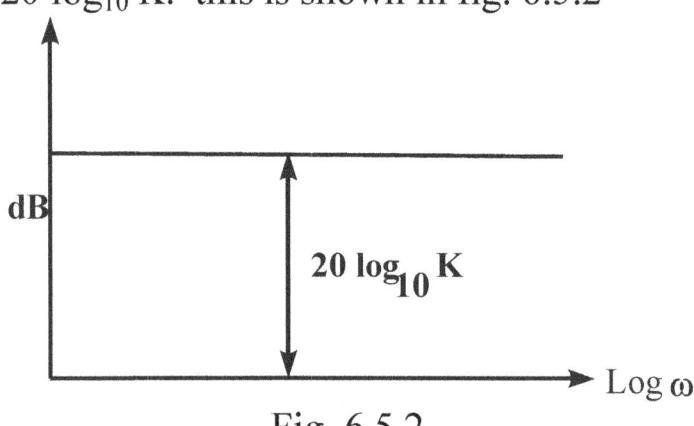

Fig. 6.5.2

2. Graph of $+20\log_{10}\left(\sqrt{1+(wT)^2}\right)$. This has two asymptotes

   i. At l.f when wT is small

   ii. At h.f when $wT > 1$

   (a) When $wT \ll 1$ at i.f i.e $w \ll \dfrac{1}{T}$

   $$20\log_{10}\left(\sqrt{1+(wT)^2}\right) = 20\log_{10} 1 = 0$$

   (b) At the corner or turnover, or breakpoint frequency

   $$wT = 1 \quad \text{or} \quad w = \dfrac{1}{T}$$

   $$20\log_{10}\left(\sqrt{1+(wT)^2}\right) = 20\log_{10} 2 = 3.0016 dB$$

(c) When $wT \gg 1$ i.e. $w = \frac{1}{T}$

$$20\log_{10}\left(\sqrt{1+(wT)^2}\right) = 20\log_{10}(wT)$$

If the frequency is doubled the dB change is also doubled

i.e. $w_2 = 2w_1$ dB change ratio $= 20\log_{10}\dfrac{w_2 T}{w_1 T} = 20\log_{10} 2$

$= 6.02$dB every time frequency is double (octave)

Thus the slope of the curve in (c) is 6db/octave or 20dB/decade change in frequency – see fig. 6.5.3

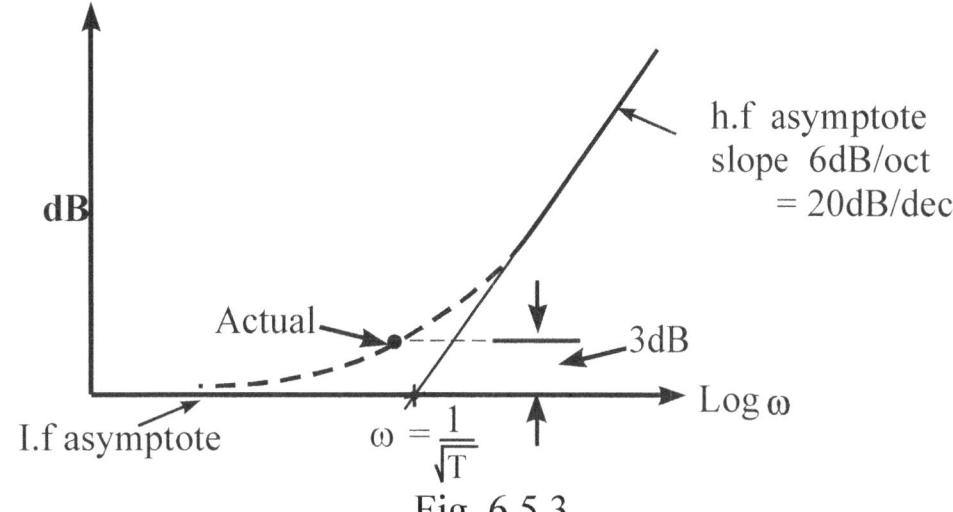

Fig. 6.5.3

Composite response obtained by subtracting graphs is shown in fig. 6.5.4

$$\left|\frac{V_O}{V_I}\right|_{dB} = 20\log_{10} K - 20\log_{10}\sqrt{1+(wT)^2}$$

Fig. 6.5.4

ELECTRONIC ENGINEERING CIRCUITS

Approximate response can be plotted from knowledge of asymptotes. Further calculation hear intersection of asymptotes reveals that

when $w = \dfrac{1}{2T}$, $= 20\log_{10}\sqrt{1+(wT)^2} = 20\log_{10}\sqrt{1.25} = 1 dB$

$w = 2T,$ $= 20\log_{10}\sqrt{5} = 7 dB$

Therefore near corner, we have values as shown in fig. 6.5.5

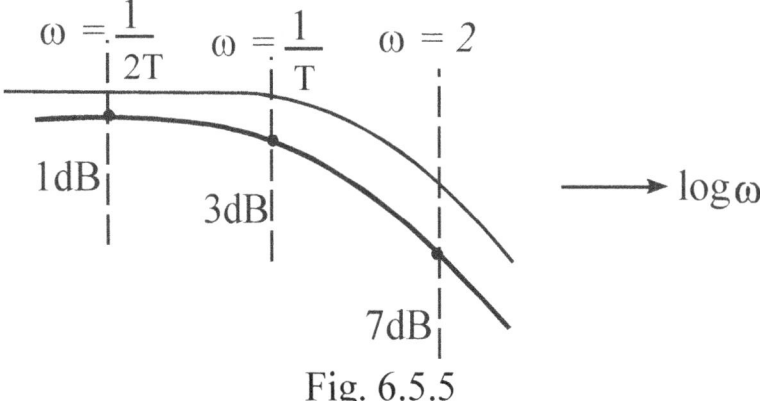

Fig. 6.5.5

Thus using these techniques complicated functions can be plotted more easily on log scale than Cartesian co-ordinates. The general form of an expression may be

$$\left|\dfrac{V_o}{V_i}\right|_{jw} = \left|\dfrac{(1+jwT_1)(1+jwT_2)(-----)}{jwT(1+jwT_2)(1+jwT_5)(-----)---(\ )}\right|$$

The technique is to plot the asymptote of each component separately and add and subtract where appropriate to achieve final result.

PHASE CHARACTERISTIC of $\dfrac{A}{1+jwT}$ **i.e.** $\phi = 1\tan^{-1} wT$

This is shown in fig. 6.5.6

(i) As $w \to 0$ $\phi \to 0$

(ii) As $w \to \infty$ $\phi \to -90^0$

(iii) $w = \dfrac{1}{T}$ $\phi = \dfrac{-\pi}{4} = -45^0$

(iv) $w = \dfrac{1}{2T}$ $\phi = -\tan^{-1}\dfrac{1}{2} = -26.50$

(v) $w = \dfrac{2}{T}$ $\phi = -\tan^{-1} 2 = 63.5^0$

ELECTRONIC ENGINEERING CIRCUITS

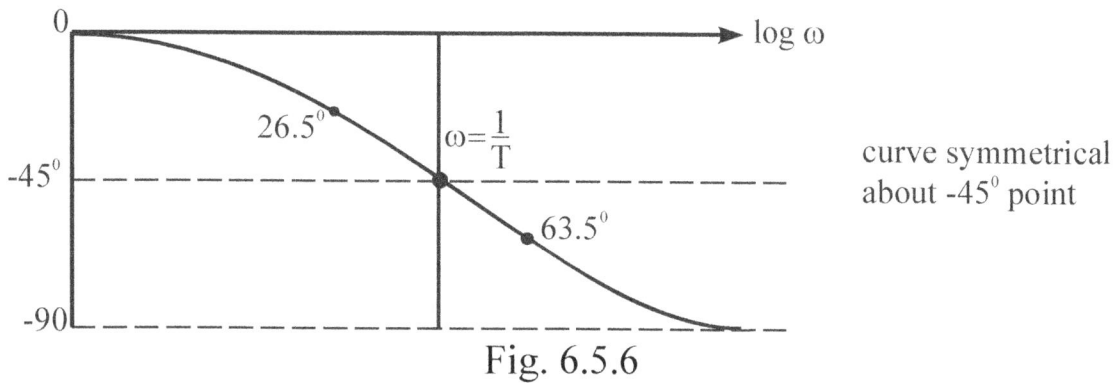

Fig. 6.5.6

Common gain characteristics

(i) $\left|\dfrac{V_O}{V_i}\right|_{jw} = |jwT|$ at w = 0, $\left|\dfrac{V_O}{V_i}\right| = 0$

$$20\log_{10}|wT| = \left|\dfrac{V_O}{V_i}\right|_{dB}$$

As frequency doubles dB change doubles. Rate is 6dB/oct i.e. 20db/dec (see fig. 6.5.7)

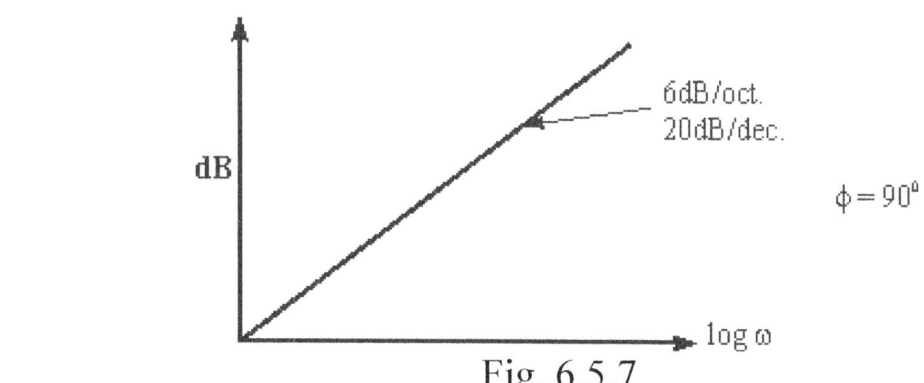

Fig. 6.5.7

(ii) $\left|\dfrac{V_O}{V_i}\right|_{jw} = \left|\dfrac{1}{(jwT)^2}\right|$ $\left|\dfrac{V_O}{V_i}\right|_{dB} = -20\log_{10}(\omega T)^2$

$$= 40\log_{10}(\omega T)$$

Rate of decrease = -12dB/octave (see fig. 6.5.8)

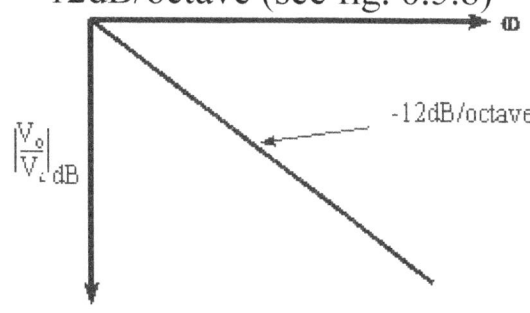

Fig. 6.5.8

Example 6.5.6

Plot of $\left|\dfrac{V_O}{V_i}\right|_{jw} = \left|\dfrac{KjwT_1}{(1+jwT_1)(1+jwT_2)}\right|$ $[T_2 > T_1]$

$\left|\dfrac{V_O}{V_i}\right|_{dB} = 20\log_{10}|KjwT_1| - 20\log_{10}\left|\sqrt{1+jwT_1}\right| - 20\log_{10}\left|\sqrt{1+jwT_2}\right|$

(a)

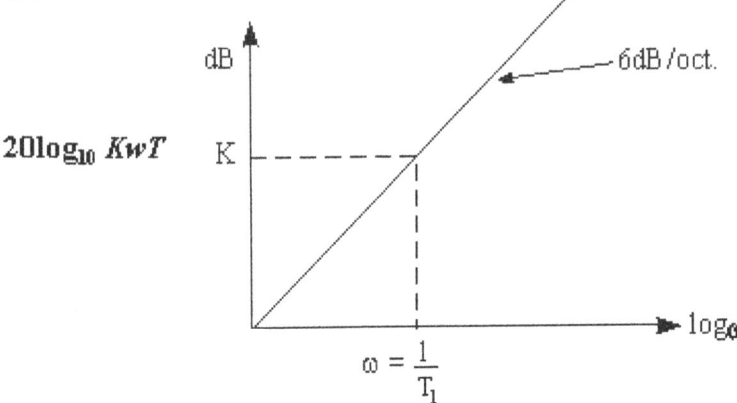

(b)

(c)

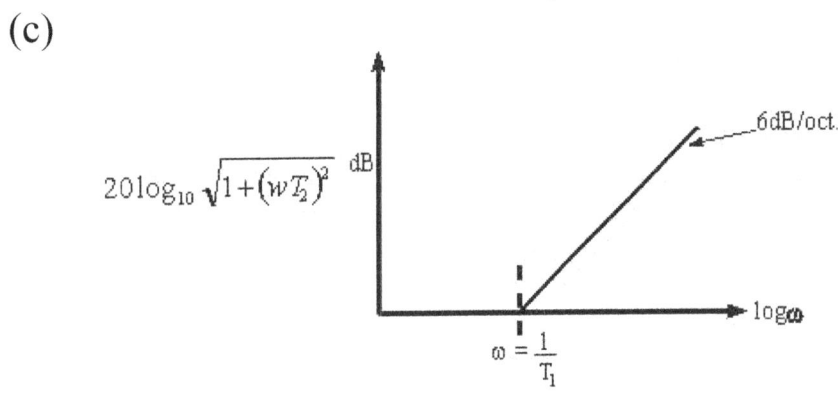

$\left|\dfrac{V_O}{V_i}\right|_{dB} = A - B - C$

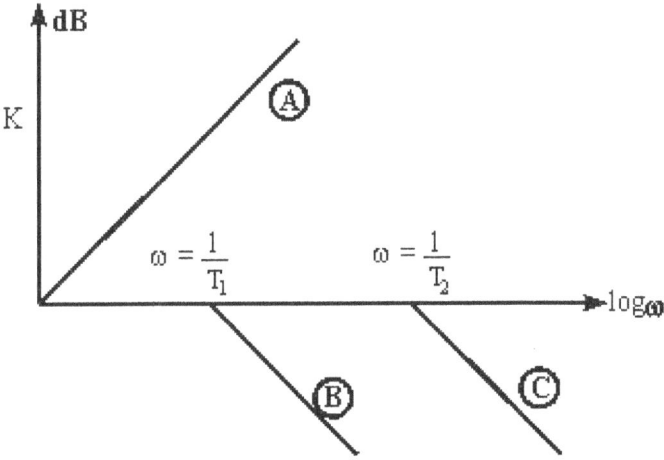

Composite characteristics

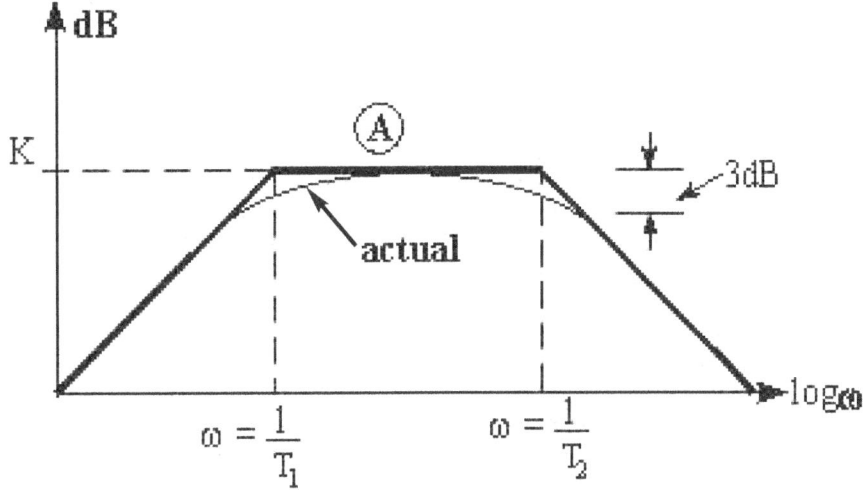

Note: These are also called Bode plots. Use is made of the fact that where graphs cut axis, $\left|\dfrac{V_o}{V_i}\right| = unity$ (see stability notes).

6.6 FIELD EFFECT TRANSISTORS (F.E.T)

These are devices which depend upon electric field to control the flow of current. Considering junction F.E.T's only their operation depends on the flow of majority carriers. It is therefore unipolar in operation compared to conventional transistor which is bipolar (hole and electron flow).

A bar of *P* type material (called the channel) is connected between the drain and course which are the terminals through which the majority carriers leave and enter the channel respectively. On the sides of the channel a heavily doped n type region is diffused and this region is termed the gate see fig. 6.6.1.

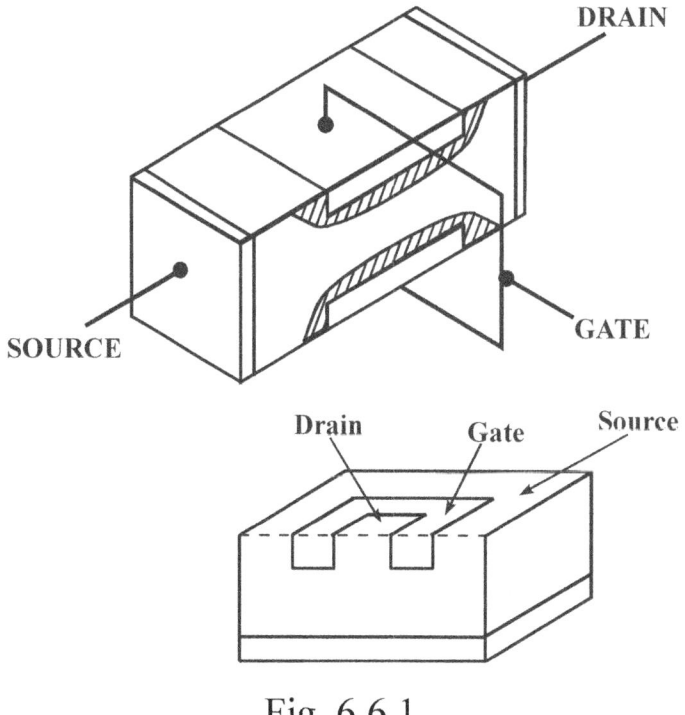

Fig. 6.6.1

A reverse bias voltage V_{gs} is applied between the gate and source. This produces a space charge region on either sides of the junction which increases with increasing reverse bias and thus reduces the effective channel width. Conduction in this region is almost zero because of the mavailiality of current carriers and the impedance between gate and course is thus extremely high. Thus the majority carriers (in the case holes) travelling from source to drain are controlled by the channel width. Hence for a fixed drain-source voltage V_{DS} the drain current will be a function of the gate-source voltage. V_{gs} controlled by the effect of its field. This type of transistor is termed a Junction Field Effect JUGFET.

It is obvious for these F.E.T's that the drain and source terminals may be interchanged because of its operation and general symmetry. Further N-channel F.E.T's can be constructed using similar techniques with semiconductors having majority carriers of opposite polarities.

6.7 CHARACTERISTICS AND OPERATION

The symbol and biasing of F.E.T transistors shown in fig. 6.7.1b and fig. 6.7.1a shows the characteristics.

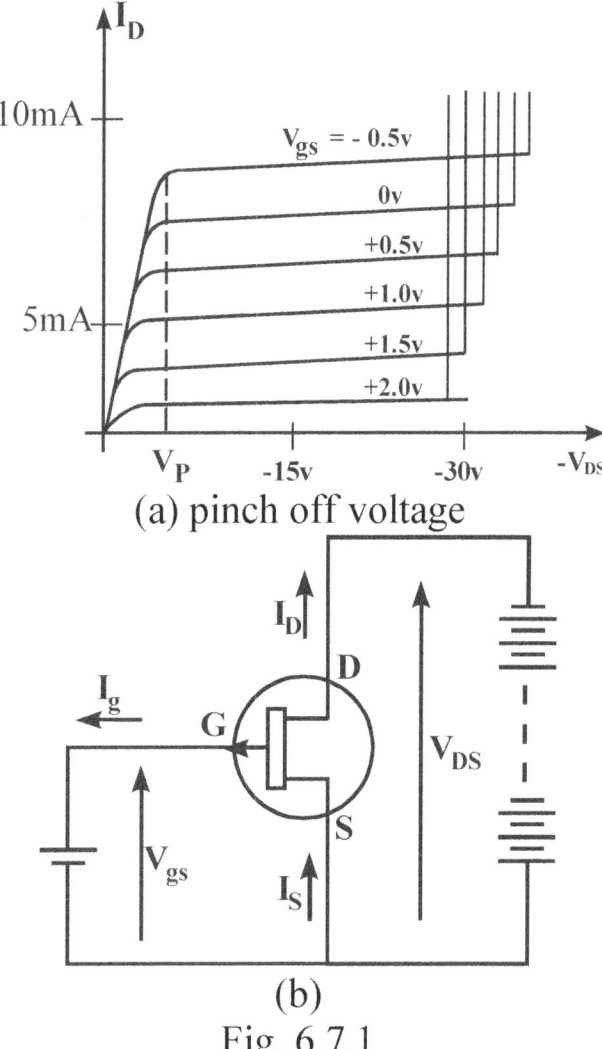

Fig. 6.7.1

At low values of V_{DS} the channel behaves as a simple semiconductor and I_D is a linear function of V_{DS} the channel being open. The voltage drop due to I_D depletion region and the source terminal reverse biases the junction and the channel begins to restrict due to the field electric. As V_{DS} is increased further a voltage V_P is reached where the channel is almost totally restricted i.e. 'pinched off'. This region is defined as non-linear because the field is not uniform in the channel. At this voltage I_D tends to a constant (approximate) value.

The channel does not close entirely (if it did I_D would be reduced to zero) because a voltage drop between the depletion region and source is necessary to reverse bias the junction and create the restricting field. As V_{DS} is increased further beyond "punched off", V_P; I_D increases only slightly until avalanche breakdown occurs between gate and source. This value decreases as the reverse bias increases since the reverse bias adds to the drain voltage. The characteristics are approximately equispaced for equal changes in gate-source voltage.

6.8 PARAMETERS

For small signal operation the F.E.T is generally operated in the linear region of its characteristics from these it is seen that

$$I_D = f(V_{DS}, V_{GS})$$

$$\delta I_D = \frac{\delta I_D}{\delta V_{DS}} \times \delta V_D + \frac{\delta I_D}{\delta V_{GS}} \times \delta V_{GS}$$

Proceeding analogous to the derivation of valve parameters
It is seen that

$$\delta I_D = \frac{1}{r_d} \delta V_D + gm \times \delta V_{GS}$$

where

$$r_d = \frac{\delta V_{DS}}{\delta I_D}\bigg|_{\delta V_{GS}=0} = \text{the drain resistance } (\sim 10 \rightarrow 100 K\Omega)$$

$$gm = \frac{\delta I_D}{\delta V_{GS}}\bigg|_{\delta V_{DS}=0} = \text{transconductance} \quad (0.1 \rightarrow 5mA/V)$$

(mutual conductance)

also donated as g_{fs} or y_{fs} *(forward transadmittance)*.

The amplification factor μ is defined as $\mu = g_m r_d$

An equivalent circuit can be derived fro small signal operation by method identical to that used for valves which gives the following equivalent circuit (fig. 6.8.1) for the output circuit.

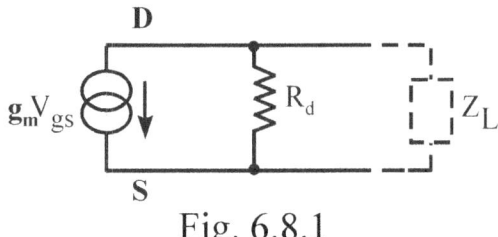

Fig. 6.8.1

Since the input is applied between gate and source junction which is reverse biased the resistance r_{gs} is very large ($\sim 10^8 - 10^{10}\Omega$) and for most purposes this can be regarded as an open circuit

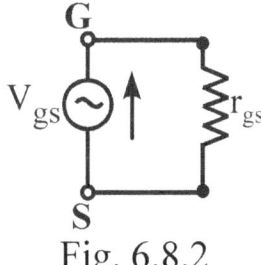

Fig. 6.8.2

Similarly the resistance between gate and drain r_{gs} (fig. 6.8.2) is of the same order of magnitude and may be omitted. The complete equivalent circuit then becomes as shown in fig. 6.8.3

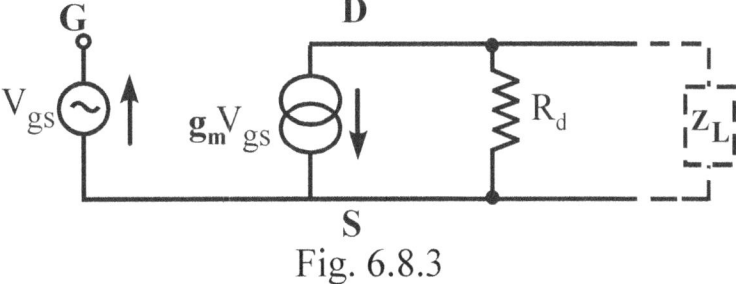

Fig. 6.8.3

Here the interelectrode capacitance C_{gs}, C_{DS}, C_{go} have been omitted. Note that striking similarity of the characteristics of the F.E.T with those of the Pentode valve amplifier. Thus analysis, biasing, coupling procedures are analogous to those used for valve circuits. Its main advantage over conventional (bipolar) transistor is its high input impedance and low noise and inherent thermal stability since the majority carrier current decreases with temperature (c. t. transistor where minority current increases with temperature). Its main disadvantages is that the gain-bandwidth product is small compared to the conventional transistor.

EXAMPLE 6.8.

It is required to

(i) Sketch the equivalent circuit of the F.E.T circuit at undband frequencies where

    Reactance of C is negligible

    Reactance of C_o is very high

    Source adequately decoupled

(ii) Find the expression for voltage gain

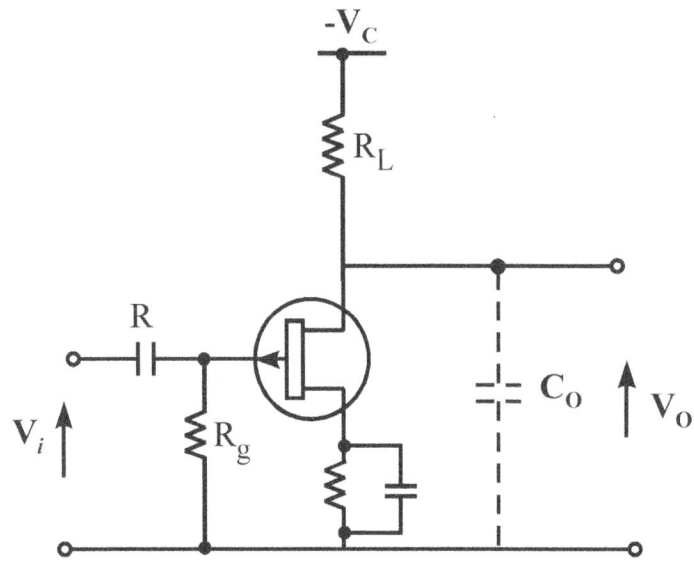

ELECTRONIC ENGINEERING CIRCUITS

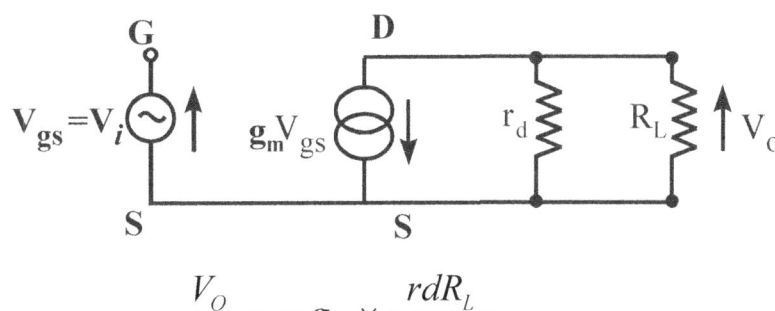

Midband voltage gain $\dfrac{V_O}{Vi} = -g_m \times \dfrac{r_d R_L}{r_d + R_L}$

CHAPTER SEVEN

7.0 POWER AMPLIFIERS

7.1 INTRODUCTION

Previous discussion of amplifiers has centred on the necessity to provide an output voltage (current) with no specific consideration of power level. It becomes important to consider both voltage and current levels where power is required to drive say a heating element, servomotor, loudspeaker etc.

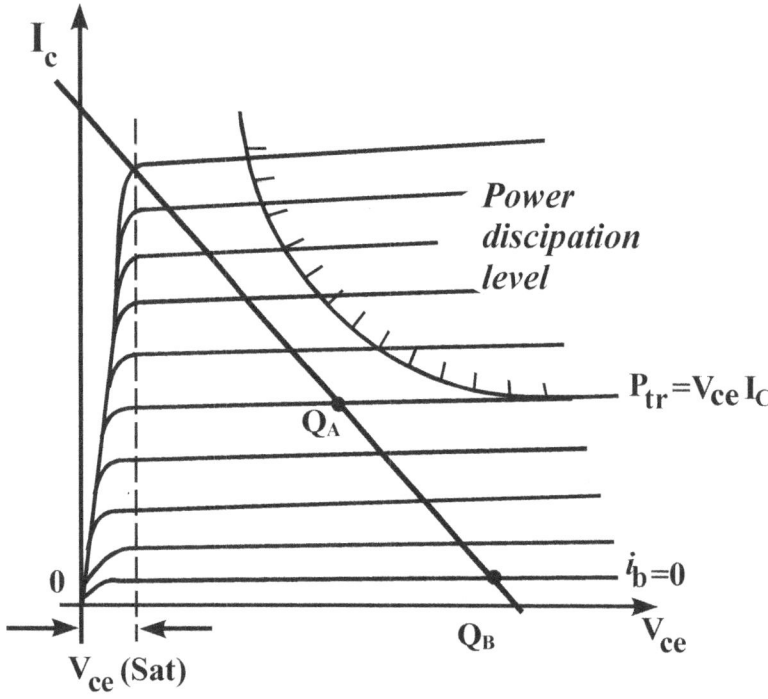

Fig. 7.1.1

Transistor must be operated below power dissipation characteristics (see fig. 7.1.1) for linear amplification. For small values of I_C and V_{Ce} distortion result due to cutoff ($i_b \sim 0$) and bottoming ($V_{Cs(sat)}$, $i_{b\ (large)}$) respectively. Operation is classified according to the method of biasing. Considering sinusoidal (or any wave shape) input with the transistor operating within its power limits:

i. **CLASS A:** is biased to the midpoint of the characteristics and thus collector current flows in the output transistor(s) for the complete cycle of input voltage (current).

ii. **CLASS B:** is biasing to collector current cutoff Q_B, hence current flows for one half of input cycle only.

iii. **CLASS AB**: is biased between Q_A and Q_B. Usually slightly above cut off. Current flows for more than ½ cycle but less than full cycle of input excursion.

iv. **CLASS C**: biased beyond cutoff therefore, current flows for less than ½ cycle. Hence cannot be directly used as a linear amplifier although it has uses in R.F amplifiers and modulators.

Design of large signal power amplifiers usually aims for maximum power output without exceeding transistor dissipation and causing excessive distortion.

Parameters used are:

i. Power output P_O = a.c power developed in load.

ii. Power output $P_{d.c}$ = $P_{d.c}$ power supplied by h.t source not that necessary dissipation = P_{tr}. Efficiency $\eta = P_O / P_{d.c}$

7.2 SINGLE ENDED AMPLIFIERS

CLASS A: Consider resistive load only driven directly or via transformer.

DIRECT: Fig. 7.2.1a shows the circuit and fig. 7.2.1b the characteristics

POWER AMPLIFIERS

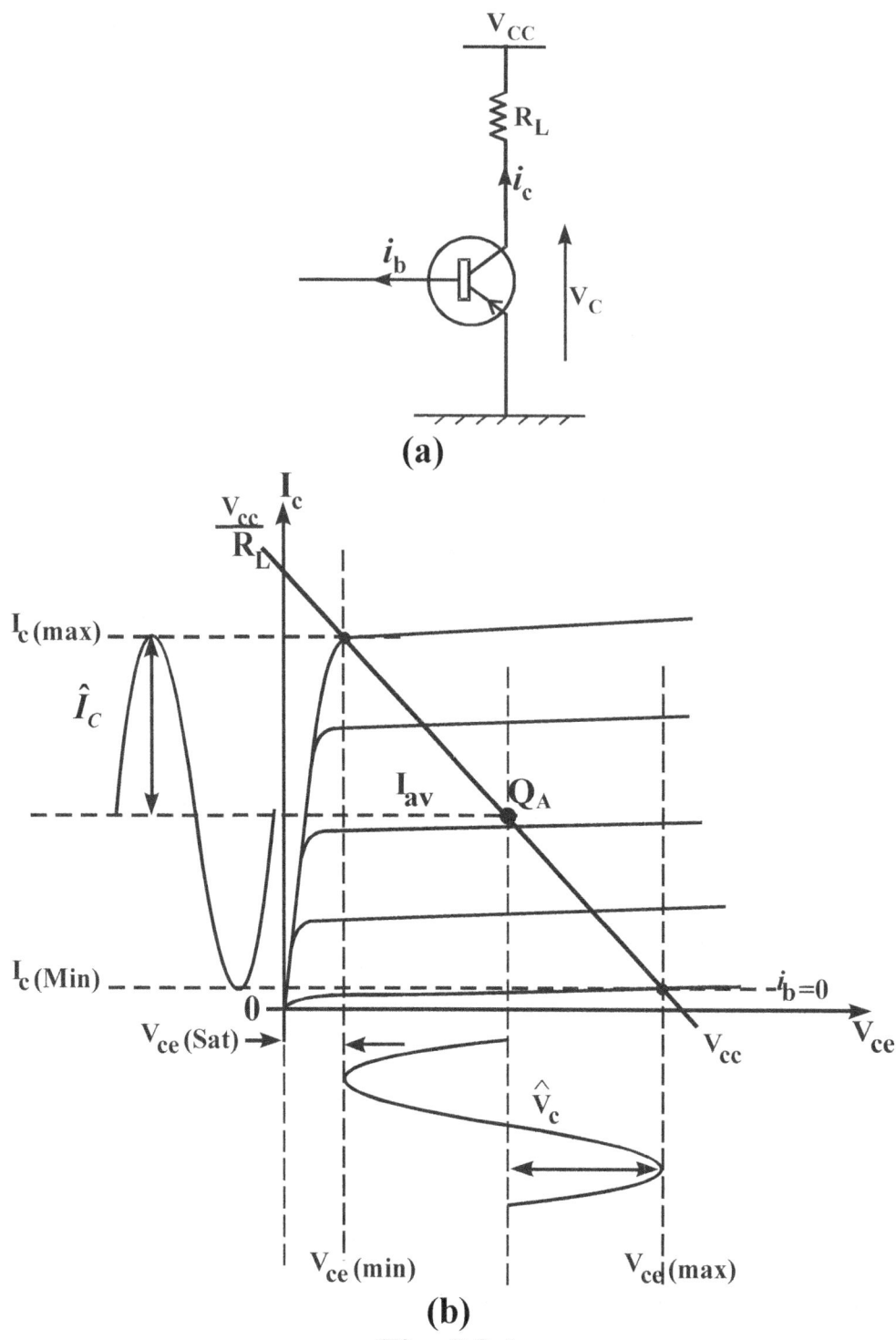

Fig. 7.2.1

$$P_K - P_K \text{ voltage} = 2V_C$$
$$= V_{C(\max)} - V_{C(\min)}$$
$$P_K - P_K \text{ current} = 2\hat{I}_C$$
$$= I_{C(\max)} - I_{C(\min)}$$

These are the maximum values which can be obtained without distortion.

Hence maximum undistortion a.c power output = $P_O = V_{r.m.s} \cdot I_{i.m.s}$

$$P_O = \frac{\hat{V}_C}{\sqrt{2}} \cdot \frac{\hat{I}_C}{\sqrt{2}} = \frac{[V_{C(max)} - V_{C(min)}][I_{C(max)} - I_{C(min)}]}{8} \text{(watts)}$$

Under normal circumstances the design is made such that

$$V_{CC} \gg V_{C(min)} = V_{Ce(sat)}$$

and that $I_{C(min)}$ is negligible (S_i transistors). These condition then give

$$V_{C(max)} = V_{CC} \ ; \ I_{C(max)} = V_{cc}/R_L$$

Therefore $P_{O(max)} = \dfrac{V_{CC}^2}{8R_L}$ watts

Power delivered by supply, $P_{dc} = I_{av} V_{CC}$. Note this is constant irrespective of signal level.

$$P_{dc} = \frac{V_{CC}}{2R_L} \cdot V_{CC} = \frac{V_{CC}^2}{2R_L}$$

Hence max^{im} efficiency $\gamma_{max} = \dfrac{P_{O(max)}}{P_{dc}} = \dfrac{1}{4} = 25\%$

N.B In the absence of input signal $V_C = V_{CC}/2$ and power dissipated in transistor for this condition is $P_{tr} = \dfrac{P_{dc}}{2} = \dfrac{V_{CC}^2}{4R_L} (watts)$

Since power from supply is constant $P_{dc} = P_O + P_{tr} + P_{RL}$ (see fig. 7.2.2). where P_{RL} is dc power expended in load = $I_{av}R_L$ assuming collector current is linear and a function of base current, $i_c \propto i_b$.

For constant supply V_{CC}

$$P_O \propto I_b^2$$

to $i_{b(max)}$ the max^{im} base current where bottoming occurs.

POWER AMPLIFIERS

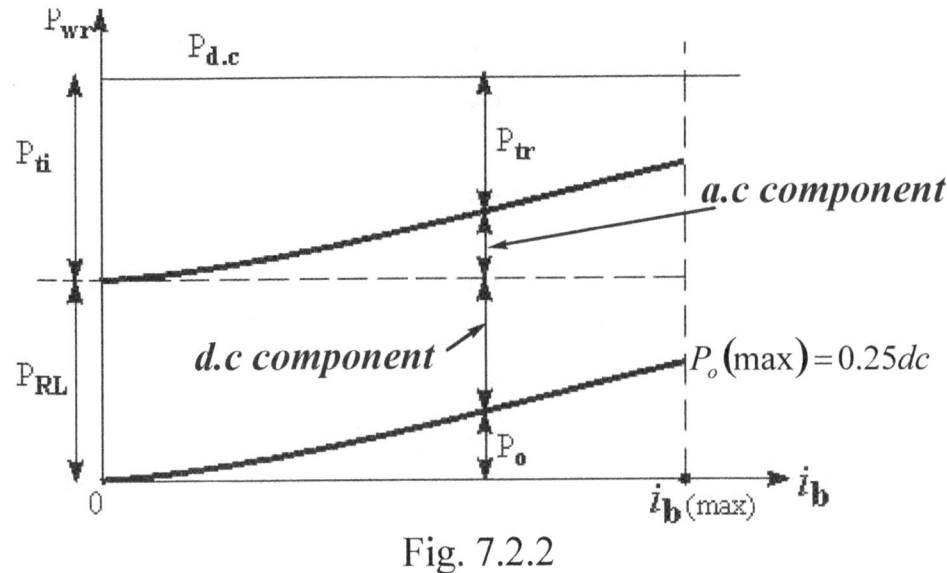

Fig. 7.2.2

7.3 TUNED CIRCUITS AND AMPLIFIERS

In the amplifiers so far discussed the general design has been to amplify over as wide a range as possible. Conversely where the requirement is amplification over a selective range of frequency only tuned circuits are employed.

Tuned Circuits Review

Parallel resonance (fig. 7.3.1)

Impedance of inductor $= R + jwL$

Impedance of capacitor $= \dfrac{1}{jwC}$

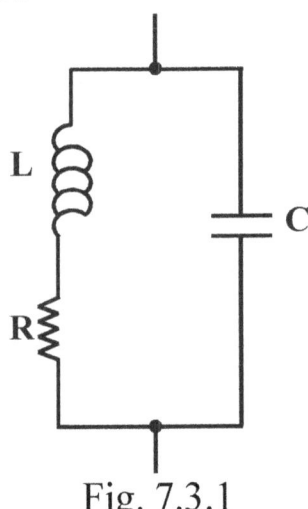

Fig. 7.3.1

POWER AMPLIFIERS

Total impedance of parallel circuit

$$Z = \frac{(R+jwL)\frac{1}{jwL}}{R+jwL+\frac{1}{jwC}} = \frac{(R+jwL)\frac{1}{jwC}}{R+j\left(wL-\frac{1}{wC}\right)}$$

In practice R is quite small (few ohms) and we may assume at the frequency concerned that R<<wL but R not << $\left(wL-\frac{1}{wC}\right)$

Since the letter may be small.

Therefore
$$Z = \frac{jwL\Big/jwC}{R+j\left(wL-\frac{1}{wC}\right)} = \frac{L/C}{R+j\left(wL-\frac{1}{wC}\right)} = \frac{L/CR}{1+j\frac{wL}{R}\left(1-\frac{1}{w^2LC}\right)}$$

at resonance $w = w_r$ and $w_r^2 LC = 1$

also $Z = \frac{L}{CR} = R_d$ a pure resistance of high value (dynamic impedance)

The Q factor (or voltage magnification) of a coil is defined as

$$\frac{Voltage\ across\ pure\ inductance}{Voltage\ across\ pure\ resistance} = \frac{I_L wL}{I_L R} = \frac{wL}{R}$$

Thus the impedance of the parallel circuit at any given frequency is

$$Z = \frac{R_d}{1+jQ\left(1-\frac{w_r^2}{w^2}\right)}$$

A plot of the impedance Z and frequency ω is shown in fig. 7.3.2

When the current is 'off tune' a fractional detuning factor δ can be defined as

$$\delta = \frac{w-w_r}{w_r}$$ i.e the fractional derivation from the resonant frequency. Therefore $(1 + \delta) = \frac{w}{w_r}$

POWER AMPLIFIERS

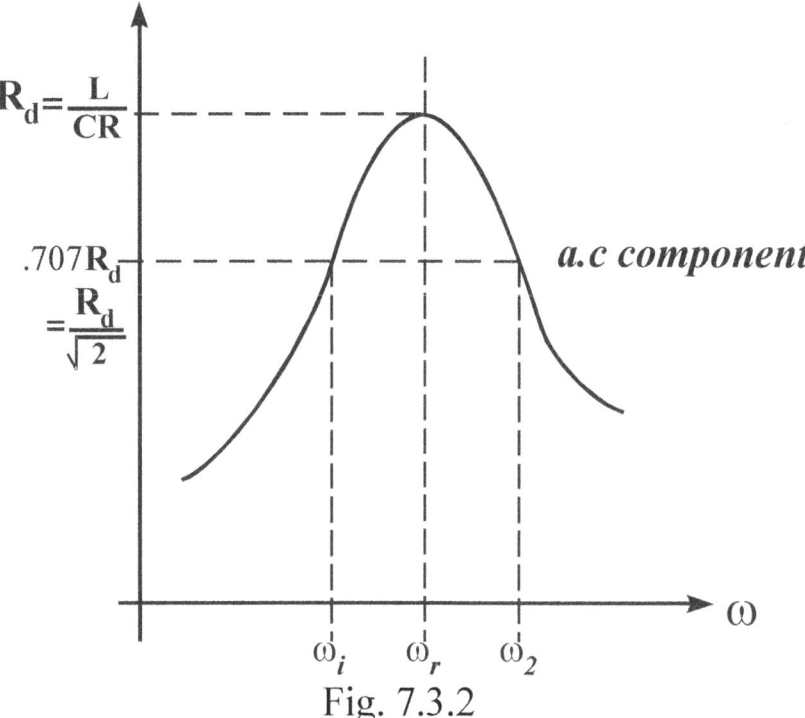

Fig. 7.3.2

Hence $(1+\delta)^{-2} = \dfrac{w_r^2}{w^2} = 1 - 2\delta$ (by Binomial theorem since δ is small)

Substituting $Z = \dfrac{R_d}{1+jQ(1-1(1-2\delta))} = \underline{\dfrac{R_d}{1+j2Q\delta}}$

When the impedance of the circuit falls to $0.707 R_d = \dfrac{R_d}{\sqrt{2}}$

$$|Z| = \dfrac{R_d}{\sqrt{1+4Q^2\delta^2}} = \dfrac{R_d}{\sqrt{2}}$$

Thus $\quad 4Q^2\delta^2 = 1$

$\quad\quad\quad 2Q\delta = 1$

Thus if the angular frequency is w_2 when Z falls to $0.707\, R_d$

$\delta = \dfrac{w_L - w_r}{w_r}$ and hence

$$Q = \dfrac{1}{2\delta} = \dfrac{w_r}{2(w_L - w_r)}$$

For the coil $Q = \dfrac{wL}{R}$

At resonance $Q_r = \dfrac{w_r L}{R}$

POWER AMPLIFIERS

but $w_r^2 LC = 1$ $\therefore w_r L = \dfrac{1}{w_r C}$ and $w_r = \dfrac{1}{\sqrt{LC}}$

Therefore $Q_r = \dfrac{w_r L}{R} = \dfrac{1}{w_r RC} = \dfrac{1}{\sqrt{LC}} \cdot \dfrac{L}{R} = \dfrac{1}{R}\sqrt{\dfrac{L}{C}}$

Also $R_d = \dfrac{L}{CR}$ $\therefore \dfrac{L}{R} = CR_d$ and $CR = \dfrac{L}{R_d}$. Substituting $Q_r = w_r CR_d = \dfrac{R_d}{w_r L}$

$Q_r = \dfrac{1}{R}\sqrt{\dfrac{L}{C}} = \dfrac{C}{L} R_d \sqrt{\dfrac{L}{C}}$ $\dfrac{1}{R} = \dfrac{C}{L} R_d$

$= R_d \sqrt{\dfrac{C}{L}}$ $\dfrac{L}{C} = RR_d$

$= \dfrac{1}{R}\sqrt{R_d R} = \sqrt{\dfrac{R_d}{R}}$ $L = CR_d R$

These are some of the expressions used to define and calculate the Q factor of a coil tuned to parallel resonance by a capacitor.

Consider the circuit of fig. 7.3.3

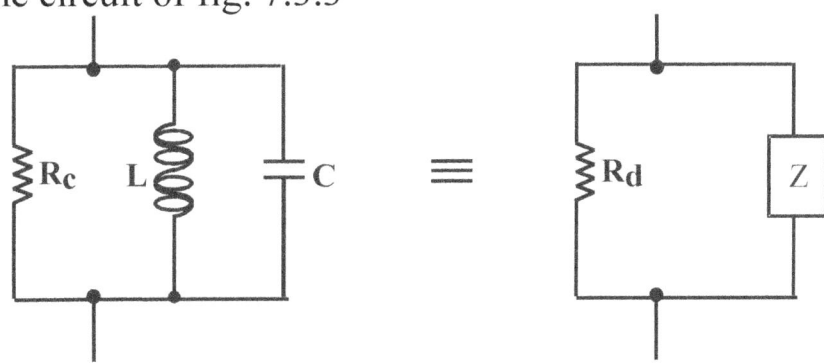

Fig. 7.3.3

$= \dfrac{jwL \cdot \dfrac{1}{jwC}}{jwL + \dfrac{1}{jwC}} = \dfrac{L/C}{j\left(wL - \dfrac{1}{wC}\right)}$

The total impedance,

$Z = \dfrac{R_d \dfrac{L/C}{j\left(wL - \dfrac{1}{wC}\right)}}{R_d + \dfrac{L/C}{j\left(wL - \dfrac{1}{wC}\right)}} = \dfrac{R_d}{1 + j\dfrac{R_d C}{L}\left(wL - \dfrac{1}{wC}\right)}$

Putting $R_d = \dfrac{L}{CR}$

$$Z = \dfrac{\dfrac{L}{CR}}{1+\dfrac{j\left(wL-\dfrac{1}{wC}\right)}{R}} = \dfrac{\dfrac{L}{CR}}{1+j\dfrac{wL}{R}\left(1-\dfrac{1}{w^2LC}\right)}$$

$$= \dfrac{R_d}{1+j2Q\delta}$$

Thus if the initial assumption (R<<wL) holds the two circuits below (fig. 7.3.4) are equivalent.

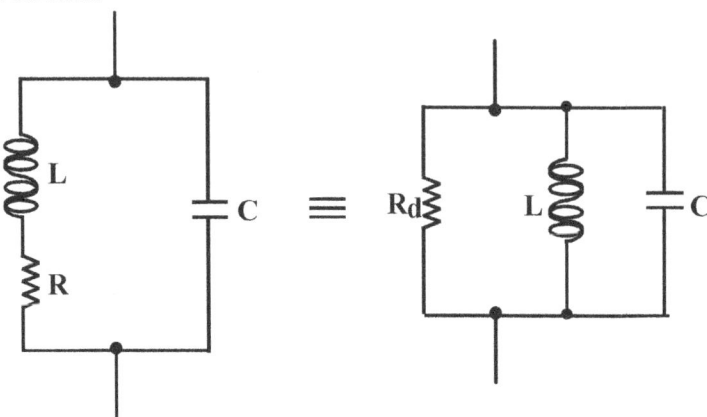

Fig. 7.3.4

Thus the practical circuit on the left may be represented by an ideal parallel L-C circuit. Shunted by a high dynamic impedance $R_d = \dfrac{L}{CR}$

7.4 F.E.T TUNED AMPLIFIERS

Fig. 7.4.1 shows the circuit diagram (of F.E.T tuned amplifier), the equivalent circuit (b) and the drain resistance r_d shunting the tuned circuit. Here the tuned circuit is shunted by the drain resistance r_d which has the result of reducing the Q factor of the tuned circuit.

In fig. 7.4.1, $R_d = \dfrac{r_d R_d}{r_d + R_d}$

POWER AMPLIFIERS

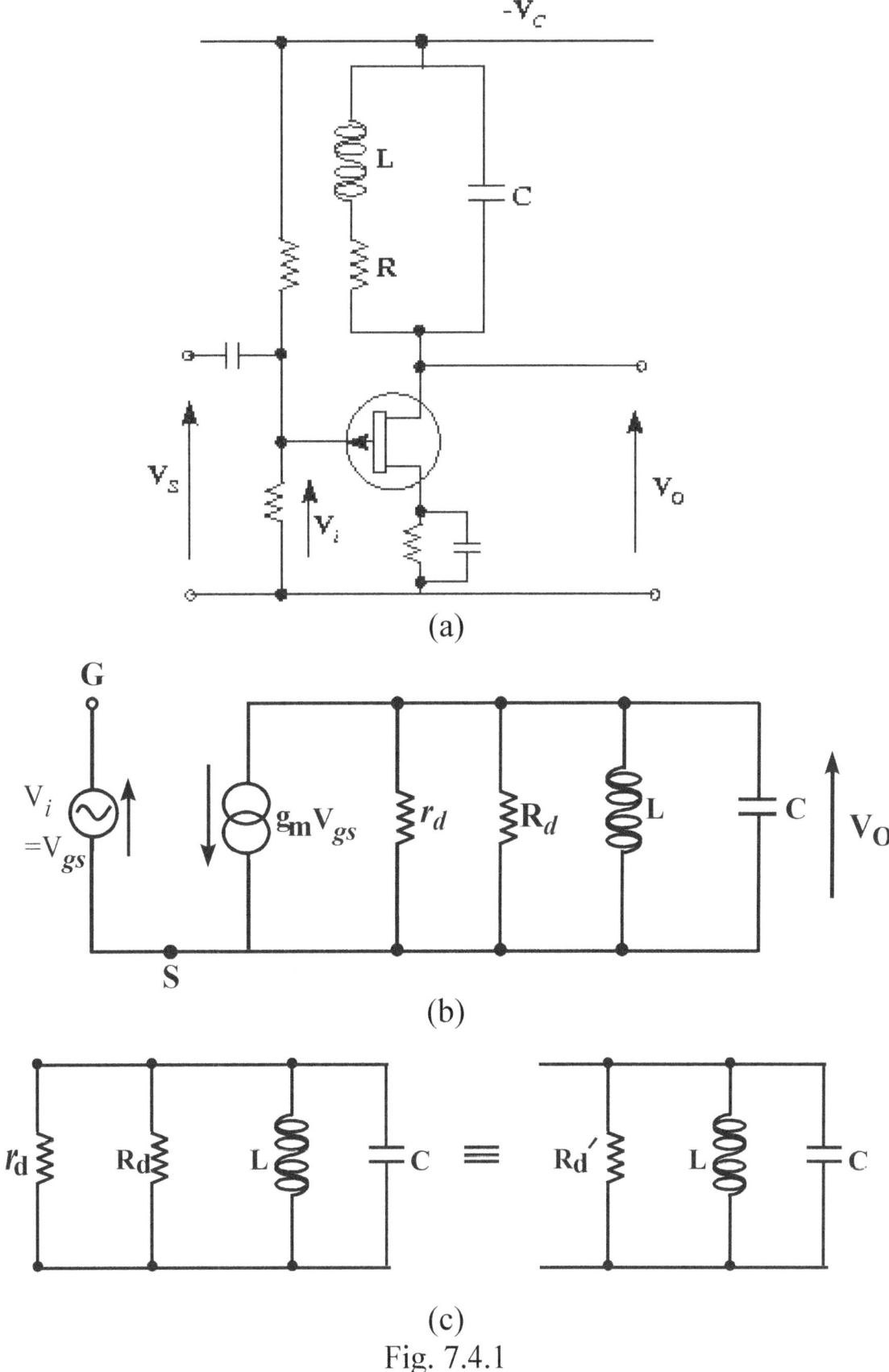

Fig. 7.4.1

Thus the loaded Q or effective Q of the tuned amplifier circuit is

$$Q_r' = w_r C R_d'$$

$$= w_r C R_d \cdot \frac{r_d}{r_d + R_d}$$

$$Q_r' = \frac{Q_r}{1 + \frac{R_d}{r_d}}$$

Therefore loaded Q factor

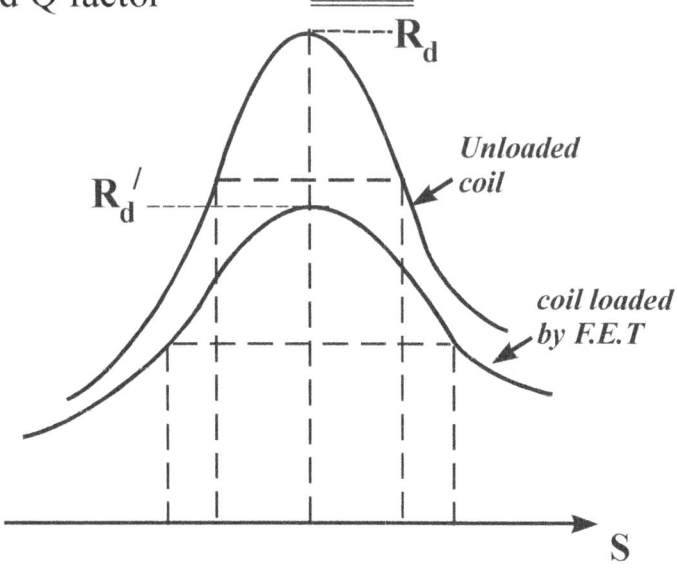

Fig. 7.4.2

This equation shows that the presence of r_d not only reduces the impedance at resonance from $R_d + R_d'$ but decreases the selectivity of the coil i.e. the ability to discriminate over a narrow band of frequency, see fig. 7.4.2. The impedance of the tuned circuit at resonance is R_d and the voltage gain of the FET tuned amplifier at resonance is given by

$$\left|\frac{V_o}{V_i}\right| = g_m r_d \cdot \frac{R_d}{r_d + R_d} = \frac{g_m R_d}{1 + \frac{R_d}{r_d}}$$

$$\left|\frac{V_o}{V_i}\right| = g_m Q_r' w_r L$$

Thus drain resistance r_d should be high so that F.E.T does not unduly load the tuned circuit.

POWER AMPLIFIERS

In valve circuits Pentode are used in preference to triodes since

r_a Pentode $\gg r_a$ triode

Since amplifier bandwidth is defined as the frequency range over which the gain does not fall below 0.707 of its maximum value (or by 3dB) see fig. 7.4.3 for sketch diagram of gain of amplifier and frequency

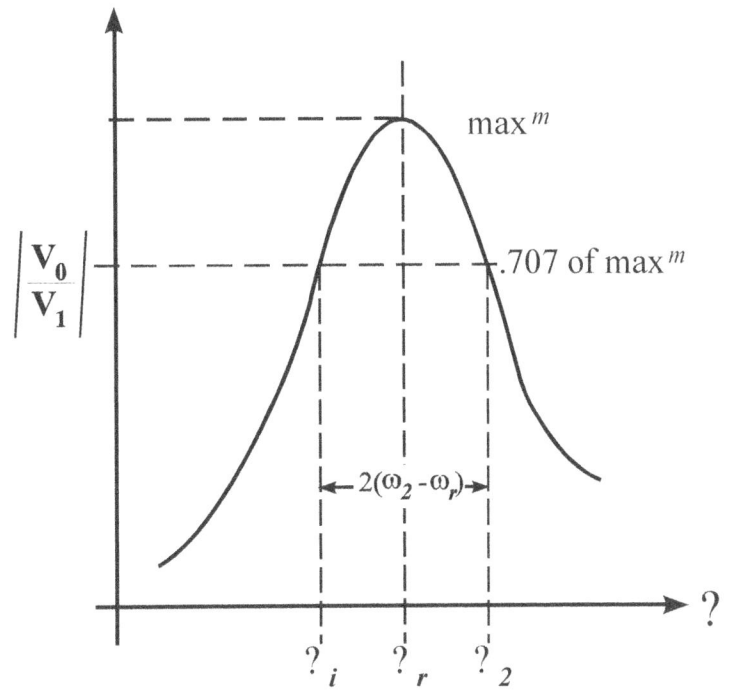

Fig. 7.4.3

$$Q_r = \frac{w_r}{2(w_2 - w_r)}$$

Bandwidth $= 2(w_2 - w_r) = \frac{1}{Q_r} \cdot w_r \, (rad/\sec)$

Thus for amplifier bandwidth $= \dfrac{1}{Q_r'} \cdot w_r$

or Bandwidth (Hz) $= \dfrac{1}{Q_r'} \times f_r$

7.5 TRANSFORMER COUPLED LOAD – CLASS A AMPLIFIER

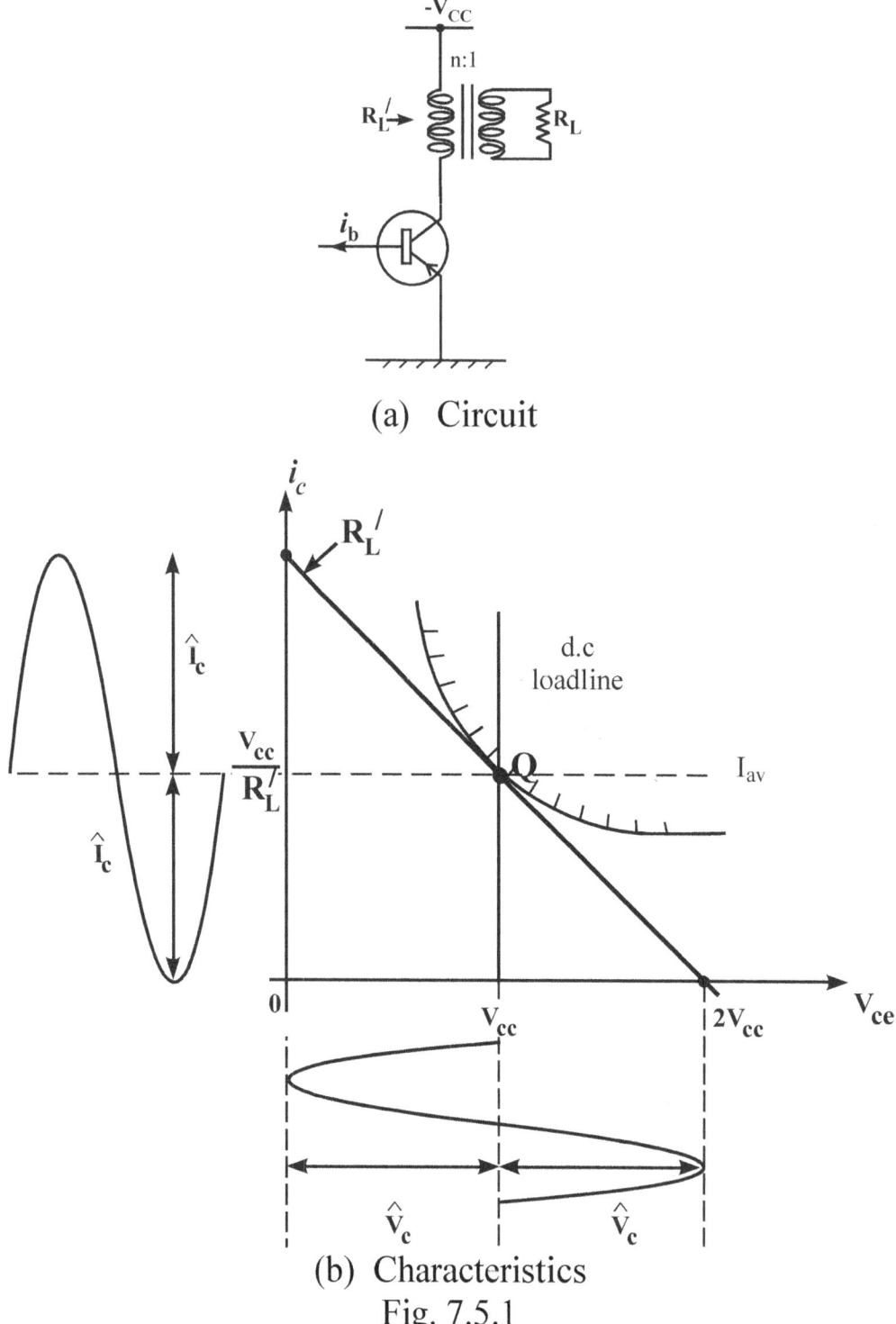

(a) Circuit

(b) Characteristics
Fig. 7.5.1

Fig. 7.5.1 shows the circuit of a transformer coupled load. An n:1 turn transformer is used to isolate the load from d.c currents and allows some flexibility in choice of effective load $R_L^/$ seen by transistor. Thus having

POWER AMPLIFIERS

no d.c in load reduces losses and increases efficiency. For an ideal transformer the effective a.c collector load is given by

$$R_L' = n^2 R_L$$

The d.c loadline is nearly vertical since the winding resistance of the transformer is usually small compared with R_L'. As before the transistor can swing from $I_c = 0$ to $V_{ce} = 0$ and biasing is at center of load line. Thus the quiescent point is at V_{cc} with a.c input signals the collector will therefore swing above and below V_{cc}. Note that maximum collector voltage is now V_{cc} which must be less that transistor breakdown voltage. Q is chosen to lie on maximum dissipation curve $P_{tr(max)}$ (hyperbola) to obtain maximum efficiency. With no input power $P_o = 0$ and $P_{dc} = I_{av} V_{cc}$. (Note again this is constant) is dissipated in the transistor.

Hence $P_{dc} = P_o + P_{tr}$ (see fig. 7.5.2

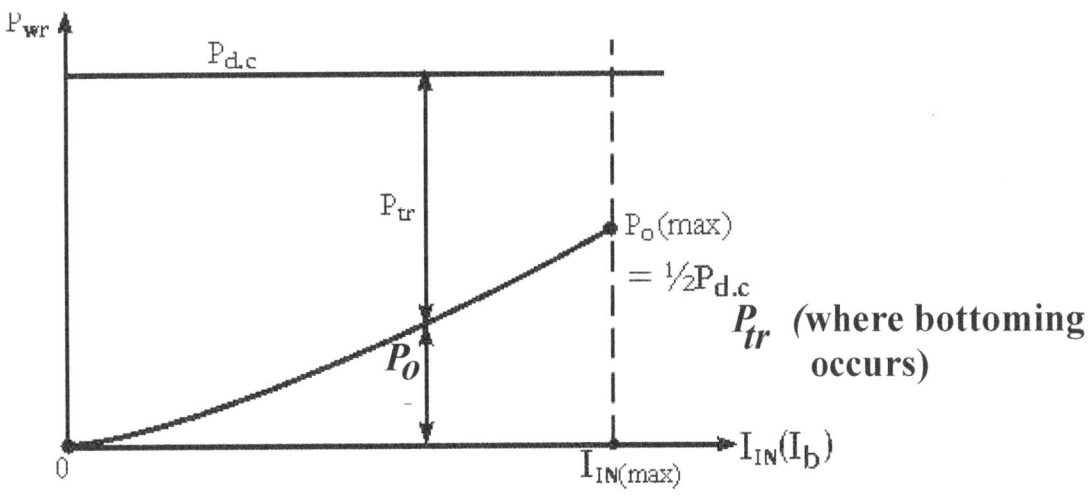

Fig. 7.5.2

Thus transistor dissipation P_{tr} decreases as P_o increases. For maximum signal input conditions the output has a peak value $\hat{V}_C = V_{CC}$

Peak value of current $\hat{I}_C = V_{CC}/R_L'$

Maximum output power $P_{O(max)} = \dfrac{\hat{V}_C}{\sqrt{2}} \cdot \dfrac{\hat{I}_C}{\sqrt{2}} = \dfrac{V_{CC}{}^2}{2R_L{}'} (watts)$

$P_{dc} = I_{av} V_{CC} = \dfrac{V_{CC}{}^2}{R_L{}'} watts.$

Hence $\eta_{max} = \dfrac{P_{O(max)}}{P_{dc}} = \dfrac{1}{2} = 50\%$

Note that $P_{tr(max)} = 2P_{O(max)}$ and power burden on transistor is exactly same as for direct load case but for a given transistor breakdown voltage the supply can only be half as large. The great advantage of this connection is that when the load is predetermined (e.g. loudspeaker) optimum matching for maximum power transfer can be achieved by adjusting the turns ratio.

Large power outputs

If a single transistor cannot supply sufficient power to a given load then the configuration of fig. 7.5.3 may be used.

1. Paralleled Transistor

Direct paralleling is uncommon because of problems giving use to thermal instability if one transistor supplies more than its fair (rated) share of load emitter resistor (unbypassed) and base resistors are included to ensure a more even distribution of power.

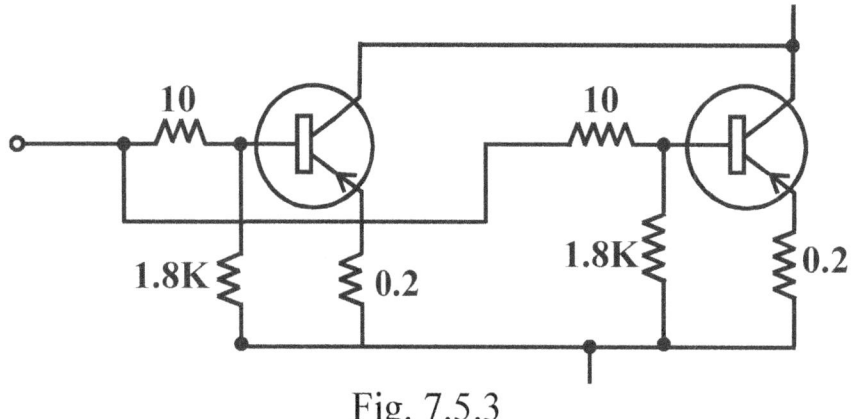

Fig. 7.5.3

(ii) Push Pull Amplifier

In the configuration shown in fig. 7.5.4 transistor must be driven in antiphase, hence the use of input transformer (or phase splitter and appropriate coupling).

If $V_1 = \hat{V} \cos wt$

$V_2 = \hat{V} \cos(wt + \pi)$

i_{b1} and i_{b2} are equal in magnitude but opposite in phase.

Assuming $i_c \propto i_b$ the collector currents produce a flux in the output transformer $\propto (i_{C1} - i_{C2})$.

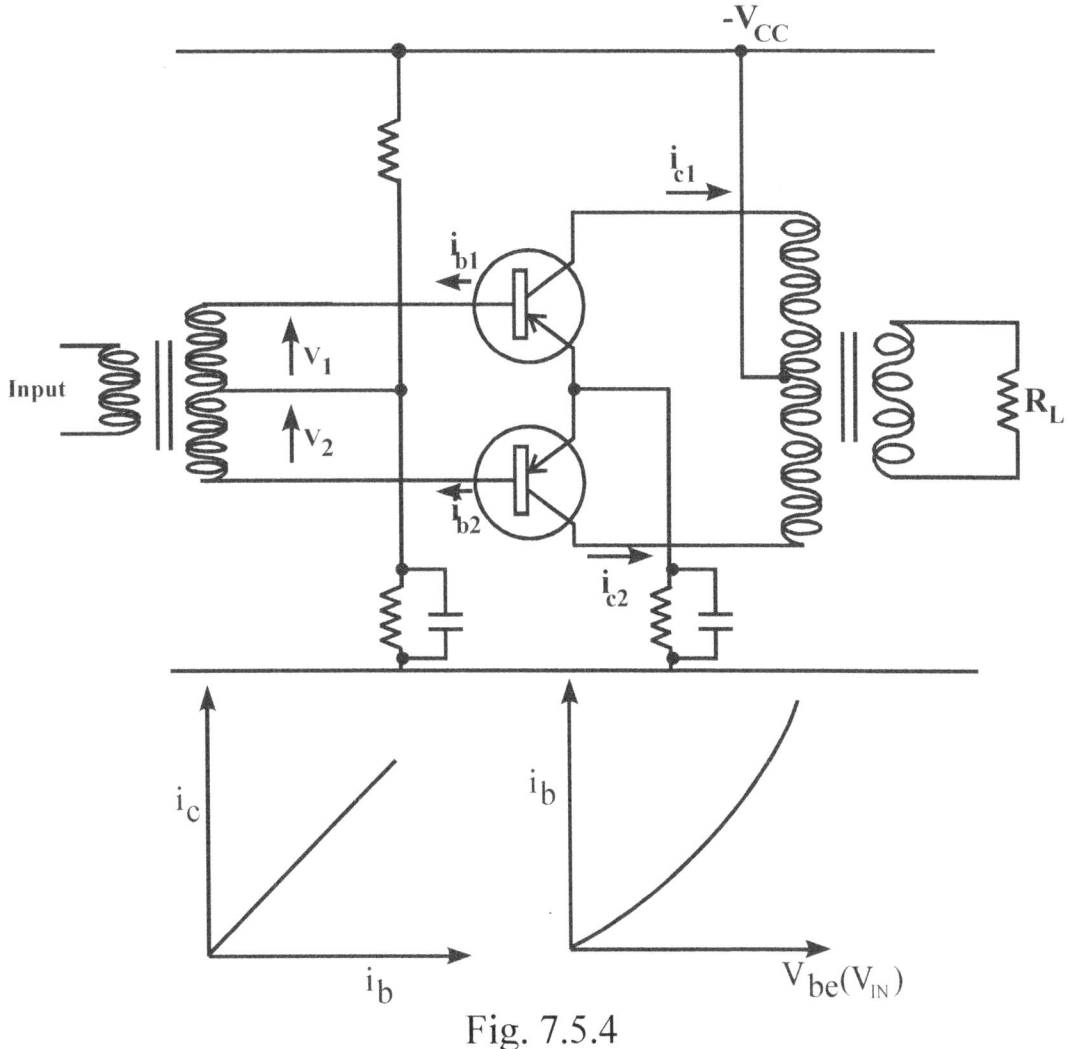

Fig. 7.5.4

POWER AMPLIFIERS

Since the input characteristic of transistor is non-linear

$$i_C \propto i_b \propto A + Bv_i + Cv_i^2 + Dv_i^3 + \text{------}$$

or $\quad i_C = a + b\hat{v}Coswt + c\hat{v}Cos\ wt + d\hat{v}Coswt + \text{----}$

This may be expanded and represented as a harmonic (Fourier series to give for each transistor.

$$i_{C1} = A_O + A_1 Cos\ wt + A_2 Cos\ 2wt + A_3 Cos\ 3wt + \text{--------------}$$
$$i_{C2} = A_O + A_1 Cos(wt + \pi) + A_2 Cos(2wt + \pi) + A_3 Cos(3wt + \pi) + \text{----}$$

Where A_O is the quiescent current in each transistor. Therefore flux

$$\propto (i_{C1} - i_{C2}) = 2A_1 Cos\ wt + 2A_3 Cos\ 3wt + 2A_5 Cos\ 5wt + \text{-----}$$

i.e.. Even harmonics cancel out and there is no net d.c magnetisation of the output transformer. Thus distortion is less than in a single ended amplifier. Which implies that a greater input signal can be applied for a given harmonic content in the output waveform. Since the dc components cancel a smaller transformer core can be used for larger ac input swings before saturation occurs.

7.6 CLASS A PUSH PULL

The circuit is in fig. 7.5.4 as with transistor biased in active regions for CLASS A. A.C currents flow in both transistors for complete input cycle (fig. 7.6.1). Putting two circuits in parallel gives both halves of transformer in operation simultaneously.

Analysis similar to that for single ended stage to give $\eta_{max} = 50\%$ but it should be noted that in this configuration both P_O and P_{dc} have doubled. Thus for a given transistor dissipation the output is doubled (with less

POWER AMPLIFIERS

distortion) for the same breakdown voltage and effective load. Fig. 7.6.2. give the equivalent circuit reduction of Class A push pull amplifier.

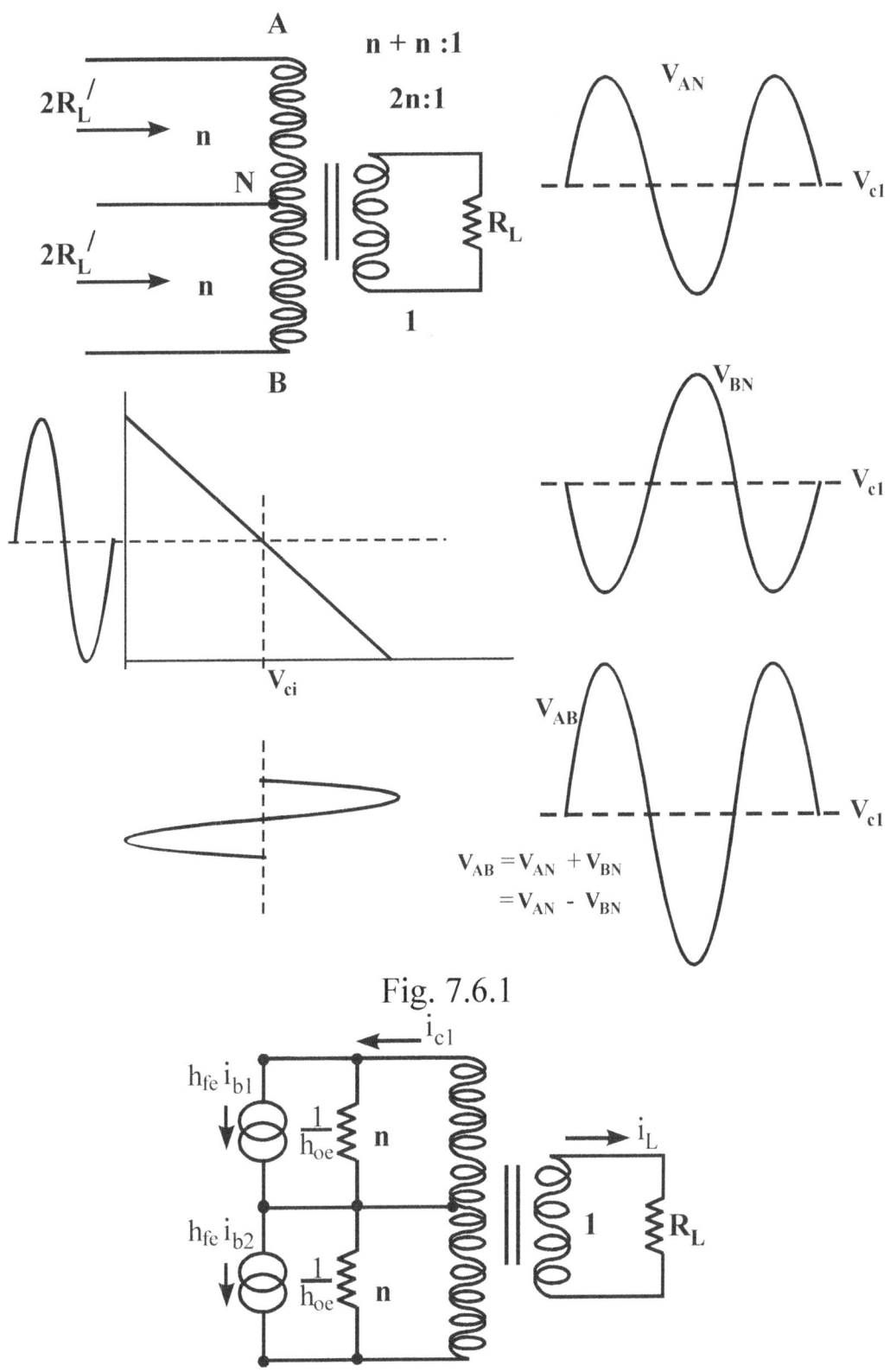

Fig. 7.6.1

Putting two circuits in parallel gives

POWER AMPLIFIERS

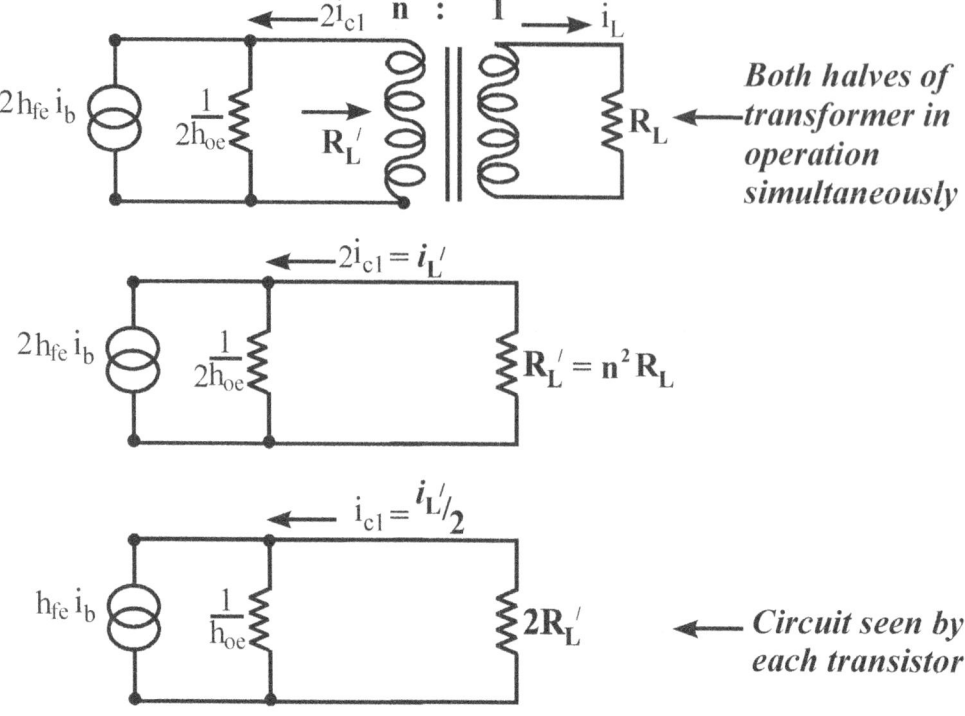

Fig. 7.6.2 Equivalent circuit reduction.

7.7 CLASS B PUSH PULL

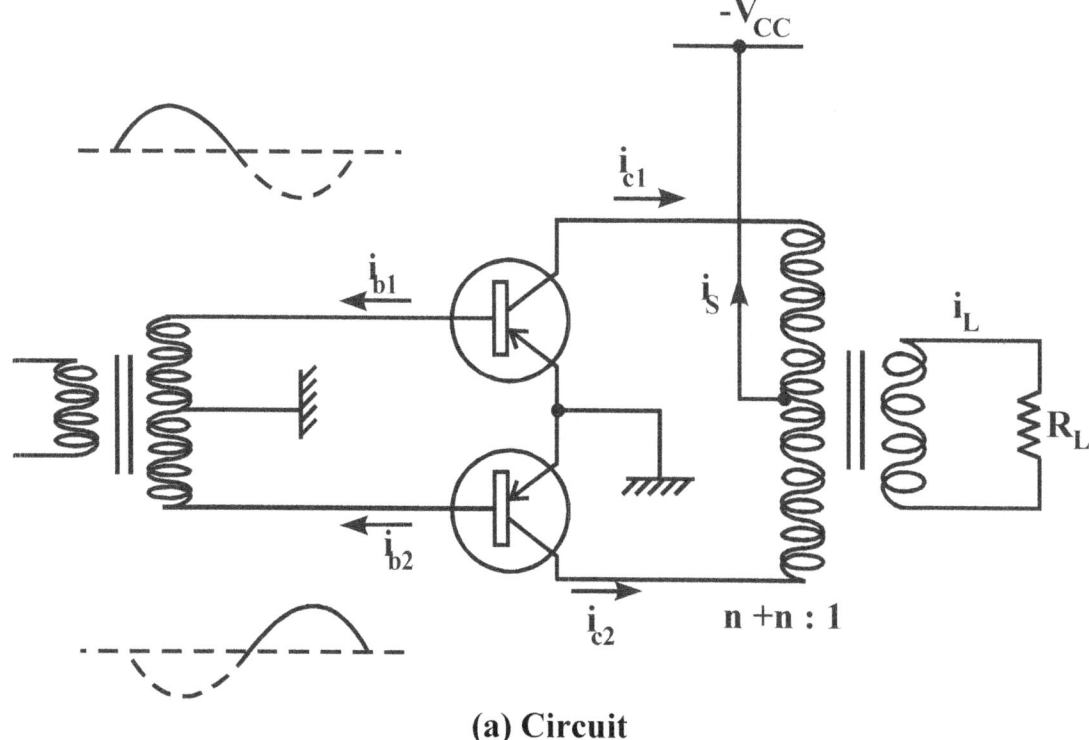

(a) Circuit

POWER AMPLIFIERS

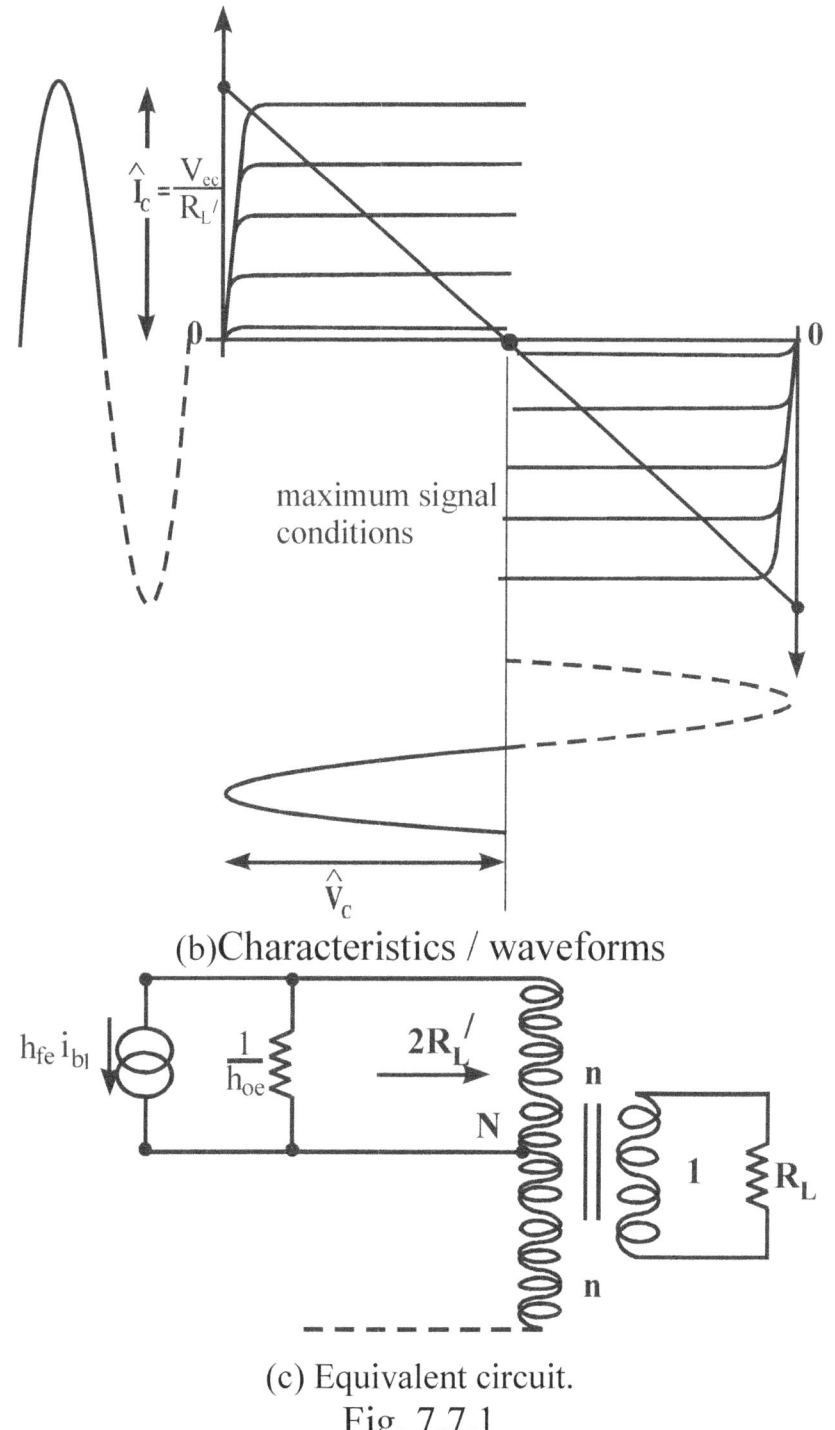

(b) Characteristics / waveforms

(c) Equivalent circuit.
Fig. 7.7.1

The circuit for class B push pull is as shown in fig. 7.7.1a with transistor biased to cut off. With this arrangement current flows in one half of transformer and the transistor only at any given time.

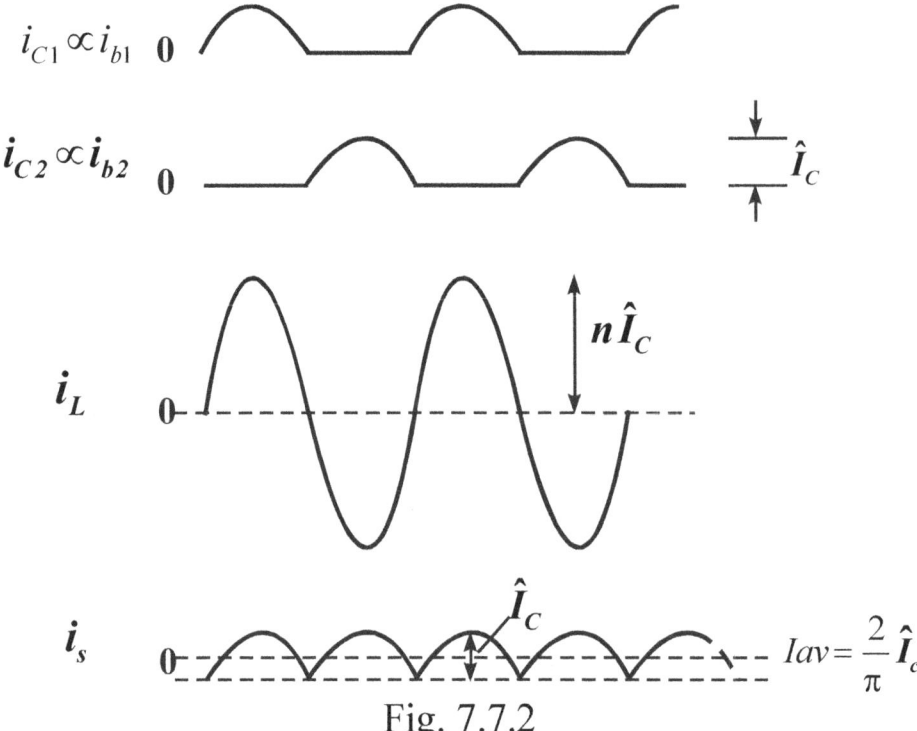
Fig. 7.7.2

Thus the effective load seen by each transistor is $R_L' = n^2 R_L$ see fig. 7.7.1c. Note that maximum base emitter voltage is $2V_{CC}$. Since during off portion voltage induced in other half of transformer (maximum of V_{CC}) is added to V_{CC}. (shown in fig. 7.7.1b)

Note in this case the d.c power is not constant but varies with input signal. Maximum dc power = $P_{dc(max)} = \dfrac{2}{\pi} \hat{I}_C V_{CC}$

Maximum collector current, $I_{C(max)} = \dfrac{V_{CC}}{R_L'}$

Maximum power output

$$= P_{O(max)} = \dfrac{\hat{v}_{C(max)}}{\sqrt{2}} \cdot \dfrac{\hat{I}_{C(max)}}{\sqrt{2}} = \dfrac{v_{CC}}{\sqrt{2}} \cdot \dfrac{v_{CC}}{\sqrt{2} R_L'} = \dfrac{v_{CC}^2}{2 R_L'} \; (watts) \qquad (\hat{v}_C = v_{CC})$$

Maximum efficiency

$$\eta_{(max)} = \dfrac{P_{O(max)}}{P_{dc(max)}} = \dfrac{V_{CC}^2}{2 R_L'} \cdot \dfrac{\pi R_L'}{2 V_{CC}^2} = \dfrac{\pi}{4} = 78.5\%$$

Note that if maximum signals are not employed the average current (and therefore power) from the supply falls.

POWER AMPLIFIERS

Since there is no d.c power in load

$$P_{dc} = P_O + 2P_{tr} \qquad \therefore P_{tr} = \frac{1}{2}(P_{dc} - P_O)$$

$$= \frac{1}{\pi}\hat{I}_C V_{CC} - \frac{\hat{I}_C^2 R_L'}{4}$$

But,

$$P_O = I_L^2 R_L = \left(\frac{\hat{I}_C}{\sqrt{2}}\right)^2 R_L = \frac{n^2 R_L \hat{I}_C^2}{2} = \frac{\hat{I}_C^2 R_L'}{2}$$

for maximum transistor dissipation $\dfrac{\delta P_{tr}}{\delta \hat{I}_C} = 0 = \dfrac{1}{\pi}V_{CC} - \dfrac{\hat{I}_C R_L'}{2}$

This occurs when $\hat{I}_C = \dfrac{2}{\pi}\dfrac{V_{CC}}{R_L'} = \dfrac{2}{\pi}\hat{I}_{C(max)}$

For maximum transistor dissipation occurs at approximately 64% of full input signal

$$P_{tr(max)} = \frac{2}{\pi^2}\frac{V_{CC}^2}{R_L'} = 0.202\, P_{O(max)}$$

At this condition $P_{dc} = \dfrac{2}{\pi} \cdot \dfrac{2}{\pi} \cdot \hat{I}_{C(max)} V_{CC}$

$$P_O = \frac{\hat{I}_C^2 R_L'}{2} = \frac{4}{\pi^2} \cdot \frac{R_L'}{2} \qquad \frac{(\propto I_b)}{(\propto I_b^2)}$$

$\eta = \underline{\underline{50\%}}$ at maximum transistor dissipation

Thus maximum power output is approximately 5 times maximum power dissipated in each transistor. Thus for same collector dissipation and CLASS B can deliver approximately 5 times more power to load than a CLASS A stage. Hence there is popular use of CLASS B for power output stages. A sketch of power and efficeiny against base current is shown in fig. 7.7.3.

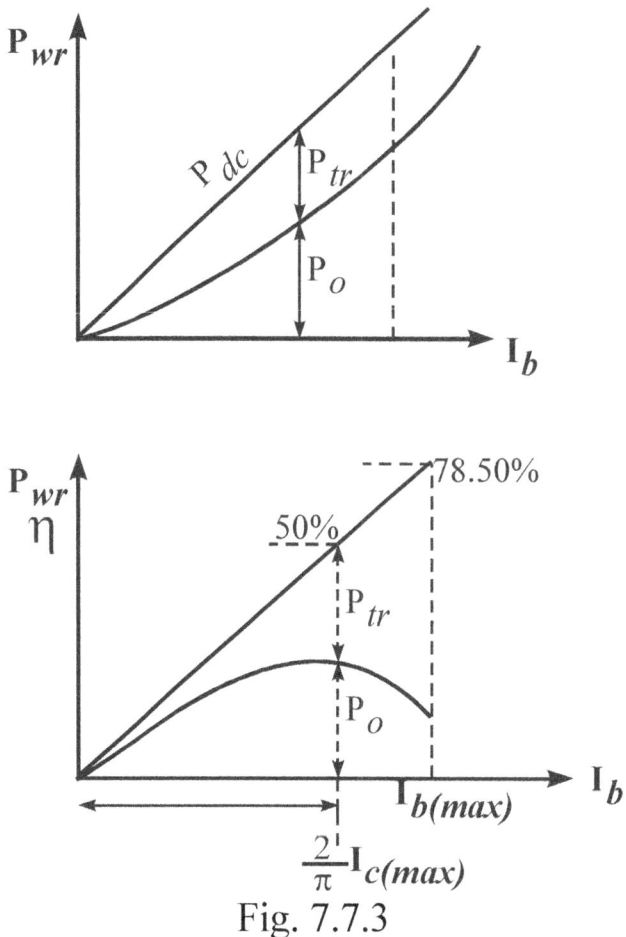

Fig. 7.7.3

7.8 CLASS AB PUSH PULL

When transistor are biased purely in CLASS B crossover distortion results during the period of time when the input signal passed through zero. Thus the transistors are slightly forward biased (CLASS AB) by R_1 and R_2 to a potential which is approximately a forward biased junction voltage.

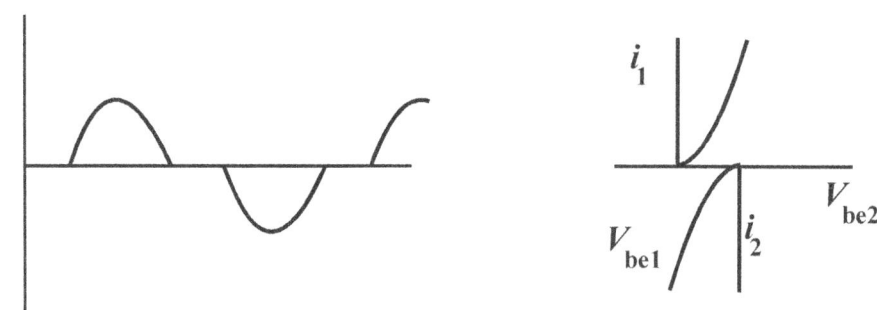

POWER AMPLIFIERS

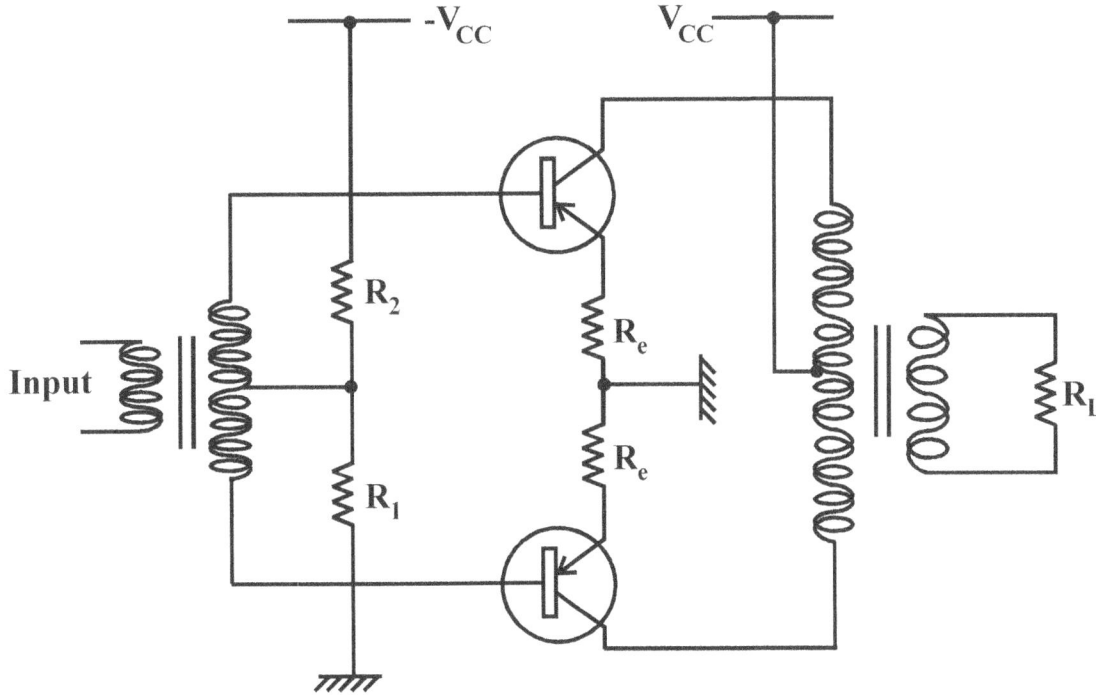

Fig. 7.8.1

For constant voltage biasing R_1 should be small (approximately few ohms) if the transistor input impedance is low (~ 30 - 50Ω). R_2 ~ 100Ω. R_1 cannot be increased significantly to reduce "bleed" otherwise;

1. Bias becomes less constant.
2. R_1 is in series with transistor input impedance and increasing it makes stage more difficult to drive.
3. Increasing R_1 and base circuit reduces d.c stability

Further it cannot be bypassed since full wave rectified currents from the two bases would produce a bias proportional to the magnitude of drive signal. An emitter resistor R_e is used to improve d.c stability. It too cannot be bypassed for reasons given above. (R_e ~ 0.1 - 1Ω) – see fig. 7.8.1)

7.9 OTHER OUTPUT CONFIGURATIONS-CLASS B

Variations in circuits are legion using similar or complementary transistors for elimination of output or input transformer or both. The output transformer can be eliminated by use of a center tapped supply see fig. 7.9.1.

Each half of the circuit is similar. The input transformer may be eliminated by using complementary transistors (matched characteristics) but off opposite polarities.

    PNP requires   -ve drive

    NPN requires   +ve drive.

Thus transistor conducts for alternate half cycles of the input waveform. If the load impedance is correctly matched the output transformer can also be eliminated. The practical circuit also eliminates the need for a phase splitter stage. Further, diodes are used in base circuit to provide bias and some temperature compensation as shown in fig. 7.9.2.

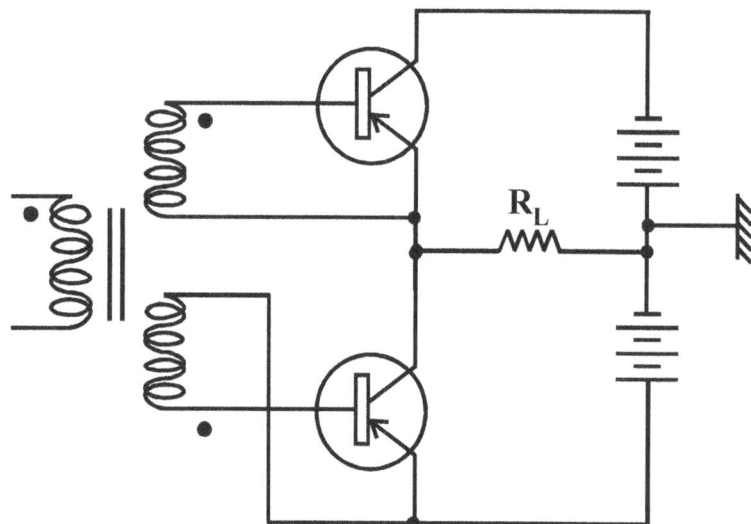

POWER AMPLIFIERS

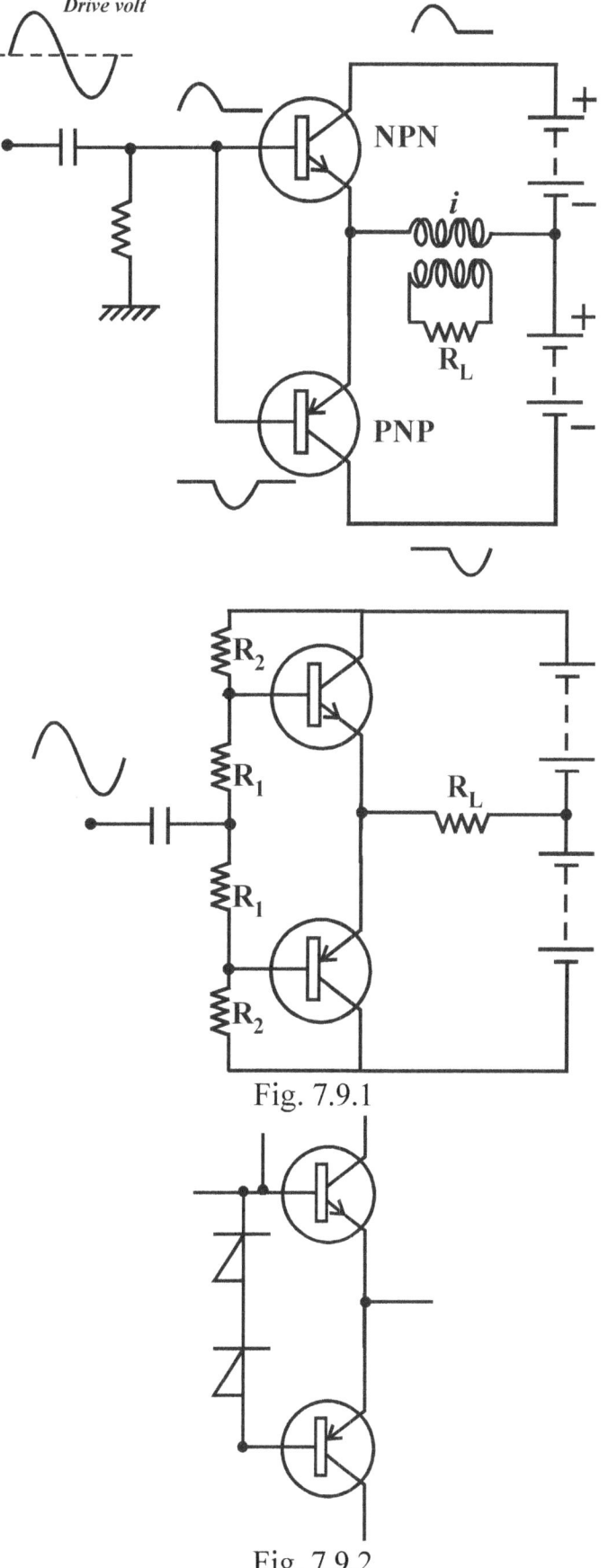

Fig. 7.9.1

Fig. 7.9.2

7.10 TRANSISTOR TUNED AMPLIFIERS

Single Tuned Stages: Referring to fig. 7.10.1, transistor coupling to the next stage is usually affected by a closely wound secondary winding or the transformer.

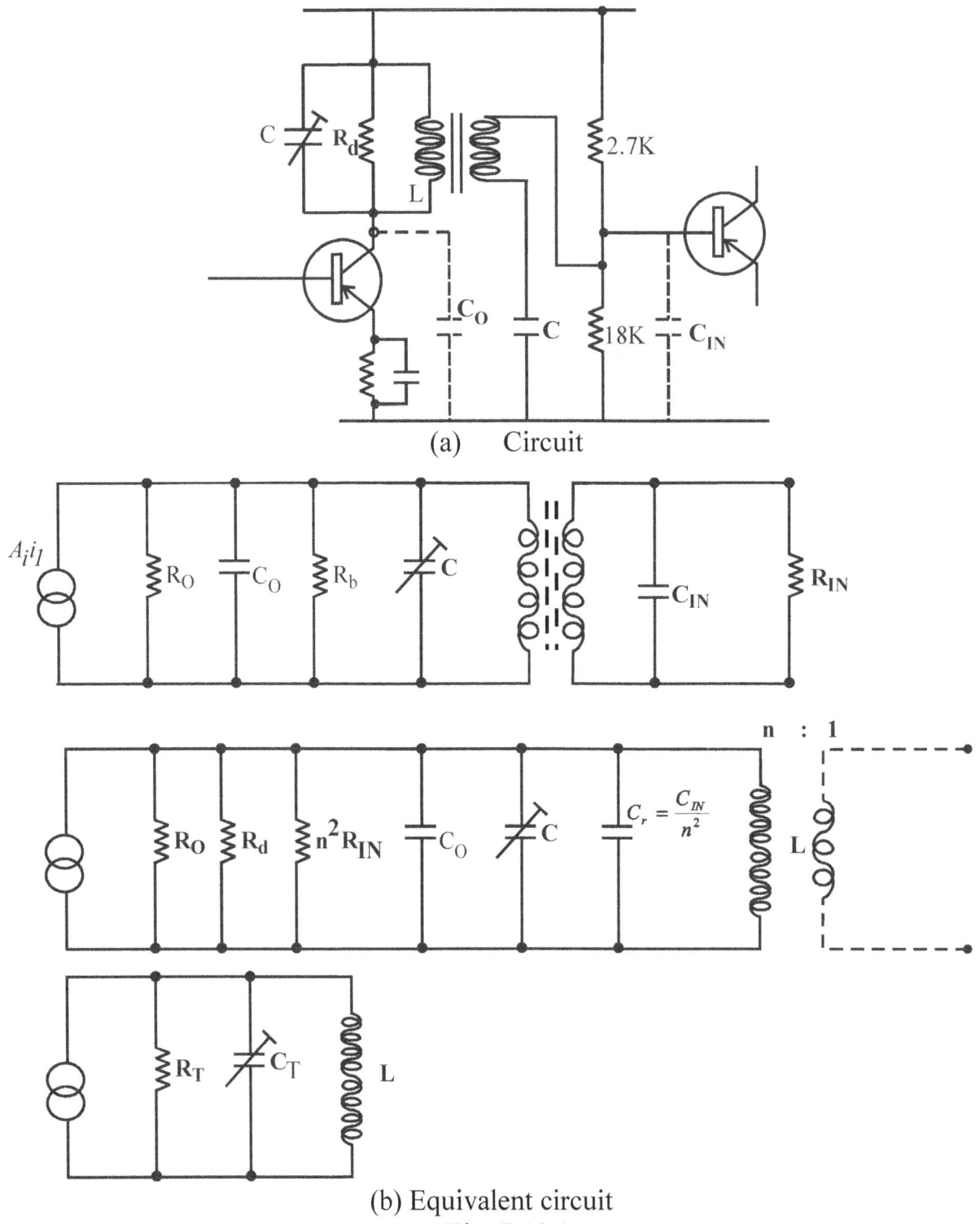

(a) Circuit

(b) Equivalent circuit
Fig. 7.10.1

The primary inductance L is tuned to resonance by the capacitor C and R_d represents the dynamic impedance of the unloaded transformer.

R_O is output resistance of transistor

A_i is current gain

C_{IN} & R_{IN} relate to driven stage. (Equivalent circuit fig. 7.10.1b) and R_T is an equivalent loaded dynamic impedance (equivalent circuit fig. 7.10.1c)

$$R_T = R_O // R_d // n^2 R_{IN}$$

$$C_T = C_O + C + \frac{C_{IN}}{n^2}$$

It is obvious from the equivalent circuits that the Q factor of the tuned circuit is affected by the loading of the transistor output impedance and the input impedance of the next stage. Thus due to the presence of stray capacitance C is adjusted such that the total capacitance C_T is tuned with L to resonance. The loaded Q factor for the complete stage can be defined as

$$Q_r' = \frac{\omega_r L}{R} = \frac{R_T}{\omega_r L} = \omega C_T R_T$$

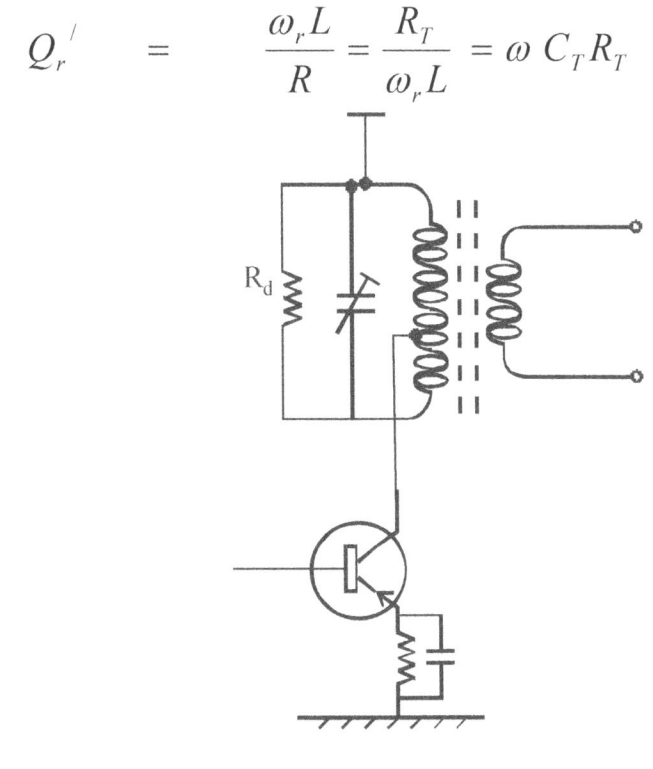

Fig. 7.10.2

The stage gain at resonance can be calculated using the value of R_T if it is known or can be calculated.

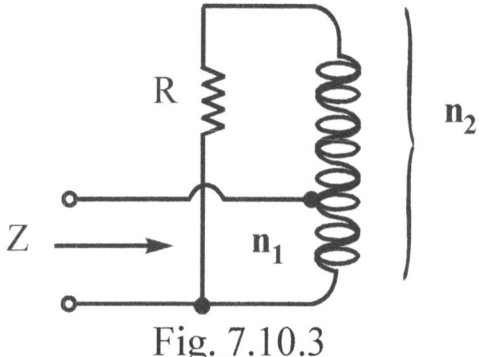

Fig. 7.10.3

The configuration shown presents some difficulties in practice to obtain optimum selectivity and therefore maximum gain due to the differing impedance levels of the tuned circuit and the transistor output resistance.

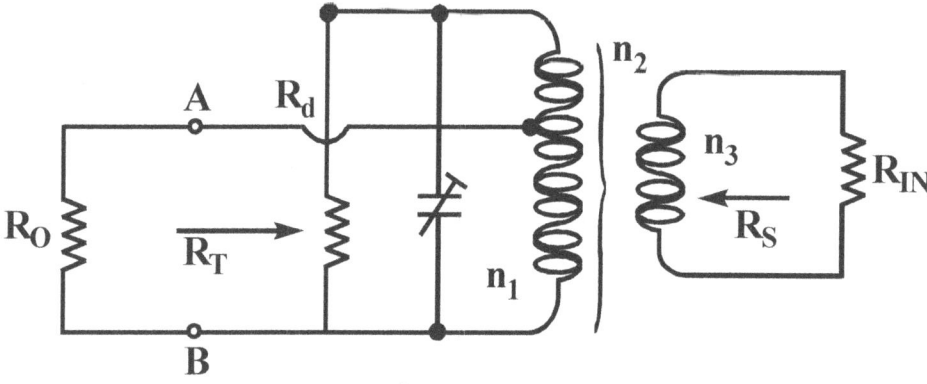

Fig. 7.10.4

Matching is usually achieved by using a tapped tuned circuit. Thus by choice of a suitable tapping see fig. 7 10 .2. The desired impendence can be presented to the transistor. Elementary circuits theory shows that the resistance measured between the top and common terminal

$$Z = R' = \left(\frac{n_1}{n_2}\right)^2 R \quad \text{(see fig. 7.10.3)}$$

Thus for a tapping half way down the coil the resistance presented to the source would be $\frac{1}{4}R$.

Thus the resistance reflected across the terminal (fig. 7.10.4) AB is given by

$$\frac{1}{R_T} = \frac{1}{R_O} + \left(\frac{n_2}{n_1}\right)^2 \frac{1}{R_d} + \left(\frac{n_3}{n_1}\right)^2 \frac{1}{R_{IN}}$$

It is this resistance which determines the bandwidth of the amplifier since the unloaded Q has restrictions on its maximum value due to size and economy. Although the primary inductance L could be varied a standard value is often used (e.g. 1mH in 475 KHz I.F amplifies) and the bandwidth is ~10KHz. Solution of these equations will give the unknown turns ratio since R_d can be calculated and R_T found from above if the resistance R_{IN} & R_O can be measured or calculated. Likewise an expression for R_s can be calculated.

7.11 DOUBLE TUNED AMPLIFIERS

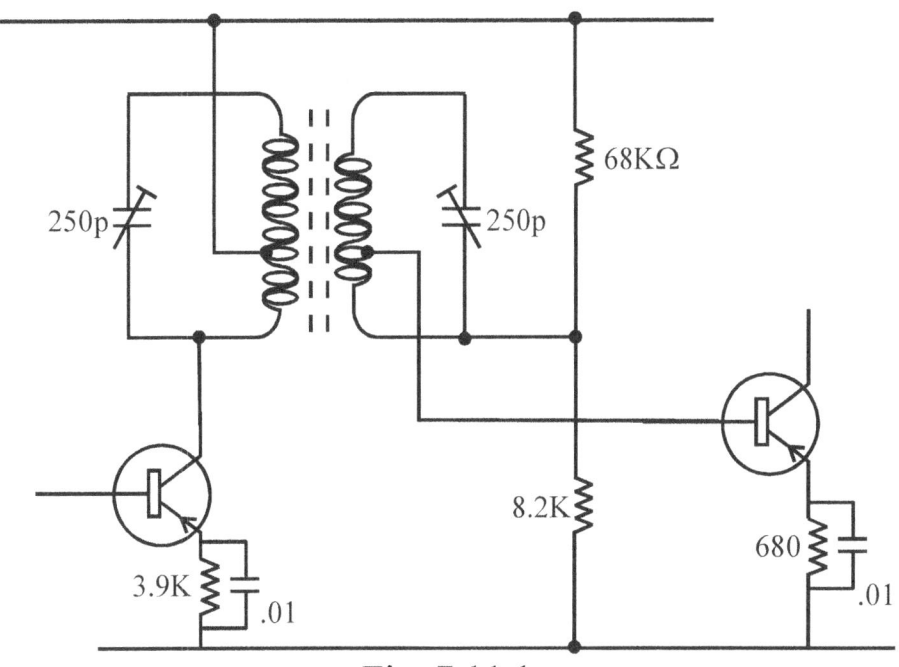

Fig. 7.11.1

The selectivity of amplifiers using single tuned transformers is often inadequate and a double-tuned circuit (fig. 7.11.1) is often used. The degree of coupling (coefficient) is defined by

$$M = \sqrt{L_1 L_2}$$

For a double tuned circuit maximum energy transfer takes place when $K = \dfrac{1}{Q}$ and this is called critical coupling –fig. 7.11.2.

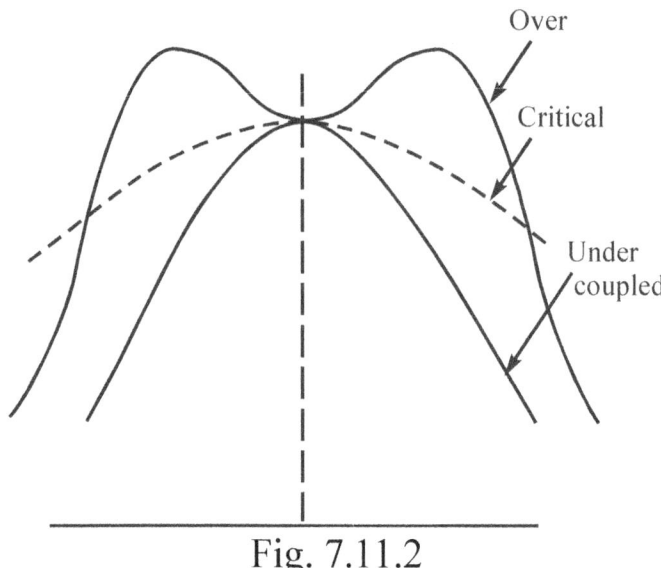

Fig. 7.11.2

POWER AMPLIFIERS

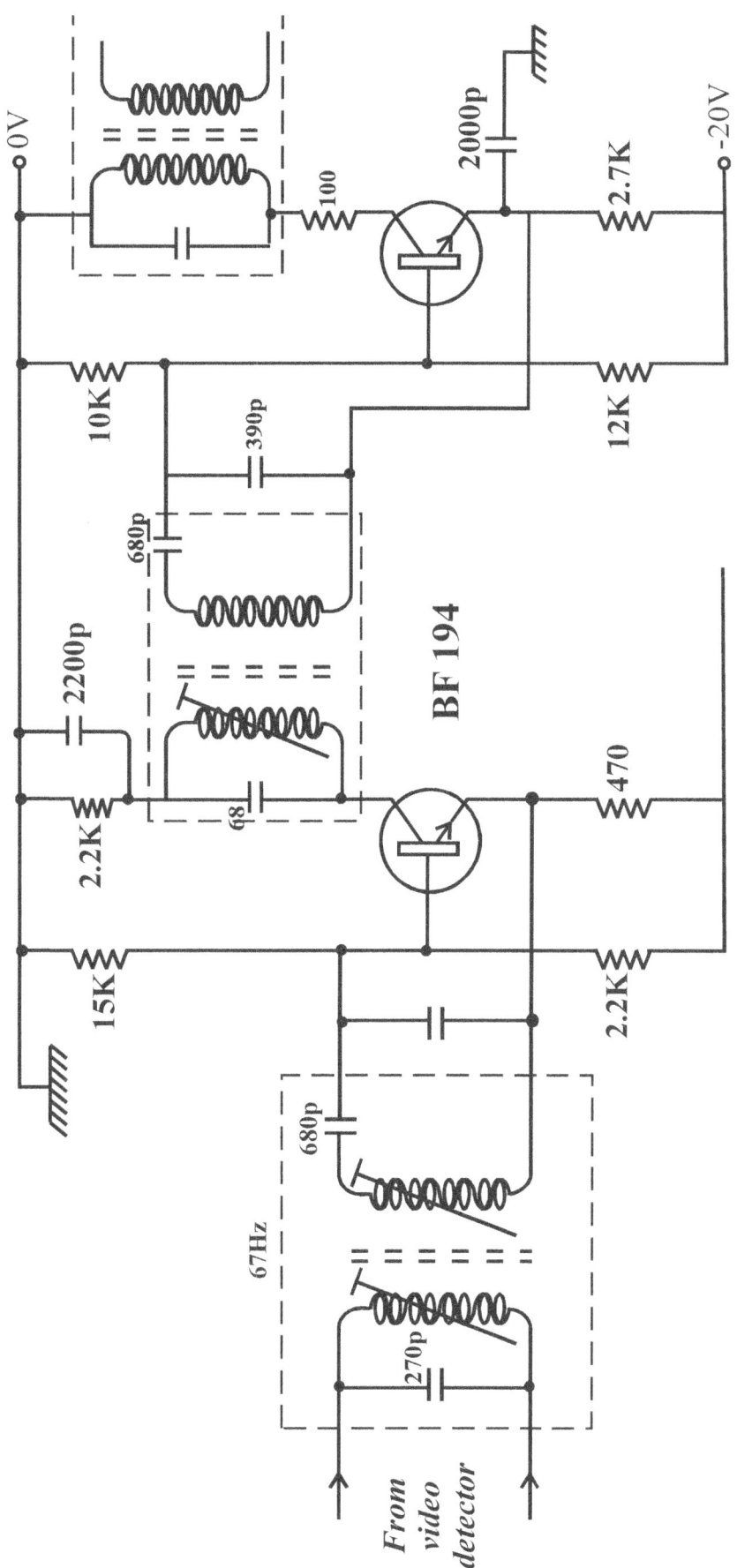

Fig. 7.11.3 Typical T.V Intercarrier sound I.F circuit.

Thus an improved bandwidth can be obtained. However, it can be shown that for a given transistor the gains is reduced compared with the case of a single tuned transformer which is only to be expected since the gain bandwidth products is a constant. The design of a multistage amplifier using double tuned circuits is fairly complicated and is perplexed by the transformers which cannot generally be regarded as loss less. Also the high output impedance of a transistor cannot reflect low impedance to the following transistor and simultaneously reflect the second transistors input impedance to load the first transistor. Fig. 7.11.1 shows the design of a multistage amplifier using double tuned amplifiers in a typical T.V intercarries sound I.F circuit.

7.12 FEEDBACK AND ITS EFFECTS

In the single and multistage amplifiers so far discussed the gain depends upon the precise values of circuit components and device parameters. The performance of such amplifiers can be modified by the application of feedback.

Definition: *A Negative Feedback Amplifier* is one in which a portion of the output voltage (or current) is added to the input voltage in a negative sense (i.e. 180^0 out of phase) thereby decreasing the effective input.

7.13 THE IMPORTANCE OF NEGATIVE FEEDBACK

The properties of amplifiers employing negative feedback enables.

1. Circuits and amplifiers be produced whose performance may be specified in terms of relatively few components.

2. Performance of circuits can be made largely independent of the precise values of device parameters thereby enabling the circuit performance to be unaffected by
   a. Changes of device (valves, transistors, F.E.T's, etc)
   b. Change of operating point due to h_{fe} changes and temperature etc.
   c. Supply voltage fluctuations
   d. Ageing where appropriate (e.g. valves g_m decreases)
3. Values of input and output impedance can be modified to obtain very high or very low values.
4. Generally reduces distortion.
5. Increases bandwidth.

7.14 VOLTAGE GAIN OF NEGATIVE FEEDBACK AMPLIFIERS

Initially some assumptions will be made to simplify the analysis. The following are not all necessary conditions and will be modified later.

1. The gain of the amplifier is $-m$ in the absence of feedback i.e.. Phase reversing with odd number of stages.

   Thus, $\frac{v_O}{v_b} = -m$ (see fig. 6.14.1)

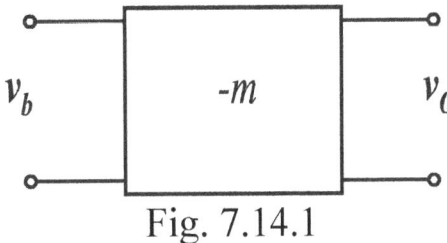

Fig. 7.14.1

2. m is real at all frequencies considered.
3. The input impedance is large in the absence of forward biased (i.e.. No current flows in the amplifier input).

4. The output impedance in the absence of feedback is very low

5. The gain of the feedback network is β so that the output of the feedback network is βV_o (fig 7.14.2)

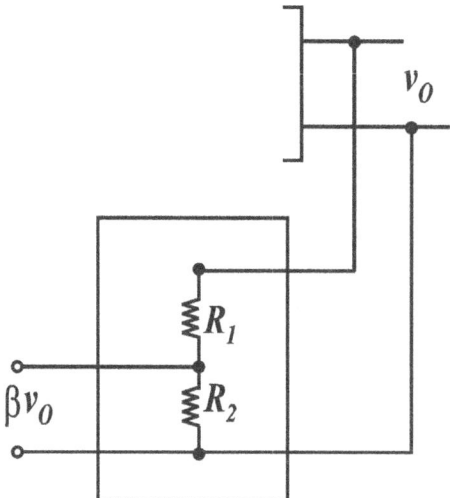

Fig. 7.14.2 Feedback Network

In the example shown $\beta = \dfrac{R_2}{R_1 + R_2}$

If v_b is input voltage and βv_o is added on the negative sense as shown in fig. 7.14.3, we have;

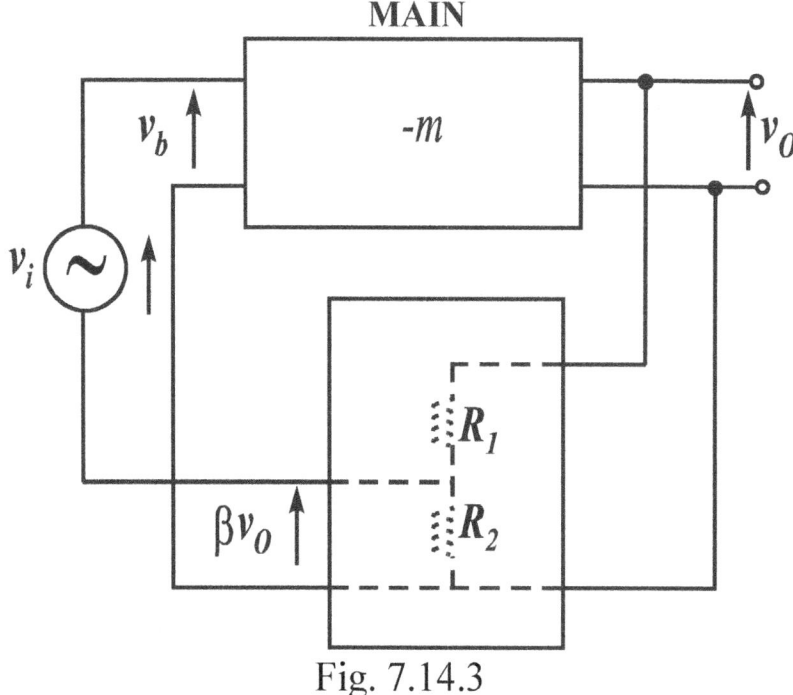

Fig. 7.14.3

POWER AMPLIFIERS

Output voltage $V_O = -mV_b$

Voltage applied to main amplifier input

$$V_b = V_i + \beta V_O$$
$$V_b = V_i - m\beta V_b$$
$$V_b (1 + m\beta) = V_i$$
$$V_b = \frac{V_i}{1+m\beta}$$

$$\therefore \quad V_O = -mV_b = \frac{-mV_i}{1+m\beta}$$

Gain of negative feedback amplifier $\dfrac{V_O}{V_i} = \dfrac{-m}{1+m\beta}$

7.15 PROPERTIES OF NEGATIVE FEEDBACK AMPLIFIERS

a. From the above $\dfrac{V_O}{V_i} = \dfrac{-m\beta}{1+m\beta} \cdot \dfrac{1}{\beta}$

   If the product of the open loop gain $m\beta >> 1$

   Then $\dfrac{V_O}{V_i} = -\dfrac{1}{\beta}$

   When the above condition applied, the overall voltage gain is not significantly dependant upon m.

   Thus the values of circuit components do not need close tolerance and the gain may be specified by relatively few components (those defining β).

b. In addition to the above if β is independent of frequency then also will V_O/V_i, provided that $m\beta >> 1$. Hence this reduces frequency distortion although the gain is reduced the frequency response is improved and the bandwidth increased at the expense of gain reduction see fig. 7.15.1

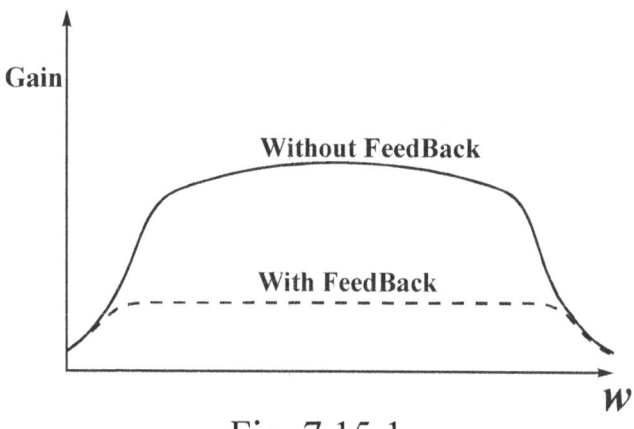

Fig. 7.15.1

c. To show that negative forward biased reduces harmonic distortion.

Assume

(i) The distortion when feedback is applied and passed through the amplifier again produces negligible further distortion- Justifying if distortion level is small <10%.

(ii) The comparison of distortion with and without feedback is effected at the same value of output voltage since distortion is function of output voltage (non-linearity of characteristics etc).

Let amplifier introduce a distortion voltage D volts.

Amplifier without feedback, $V_O = -mV_b + D$

Amplifier without feedback, $V_b = V_i + \beta V_O$
$$= V_i + \beta[-mV_b + D]$$
$$V_b = V_i - m\beta V_b + \beta D$$
$$V_b(1+m\beta) = V_i + \beta D$$
$$V_b = \frac{V_i}{1+m\beta} + \frac{\beta D}{1+m\beta}$$

Output voltage $\quad V_O = -m\left[\dfrac{V_i}{1+m\beta} + \dfrac{\beta D}{1+m\beta}\right] + D$

∴ Distortion with forward bias $\quad = D\left[1 - \dfrac{m\beta}{1+m\beta}\right] = D\left[\dfrac{1}{1+m\beta}\right]$

POWER AMPLIFIERS

It is seen that distortion is reduced in the same ratio as gain when feedback is applied.

d. To show that negative feedback reduces phase shift. Let the gain be complex $m = p + jq$.

ϕ = phase angle between input and output in absence of feedback $= \tan^{-1}\left(\dfrac{q}{p}\right)$

With feedback $\dfrac{V_O}{V_i} = \dfrac{-m}{1+m\beta} = \dfrac{-(p+jq)}{1+\beta(p+jq)}$

Note $\left[\dfrac{X_1 \angle \theta_1}{X_2 \angle \theta_2} = \dfrac{X_1}{X_2} \angle \theta_1 - \theta_2\right]$

Phase angle with feedback $= \tan^{-1}\left(\dfrac{q}{p}\right) - \tan^{-1}\left(\dfrac{q\beta}{1+p\beta}\right)$ (negatively 180^0 phase angle). Thus phase angle is produced by forward bias.

7.16 METHODS OF APPLYING NEGATIVE FEEDBACK

It is important to note that;

1. When the voltage feedback in a portion of the output voltage across the load voltage feedback is said to have been applied.

2. When the voltage feedback is proportional to current in the load current feedback is said to have been applied.

In general an odd number of phase reversing stages must be used. It should be noted that although the gain of the amplifier with negative feedback is reduced the advantages far outweigh thus factor which can be compensated for by cascading feedback amplifiers. The methods of applying negative feedback are.

a. **Series Voltage Feedback**. Fig. 7.16.1 shows arrangement for series feedback.

POWER AMPLIFIERS

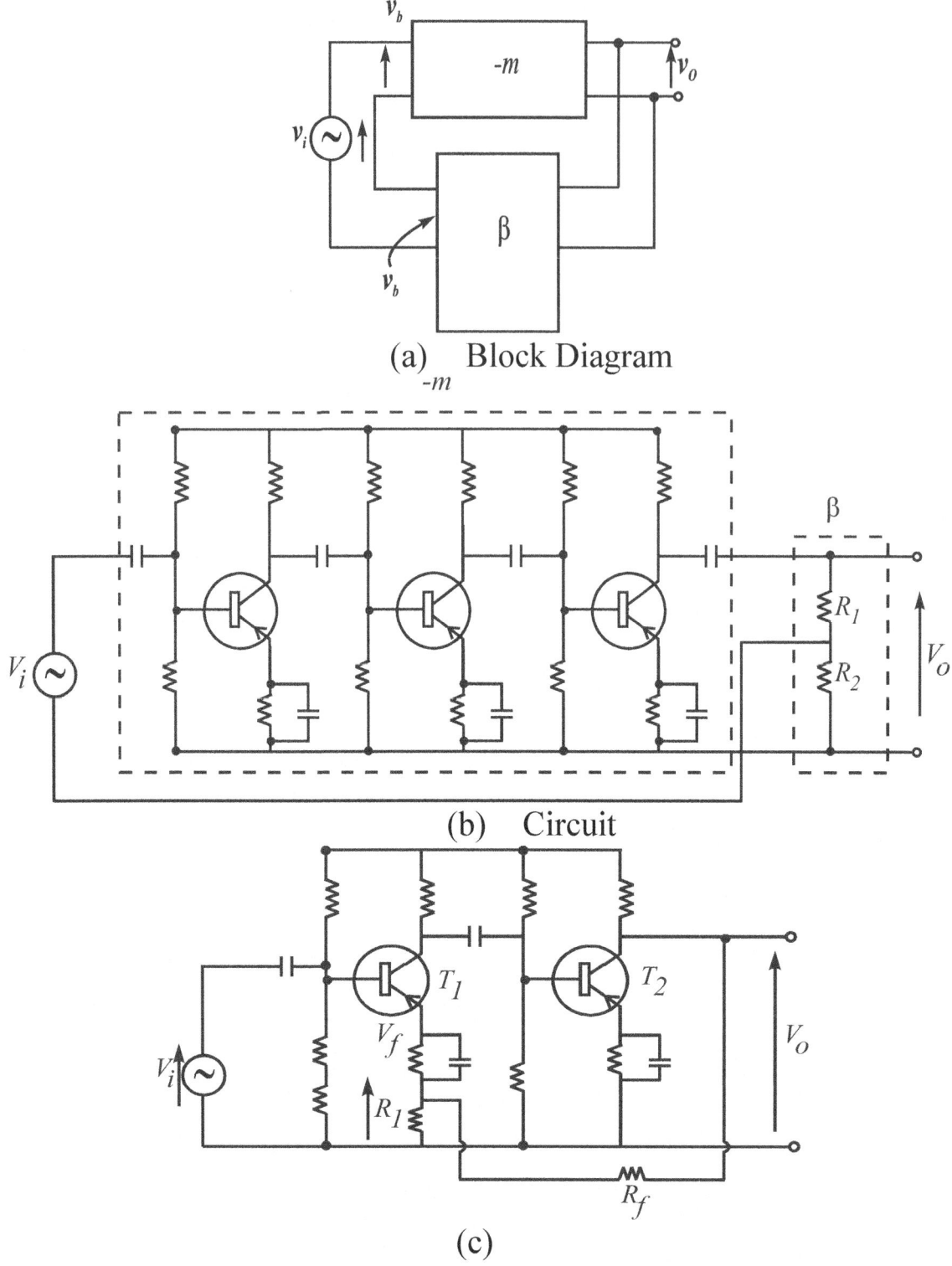

Fig. 7.16.1 Series voltage Feedback

For fig. 7.16.1c feedback is negative since one polarity reversal occurs between emitter of T_1 and collector of T_2 where feedback voltage is obtained.

b. **Shunt Voltage feedback.** The arrangement of fig. 7.16.2 shows a shunt voltage feedback.

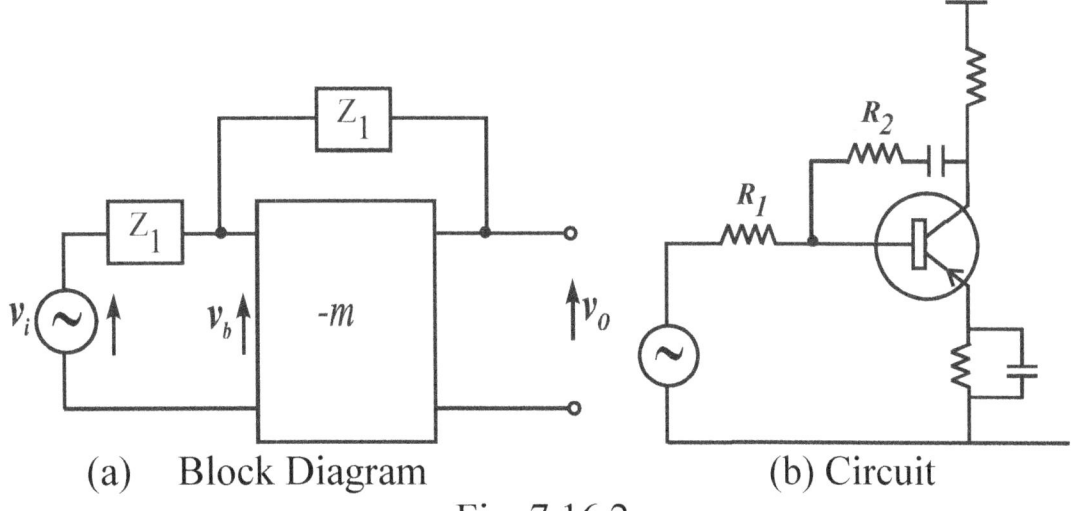

(a) Block Diagram (b) Circuit

Fig. 7.16.2

c. **Series Current Feedback**: The block diagram of fig 7.16. 3 shows a series current feedback.

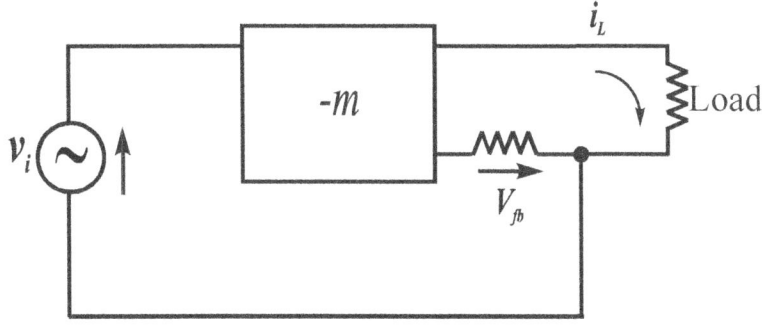

Fig 7.16.2
Feedback Voltage $V_{fb} = i_L R_S$
$\propto i_L$

d. **Shunt Current Feedback**: Fig. 7.16.3 gives the block diagram and circuit for shunt current feedback.

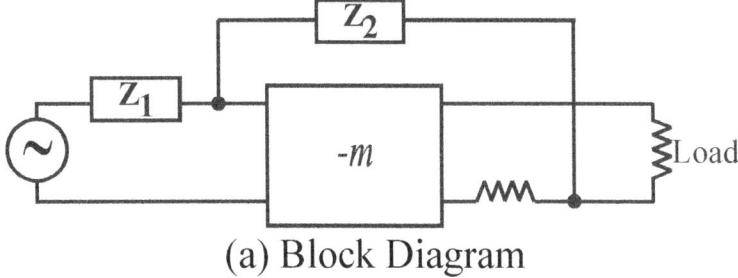

(a) Block Diagram

POWER AMPLIFIERS

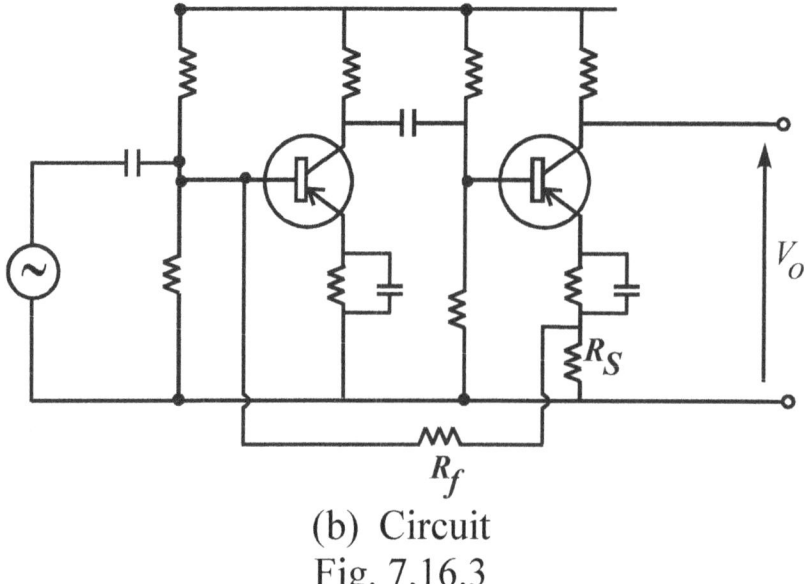

(b) Circuit

Fig. 7.16.3

7.17 EFFECT OF NEGATIVE FEEDBACK ON INPUT AND OUTPUT IMPEDANCE

e.g. series voltage feedback.

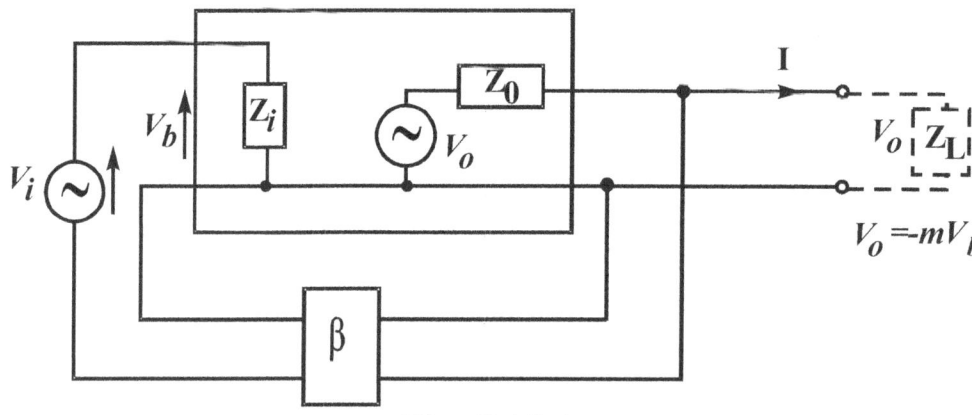

Fig. 7.17.1

Let Z_i be input impedance in absence of feedback with feedback the input voltage to amplifier, $V_b = V_i + \beta V_o$

$$V_b = V_i - m\beta V_b$$

$$V_b = \frac{V_i}{1+m\beta}$$

Thus input current to complete feedback amplifier = V_b/Z_i

$$= \frac{V_i}{Z_i(1+m\beta)}$$

Hence input impedance has been increased to $Z_i(1+m\beta)$

POWER AMPLIFIERS

Output Impedance: If current is taken from output terminals then V_O will fall (due to drop across Z_O) and the voltage feedback to the input ($V_i + \beta V_O$) will be increased thus tending to maintain the same level of output.

Let Z_O be output impedance in absence of feedback.

Let V_O be output voltage when current $I = 0$

$$\therefore \quad V_O = V_O - IZ_O \text{-----------------(A)} \quad (V_O = -mV_b)$$
$$\therefore \quad V_b = V_i + \beta(V_O - IZ_O)$$

Hence $V_O = -m[V_i + \beta(V_O - IZ_O)]$

$$V_O = \frac{-mV_i}{1+m\beta} + \frac{m\beta Z_O I}{1+m\beta} \text{-----------------(B)}$$

Substituting the value of V_O (B) in equation (A)

We have, $V_O = V_O - IZ_O = \dfrac{-mV_i}{1+m\beta} - \dfrac{Z_O}{1+m\beta} \cdot I$

Thus, amplifier behaves as a generator having an O/C potential difference of $V_O = \dfrac{-mV_i}{1+m\beta}$ and an internal impedance of $\dfrac{Z_O}{1+m\beta}$

$\left(\text{or} \Rightarrow \dfrac{Z_O}{m\beta} \text{ for } m\beta \gg 1\right)$.

The advantages of increased input impedance and low output impedance are obvious. Other feedback configurations modify input and output impedance as summarized below.

Feedback	Effect on input impedance	Effect on output impedance
Series Voltage	Raised	Lowered
Shunt Voltage	Lowered	Lowered
Series Current	Raised	Raised
Shunt Current	Lowered	Raised.

CHAPTER EIGHT

8.0 STABILITY OF FEEDBACK AMPLIFIERS

8.1 INTRODUCTION

The gain of a negative feedback has been derived as $\frac{-m}{1+m\beta}$. The system stability may be described according to whether the roots of the equation $1 + m\beta = 0$; determining the gain cause the output to decrease to zero in the absence of an input signal or to increase without limit.

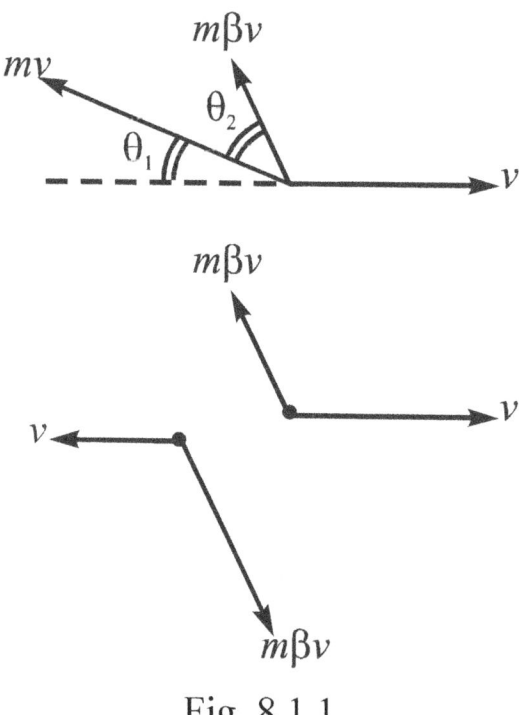

Fig. 8.1.1

In general gain and feedback functions may be complex. For example, at a particular frequency, ω, these may be $m\angle\theta_1$ and $\beta\angle\theta_2$ respectively. If v is the input reference vector, feedback voltage = $m\beta v$. When plotting the response of $m\beta$ to fixed amplitude, varying frequency input the vector diagram usually rotated by 180^0. Thus for any value of ω in the range $0 \rightarrow \infty$ a point may be plotted in the complex plane specifying $[m\beta v]_{j\omega}$.

STABILITY OF FEEDBACK AMPLIFIERS

The line joining these points obtained with different values of w is called a **NYQUIST DIAGRAM** (see fig. 8.1.1. and 8.2.1).

8.2 NYQUIST CRITERION OF STABILITY

The vector locus is usually normalized plotting $[m\beta]_{jw}$ with the point $(-1 + jo)$ as the reference position.

If the locus $[m\beta]_{jw}$ in the complex plane corresponding to all values of w from 0 to $\pm\infty$ encircles the point $(-1 + jo)$ the system will be unstable as a feedback amplifier.

If the locus passes through the point $(-1 + jo)$ the system is without damping.

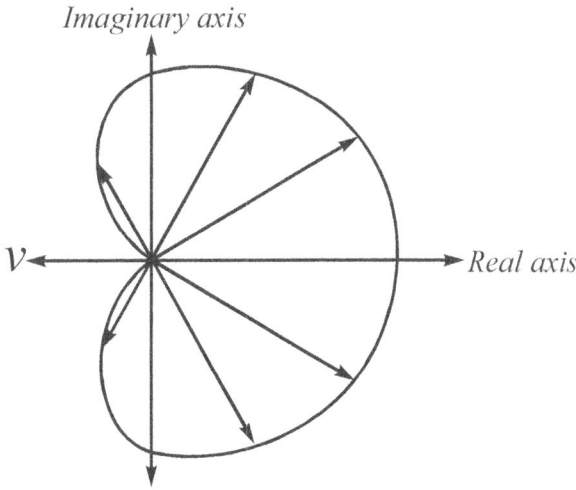

Fig. 8.2.1

If the locus neither passes through nor encircles the point $(-1 + jo)$ the system will be stable as a feedback amplifier.

Let us consider a 3 stage amplifier of fig. 8.2.2

C_{S1} = $C_{ce1} + C_{be2} + (1 + A_2)C_{cb2}$ + starys.

C_{S2} = $C_{ce1} + C_{be2} + (1 + A_2)C_{cb2}$ + starys.

C_{S3} = C_{ce3} + starys

C_{S4} = Input capacitance of 1st stage = $C_{be1} + (1 + A_1)C_{cb1}$ + starys.

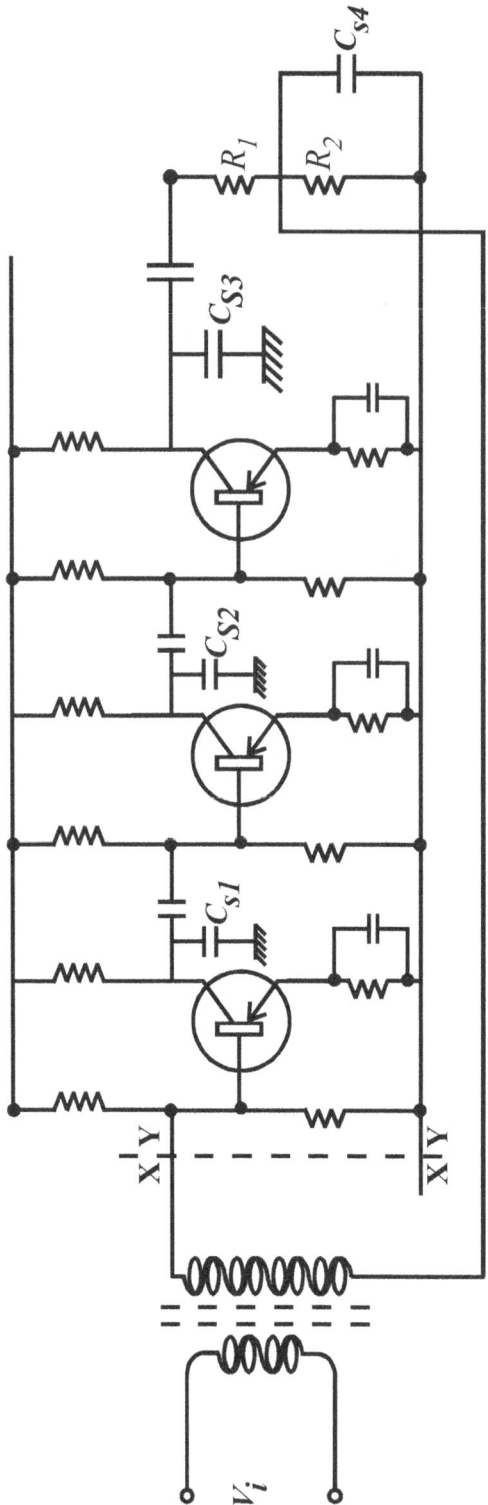

Fig 8.2.2 Three Stage Amplifier

At low frequencies (fig. 8.2.3), the vector locus moves in three quadrants since the 3 coupling capacitors give rise to phase loads ($\rightarrow 90^0$ maximum). At high frequency the locus moves in four quadrants due to the 4 lags (each $\rightarrow 90^0$) because of the presence of C_{S1} - C_{S4}

STABILITY OF FEEDBACK AMPLIFIERS

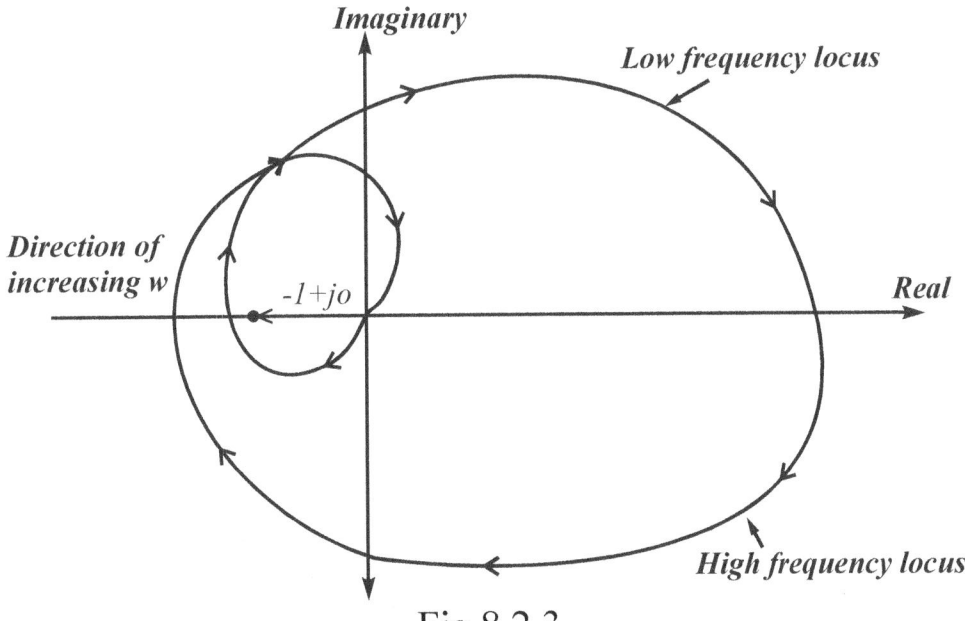

Fig 8.2.3

This vector locus of the open loop gain $m\beta$ encircles the point $-1+jo$ and of connected as a closed loop amplifier the system will be unstable. Here the loop gains $m\beta > 1$ and when **xx** is joined to **yy** (fig. 8.2.2) the amplifier can provide its own input in the absence of input signal. This is obviously an unsatisfactory condition for linear amplification.

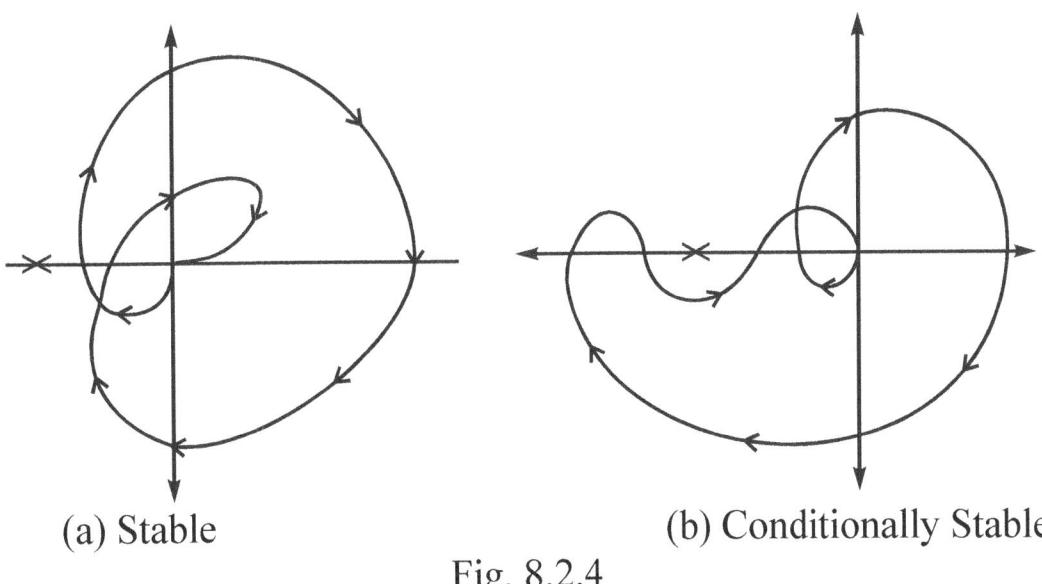

(a) Stable   (b) Conditionally Stable

Fig. 8.2.4

Thus it is possible to predict the closed loop behaviour of a feedback system from its open loop response. Fig. 8.2.4 shows the case when the vector locus of open loop gain $m\beta$ do not encircle the point $(-1+jo)$

8.3 STABILISATION OF FEEDBACK AMPLIFIERS

In order to obtain a stable system the nyquest diagram must be shaped such that the locus does not encircle the $(-1 + jo)$ point.

This may be achieved by:

i. Reducing the values of $m\beta$ thereby collapsing the diagram until it passes inside the $(-1 + jo)$ point. However using this method the value of $m\beta$ may then be too small to be useful.

ii. Stabilisation by increasing the target lag. Here the target time constant is supplemented by additional capacitor (see fig. 8.3.1) to make it mush larger. (Note that additional of extra capacitors does not introduce an extra lag.) The value of the increased time constant is chosen such that by posing gain at the rate of 6dB/Oct (20dB/Decade) zero dB (unity) loop gain is reached before the loop phase shift reaches 360^0.

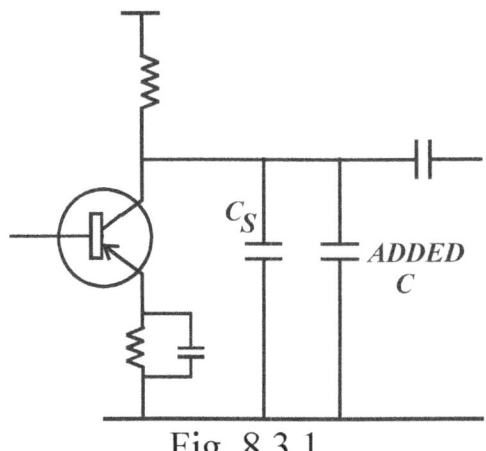

Fig. 8.3.1

The stabilizing lag introduces a [phase angle of 90^0 (approximately) over a wide frequency range for that stage. Only at frequency near to where loop gain is zero dB (unity) do the other lags introduce appreciable phase shift. This method however severely reduces the bandwidth before feedback is applied.

iii. **Lead Stabilisation.** A similar method to above is applicable at low frequency. Where one of the coupling time constants is made much smaller than the others.

iv. **Network Stabilisation.** Many networks exist which may be introduced into an amplifier or feedback path which may be designed to severely reduces the loop gain when the loop phase shift is approaching 360^0 (see fig. 8.3.2).

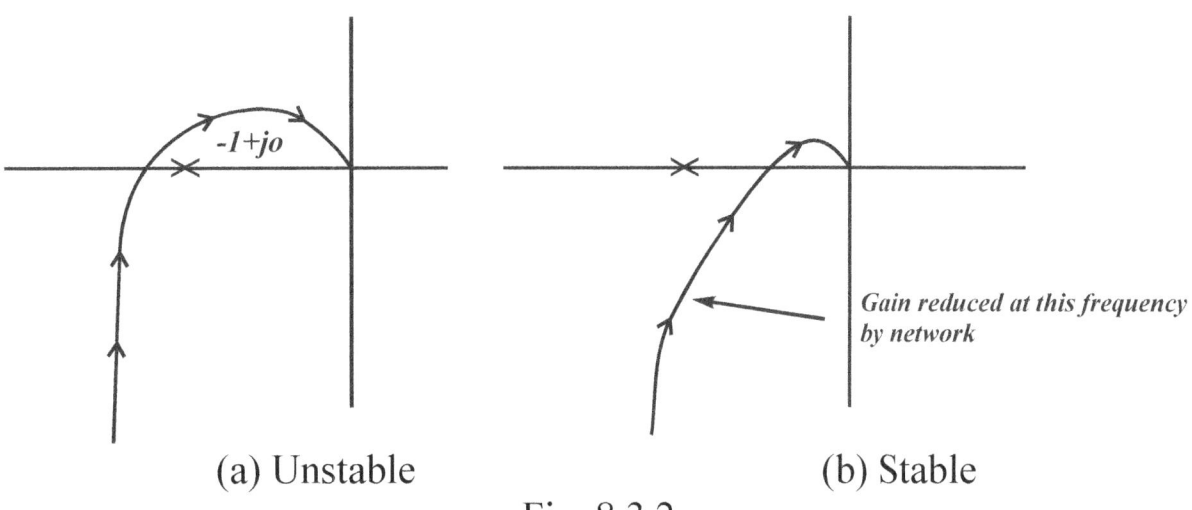

(a) Unstable   (b) Stable

Fig. 8.3.2

The shaping of the frequency response locus is governed by many factors, for instance is the amplifier (or system) to respond to pulse or step input. If so transient response is important and the design must cater for an adequate gain and phase margin (see text book on control systems)

8.4 OSCILLATORS

An oscillator is a feedback system which is deliberately designed to be unstable. The usual arrangement consist of amplifier and phase shift network connected to the output of the amplifier as represented in the block diagram of fig. 8.4.1.

STABILITY OF FEEDBACK AMPLIFIERS

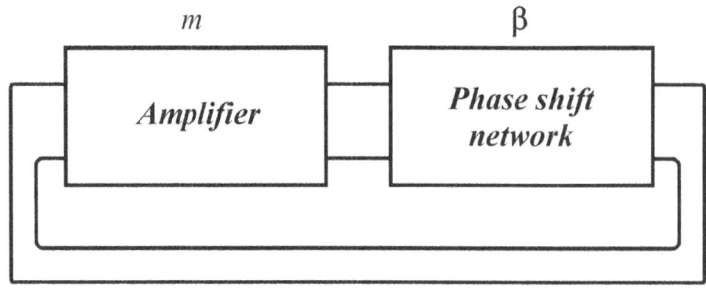

Fig. 8.4.1

Broadly speaking two main classes exist;

i. Sine wave: (a) Using LC circuits (b) Using RC circuits

ii. Non Sinusoidal-Relaxation oscillators – Multivibrators circuits requirements

1. Obtain relationship between circuit constants and device parameters which must be satisfied to produce a loop gain of $\geq$ unity to ensure oscillation.
2. Frequency of Oscillation if the above (maintenance) condition is satisfied.

Nyquist's criterion states that the system will be unstable as a feedback amplifier if the frequency response locus of $m\beta$ encircles $(-1+jo)$ and if it passes through $(-1+jo)$ point the system will be without damping. This latter condition represents the limiting dace of stability which corresponds to the maintenance condition.

8.5 PROCEDURE FOR ACHIEVING OSCILLATION

i. Break loop at convenient point xx (fig. 8.5.1)

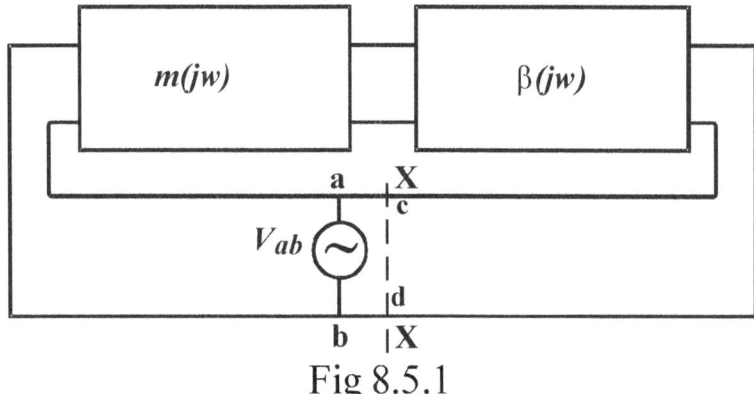

Fig 8.5.1

STABILITY OF FEEDBACK AMPLIFIERS

ii. Connect generator V_{ab} to a and b to give $\dfrac{V_{cd}}{V_{ab}} = m\beta(jw)$

iii. If rector locus passes through $(-1 + jo)$ the $\dfrac{V_{cd}}{V_{ab}} = 1$. (Note that m contains phase reversal sign for amplifier with odd number of stages).

iv. To ensure passage through $-1 + jo$ loop phase shift must be $2n\pi$ (n usually $\pm r\, 0$). This imaginary part of $m\beta(jw)$ must be zero which gives the frequency of oscillation.

v. Equate the real part of $m\beta(jw)$ to unity to find the minimum maintenance condition. Fig. 8.5.2 gives the block diagram of the oscillator.

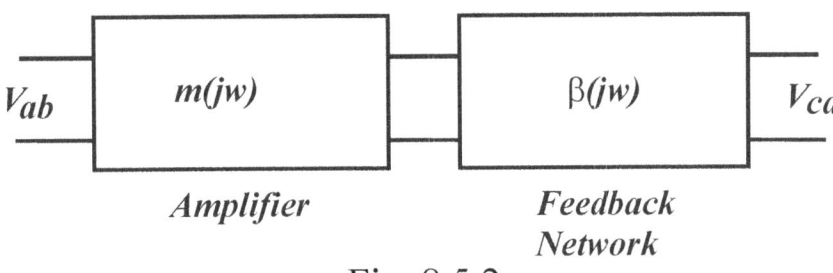

Fig. 8.5.2

From the foregoing brief explanation of nyquist's criterion the open loop gain $m\beta(jw)$ must be unity for oscillation to take place. This may be achieved in a number of ways. Usually it is possible to design amplifier with a phase shift 180^0 (odd number of stages) or 360^0 (even number of stages) and have reasonable amount of gain quite easily. Thus the feedback network is usually frequency sensitive and arranged such that the total loop phase shift is 0^0 or 360^0 which with the above amplifiers taken as an example must have a network phase shift of 180^0 and 0^0 respectively.

STABILITY OF FEEDBACK AMPLIFIERS

Types of Sine Wave Oscillators (shown in fig. 8.5.3)

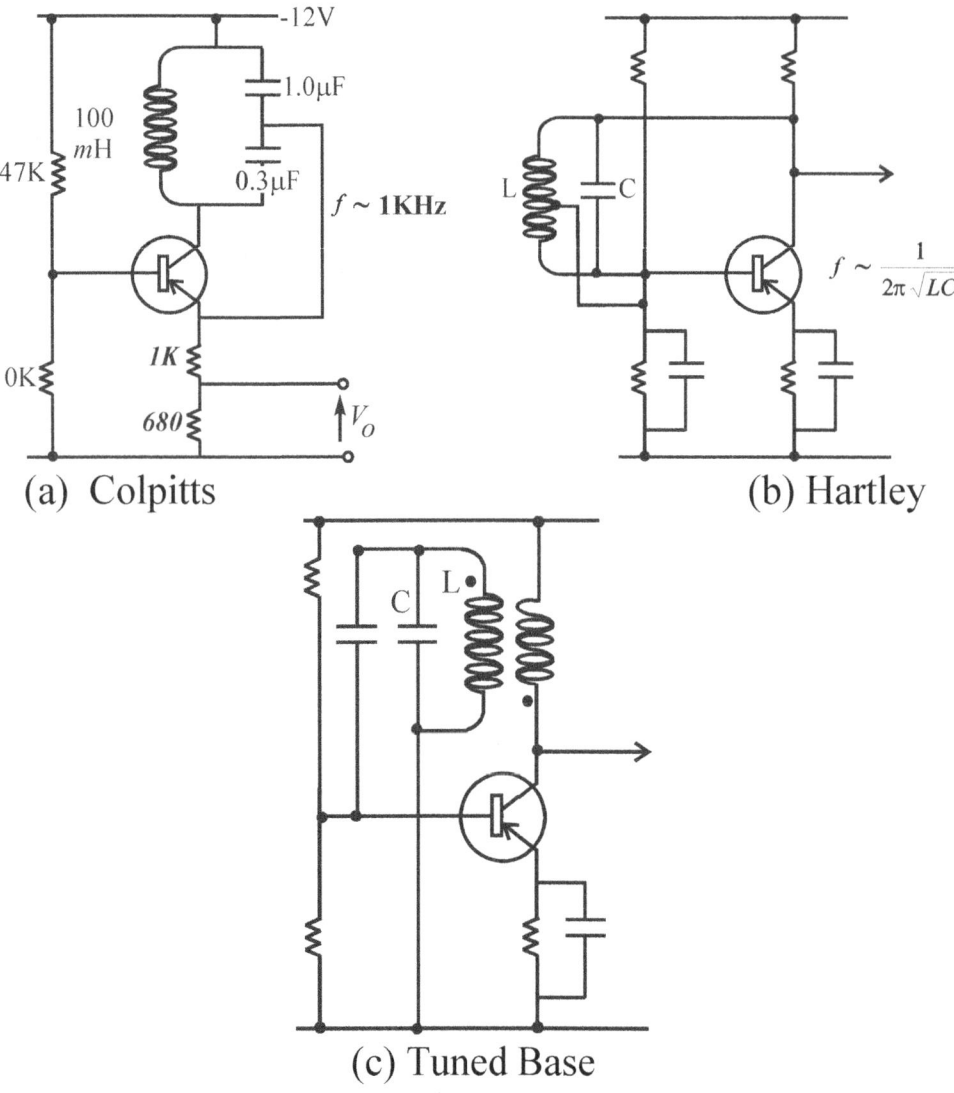

(a) Colpitts

(b) Hartley $\quad f \sim \dfrac{1}{2\pi\sqrt{LC}}$

(c) Tuned Base

Analysis is usually tedious and with approximations, frequency expressions are generally reduced to that of the tuned circuit.

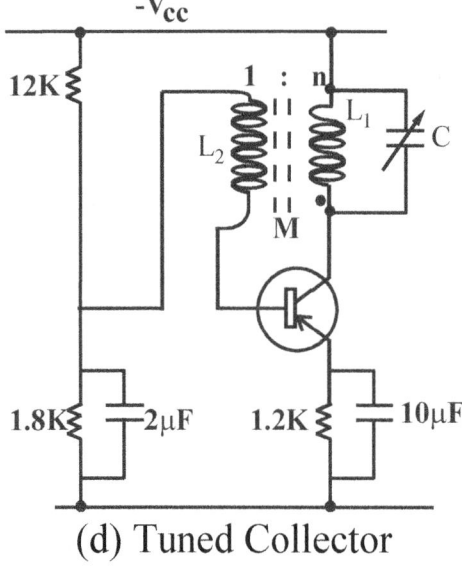

(d) Tuned Collector

229

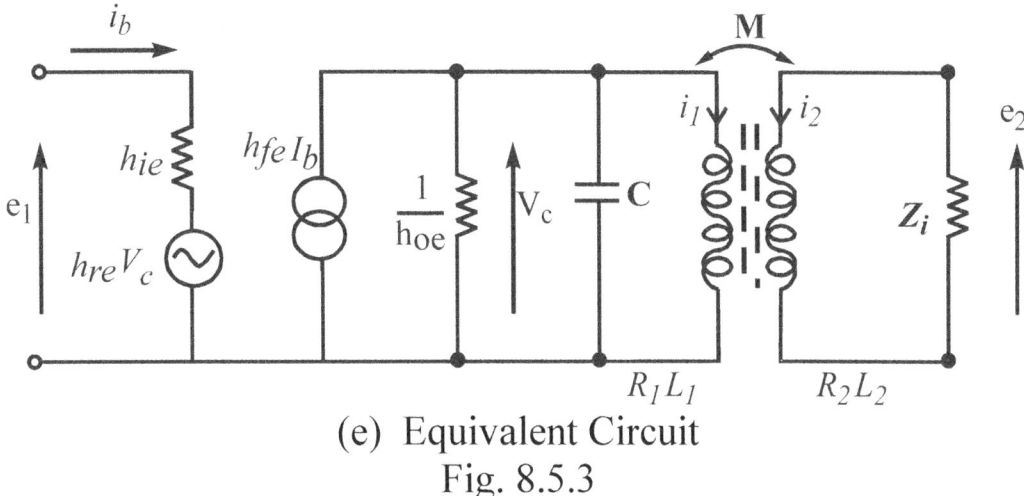

(e) Equivalent Circuit
Fig. 8.5.3

Z_i is the input impedance of the resistor ie. For a.c it is h_{ie} // 12K ~ h_{ie}. Solving the 3 simultaneous equations for the above circuit in terms of B_2 / e_i and equating imaginary part to zero gives the frequency of oscillation.

$$f = \frac{1}{2\pi}\sqrt{\frac{h_{ie} + R_2 + R_1 \Delta_h \left(\frac{R_2}{h_{ie}} - 1\right)}{\frac{\Delta_h}{h_{ie}}(L_1 L_2 - M) + C(h_{ie}L_1 + R_1 L_2 + R_2 L_1)}}$$

In practice R_1, R_2 are small compared with the reactance of $L_1 L_2$ and $M = \sqrt{L_1 L_2}$ (lose coupling with K = 1) and above approximates to

$$f = \frac{1}{2\pi\sqrt{LC}} \qquad (note, \Delta_h = h_{ie}h_{re} - h_{fe}h_{oe})$$

Fig. 8.5.4

STABILITY OF FEEDBACK AMPLIFIERS

Similarly equating real parts gives maintenance condition,

$$M h_{fe} = -\frac{1}{L_1 C}(L_1 L_2 - M^2)C + L_2\left(-R_1 \frac{\Delta_h}{h_{ie}} + 1\right)$$

Approximating as before to give $h_{fe} = \frac{M}{L_1} = \sqrt{\frac{L_2}{L_1}}$

There are many variations using tuned circuit in tapped tuned circuits. e.g. Hartley circuit of fig. 8.5.4

8.6 R-C SINE WAVE OSCILLATORS

1. **Bridge Networks** E.g Wein
   Here amplifier in (fig. 8.6.1) is assumed to have an even number of stage ie amplifier phase shift 0^0 or 360^0

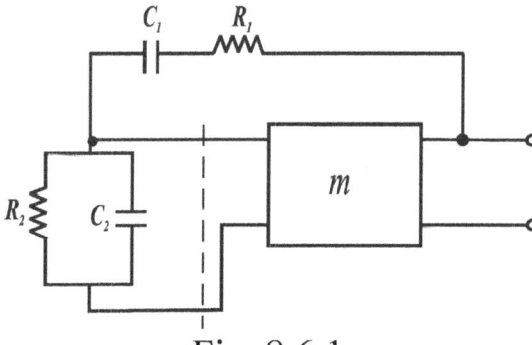

Fig. 8.6.1

2. **Feedback Network** is represented in fig. 8.6.2a

   i. At low frequency looks like (b) (dormant impedance shown)

   ii. At high frequency look like (c)

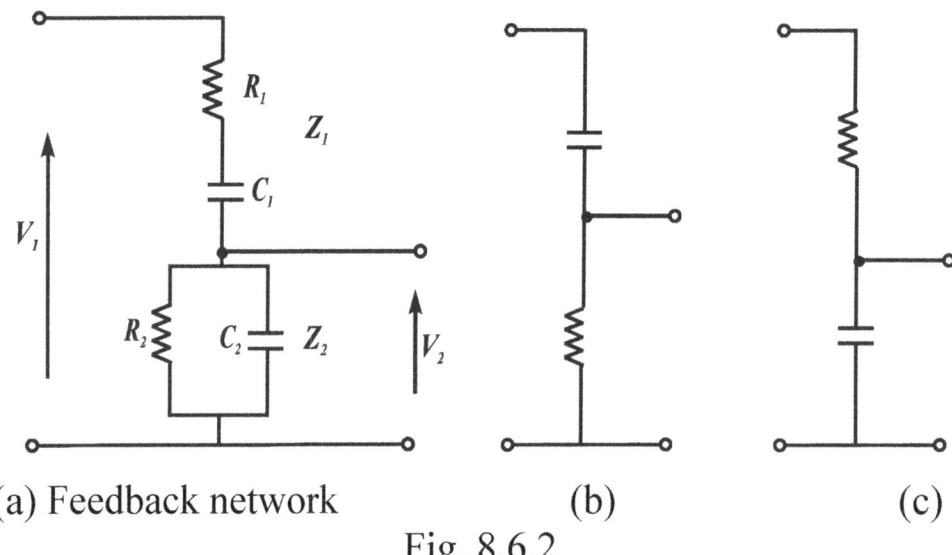

(a) Feedback network   (b)   (c)

Fig. 8.6.2

STABILITY OF FEEDBACK AMPLIFIERS

There must be some frequency where phase angle changes from lagging to loading (ie zero phase angles for the network. Thus at this frequency the amplifier output and input are exactly in phase and of the loop gain is ≥ 1 then oscillation takes place. The output voltage V_2 is given by;

$$V_2 = \frac{Z_2}{Z_1 + Z_2} V_1, \quad Z_1 = R_1 + \frac{1}{SC_1} = \frac{1 + SC_1 R_1}{SC_1}$$

$$Z_2 = \frac{R_2}{1 + SC_2 R_2} = \frac{R_2/SC_2}{R_2 + 1/SC_2}$$

$$\left.\frac{V_2}{V_1}\right|_S = \frac{R_2/(1+SC_2 R_2)}{(1+SC_1 R_1)/SC_1 + R_2/(1+SC_2 R_2)}$$

$$= \frac{SC_1 R_2}{SC_1 R_2 + (1+SC_1 R_1)(1+SC_2 R_2)}$$

$$= \frac{SC_1 R_2}{1 + S(C_1 R_2 + C_1 R_1 + C_2 R_2) + S^2 C_1 R_1 C_2 R_2}$$

For sine wave $s = jw$

$$\beta = \left.\frac{V_2}{V_1}\right|_{jw} = \frac{jwC_1 R_2}{1 - w^2 C_1 R_1 C_2 R_2 + jw(C_1 R_1 + C_2 R_2 + C_1 R_2)}$$

For zero shift at frequency of oscillation

$$1 - w_O^2 C_1 R_1 C_2 R_2 = 0$$

$$\underline{\omega_O = \frac{1}{\sqrt{C_1 R_1 C_2 R_2}} = \textit{Frequency of oscillation}}$$

For unity loop gain at frequency of oscillation

$$m\left(\frac{V_2}{V_1}\right)_{jw} = \frac{m w_O C_1 R_2}{w_O (C_1 R_2 + C_2 R_2 + C_1 R_1)} = 1$$

$$m = \frac{C_1 R_2 + C_1 R_1 + C_2 R_2}{C_1 R_2}$$

$$\underline{\underline{m = 1 + \frac{R_1}{R_2} + \frac{C_2}{C_1}}}$$

If $R_1 = R_2 = R$ and $C_1 = C_2 = C$

$$w_O = \frac{1}{CR} \quad m = 3$$

Such a circuit forms the basis of many laboratory oscillators for low frequency (~ 100KHz).

Range change usually accommodated by switched capacitance and variable frequency by variable resistance (fig. 8.6.3).

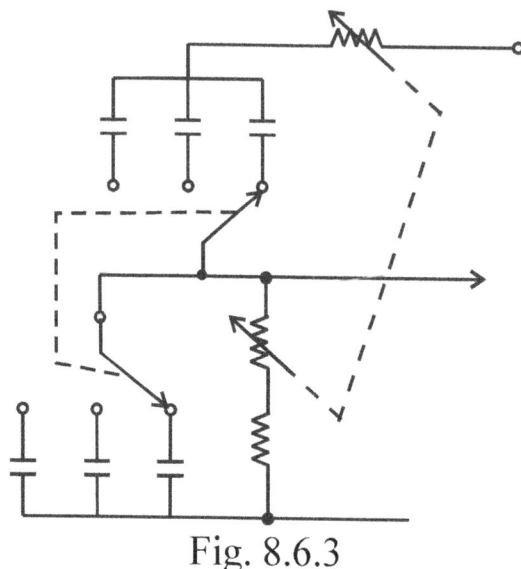

Fig. 8.6.3

Note that for fixed C frequency varies inversely with R (c.f LC oscillation varies $\propto \frac{1}{\sqrt{R}}$). Such circuits require amplitude stabilisation (usually with thermistors).

8.7 PHASE SHIFT OSCILLATORS – RC ladder networks (fig. 8.7.1)

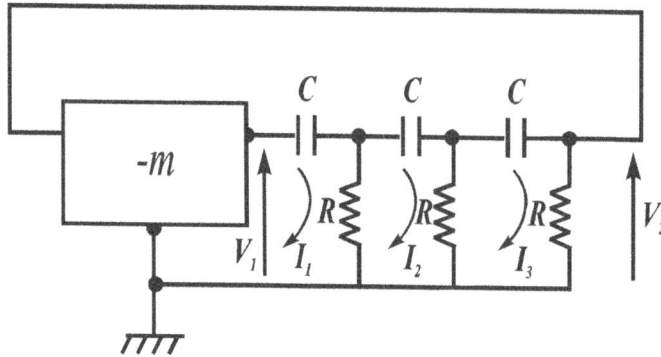

Fig. 8.7.1

Here amplification phase shift is 180^0. Therefore network must provide another 180^0 at least there RC sections are needed to provide 180^0 shift.

STABILITY OF FEEDBACK AMPLIFIERS

$$V_1 = \left(R + \frac{1}{SC}\right)I_1 - RI_2$$

$$0 = -RI_1 + \left(2R + \frac{1}{SC}\right)I_2 - RI_3$$

$$0 = -RI_2 + \left(2R + \frac{1}{SC}\right)I_3$$

Network output $V_Z = RI_S$

$$\Delta = \begin{vmatrix} R + \frac{1}{SC} & -R & 0 \\ -R & 2R + \frac{1}{SC} & -R \\ 0 & -R & 2R + \frac{1}{SC} \end{vmatrix} \qquad I_3 = \frac{\Delta_{13}}{\Delta}$$

$$\Delta = R_3 + \frac{6R^2}{SC} + \frac{5R}{S^2C^2} + \frac{1}{S^3C^3} \qquad \Delta_{13} = V_1 \begin{vmatrix} -R & 2R + \frac{1}{SC} \\ 0 & -R \end{vmatrix}$$

$$\therefore V_2 = \frac{V_1 R^3}{\Delta} \qquad\qquad = V_1 R^2$$

$$\frac{V_2}{V_{11}} = \frac{R^3}{R^3 + \frac{6R^2}{SC} + \frac{5R}{S^2C^2} + \frac{1}{S^3C^3}} = \frac{S^3 R^3 C^3}{S^3 R^3 C^3 + 6S^2 C^2 R^2 + 5SCR + 1}$$

For sine waves $s = jw$

$$\left.\frac{V_2}{V_1}\right|_{jw} = \frac{-jw^3 R^3 C^3}{-jw^3 R^3 C^3 + 5jwCR + 1 - 6w^2 C^2 R^2}$$

For 180^0 phase shift in network (-ve sign)

$$1 - 6w^2 C^2 R^2 = 0$$

$$\underline{\underline{w = \frac{1}{\sqrt{6}CR}}}$$

CHAPTER NINE

9.0 PRINCIPLES OF COMMUNICATION SYSTEMS

9.1 INTRODUCTION

Communication is self explanatory and implies the transmission of information usually between two remote points. Information may be transmitted

(a) Using conductors physically connecting the two points
(b) Using electromagnetic radiation.

Systems operating under (a) are commonly found in telegraphy or telephony applications. Those under (b) can be that of radio communication.

Electromagnetic Fields

It is a well known fact that when a current flows in a conductor two fields are created.

i. Induction field
ii. Radiation field

At frequencies below 20 KHz the magnitude of the radiation field is negligible compared with the induction field. For example in an inductor one can safely assume that below 20 KHz all of the energy goes into the induction field and very little into the radiation field, thus giving total energy absorption over one cycle of practically zero.

At frequencies above 20 KHz an increasing proportion of energy is supplied to the radiation field and transmitted into space in the form of electromagnetic energy.

From principles established in field theory the rate of decay of induction field with distance from the conductors is fairly rapid but this is not the case with the radiation field.

Hence if a second conductor (aerial) is placed some distance away from the 'radiating' conductor (transmitter) will have an e.m.f induced into it due to the electromagnetic radiation. Hence this phenomenon may be used as a method of communication without the use of a direct electrical connection. The magnitude of the e.m.f induced in the aerial will depend upon its design and the field strength of the radiation field in which it is positioned and may vary from a few *μV* to a few *mV*.

9.2 MODULATION AND FREQUENCY TRANSLATION

In normal everyday life communication of *"information"* by direct methods is at relatively low frequencies e.g. speech/music ≤ 20 KHz, measurements of physical quantities few Hz to 5 KHz.

For the transmission of this information to a remote point some form of modulation is necessary. In the process of modulation a parameter of the basic electromagnetic wave (called the carrier wave) is made to vary in sympathy with the information it is descried to transmit.

The nature of the fundamental methods of modulating waveforms may be divided into two broad classes:- analogue techniques and digital techniques. Initially emphasis in this notes will be given to analogue techniques which are in wide usage, although digital techniques are finding increasing applications in broadcast techniques as well as their more well known modes for 'data' transmission.

A simple example of a digital modulation is HORSE code whereby the carrier signal is switched ON and OFF according to some predetermined rules. This is essentially a CODE system and the receiver must decode the signal on reception.

However, the main forms associated with analogue methods of information transmission are carrier modulation adinered by varying

i. Its peak amplitude   -Amplitude Modulation (AM)
ii. Its instantaneous frequency - Frequency Modulation (FM)
iii. Its phase   - Phase Modulation (PM)

Consider, initially that it is desirable to transmit information about an audio signal (say a single tone of 1KHz) to a remote location. Assuming that physical wiring is impracticable (e.g. it may be to an aircraft etc), direct transmission is not possible for two good reasons and frequency translation to a higher frequency range is necessary. First electromagnetic radiation at low (say audio) frequencies is not possible as slated in the foregoing and secondly even if it were possible the aerial lengths required would then be impractical.

Since the velocity of electromagnetic waves is free space (3×10^8 m/s- velocity of light) is related to frequency and wavelength (λ) by

$$V = f\lambda$$

It follows that (if it were possible) the transmission of a 1KHz signal would require an aerial length of 3×10^5 meters !!!!!

Hence the modulations process usually incorporates a frequency translation to a higher frequency range for ease and economy of transmission, i.e.

Low	L.F	30 – 300KHz	Marine, Navigation
Medium	M.F	300-3,000KHz	Broadcast
High	H.F	3-30MHz	Long Distance Communications
Very High	V.H.F	30-300MHz	T.V, F.M. Radio
Ultra High	U.H.F	300-3,000MHz	T.V, Radar
Super High	S.H.F	3,000 – 30,000MHz	Radar, Space Communications

Methods and modes of modulation vary considerably from AM techniques at L.F through AM/FM techniques at HF/VHF to PM and digital (pulse) techniques at S.H.F.

9.3 THE FUNDAMENTAL MODULATION PROCESSES

In most practical applications the CARRIER or BASE BAND signal is usually sinusoidal in nature.

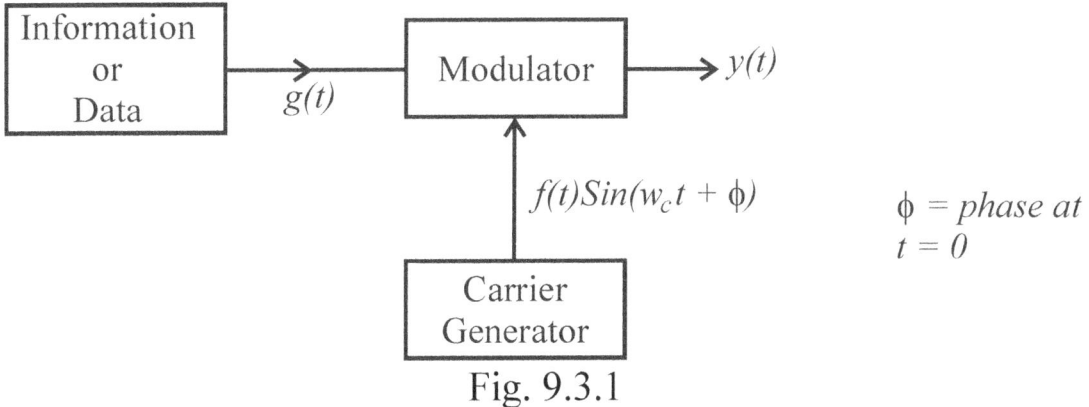

Fig. 9.3.1

Some parameter of the carrier waveform is varied 9amplitude, frequency, phase) by the modulator prior to its power amplification and radiation-see fig. 9.4.1 for block diagram of process.

9.4 AMPLITUDE MODULATION (AM)

Here the peak amplitude of the carrier is made to vary according to the information or data. For simplicity consider s single tone modulation by a sine wave of frequency p fig. 9.4.1 and let $\phi = 0$.

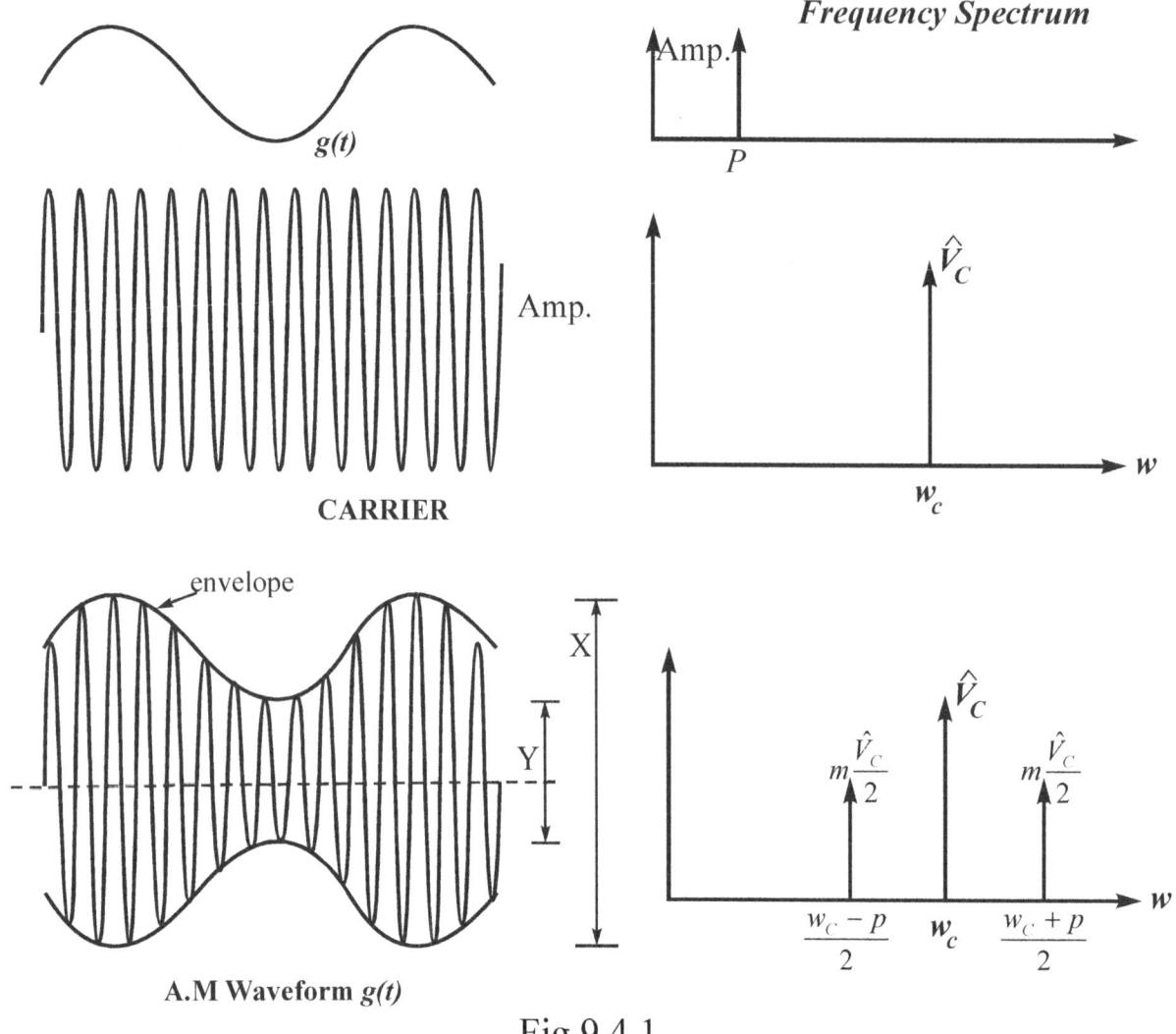

Fig 9.4.1

The peak amplitude of the carrier can be written as

$$f(t) = \hat{V}_C (1 + mg(t))$$

Where $g(t) = \sin pt$ the modulation function and m is called the depth of modulation or modulation index.

In this case where $\phi = 0$ the modulated output is given by

$$\begin{aligned} y(t) &= f(t) \sin w_C t = V_C(1 + m \sin pt) \sin w_C t \\ &= V_C \sin w_C t + m\hat{V}_C \sin pt \sin w_C t \\ &= V_C \sin w_C t + \frac{mV_C}{2} \cos(w_C - p)t - \frac{mV_C}{2} \cos(w_C + p)t \end{aligned}$$

= Unmodulated + Two information carrying sidebands
    CARRIER   LOWER LSB.   UPPER USB

Note that the 'information' about the original signal is contained in the sidebands one above and one below the carrier, and the carrier is merely used as a vehicle for radiation.

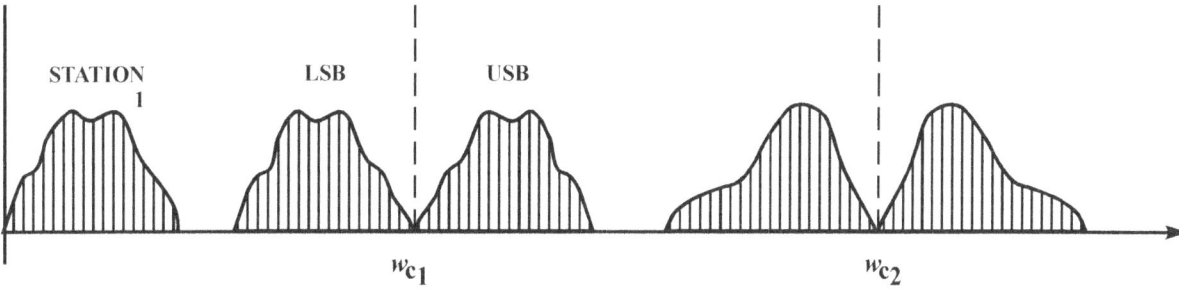

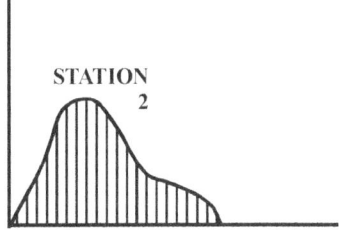

Fig.9.4.2

The modulation depth or index m can vary between 0 and 1 and is given by $m = \frac{X-Y}{X+Y}$ in practice this usually has a maximum value ~ 0.9.

In general the information is not just a single tone but a complex of constituent frequencies of various amplitudes. Thus the sidebands are not single frequencies above and below the carrier but a band of frequencies as represented in fig. 9.4.2. Since the carrier may be placed at any frequency suitable for transmission the above shows a second carrier frequency indicating that several stations may transmit simultaneously without interference providing that the sidebands do not overlap. Because of pressure on the available "space" in a given frequency range e.g. (M.F) the bandwidth used on A>M radio broadcast is about 5KHz above and below the carrier (Note, this is not very good for true music reproduction ~ 15-20KHz but sufficient for most needs.

9.5 POWER CONSIDERATIONS

Consider an e.m.f due to an AM waveform to be of the form,

$$V = \hat{V}_C (1 + \sin pt) \sin w_c t$$

If this is applied to a transmitting aerial of effective radiation resistance R a current i would flow and is given by,

$$i = \frac{V}{R} = \frac{\hat{V}_C}{R}(1 + \sin pt)\sin w_c t$$

The instantaneous power in R is given by $i^2 R$

$$= \frac{V^2}{R} = \frac{(\hat{V}_C (1 + \sin pt)\sin w_c t)^2}{R}$$

The total power dissipated in R will be of the form

$$P_t = \frac{\hat{V}_C^2}{2R} + \frac{m\hat{V}_C^2}{\delta R} + \frac{m\hat{V}_C^2}{\delta R} \qquad \left[\begin{array}{l}\text{Note, mean value of } \sin^2\theta, \cos^2\theta \\ \text{over one period} = \frac{1}{2}\end{array}\right]$$

$$= \frac{\hat{V}_C^2}{2R} + \frac{m\hat{V}_C^2}{4R}$$

$$= \frac{\hat{V}_C^2}{2R}\left[1 + \frac{m^2}{2}\right]$$

If the carrier is unmodulated ($m = 0$) the power in the carrier waveform is

$$P_C = \frac{\hat{V}_C^2}{2R}$$

Thus the power contained in the two side bands is $\frac{m^2}{2} P_C$

Total Power $P_t = P_C\left[1 + \frac{m^2}{2}\right]$

Therefore the ratio

$$\frac{\text{Sideband power}}{\text{Total power}} = \frac{m^2/2 \, P_C}{P_C\left[1 + \frac{m^2}{2}\right]} = \frac{m^2}{2 + m^2}$$

Thus for $M = 1.0$ ratio is $\frac{1}{3}$ and for $M = 0.1$ ratio is $\frac{1}{20}$. Hence the reason for a high percentage modulation.

9.6 SIDEBAND OPERATION

From the foregoing it is seen that all of the information is contained in the sidebands and none in the carrier and hence the carrier may be suppressed with a considerable saving in power output.

In such a situation both sidebands are present and operation is called Double-sideband suppressed carrier DSBSC operation.

However since both of the sidebands convey information of the original signal one of these may also be suppressed with a further saving in power and a bandwidth reduction by ½. This operation is called single sideband suppressed carrier SSBSC.

SSBSC operation is therefore attractive where power output is at a premium and the reduced bandwidth allows a greater number of transmitters to use the same band of frequencies than would be the case with conventional transmission of carrier plus both sidebands.

Examples in this area include mobile telecommunication, radio telephony and amateur transmissions.

$$\frac{Mean\ SSBSC\ Power\ Output}{Conventional\ AM\ Total\ Power\ Output} = \frac{m^2/4 \times P_C}{P_C\left[1 + \frac{m^2}{2}\right]} = \frac{m^2}{4 + 2m^2}$$

Thus for $m = 1$ ratio is $\frac{1}{6}$ – a considerable saving. The price paid for this saving in power and bandwidth is an increased complexity in both the transmission and reception apparatus. It is therefore common practice for normal broadcasting to be done with conventional A.M i.e. Transmission of carrier and both sidebands with the consequent simplicity in receiver design, commonly seen in the domestic radio.

The SSBSC receivers are generally used by professionals and enthusiastic amateurs.

9.7 VESTIGAL SIDEBAND TRANSMISSION (VSB)

This is a special case of double sideband transmission in which one complete sideband is transmitted and a small portion or vestige of the other sideband is transmitted –see fig. 9.7.1.

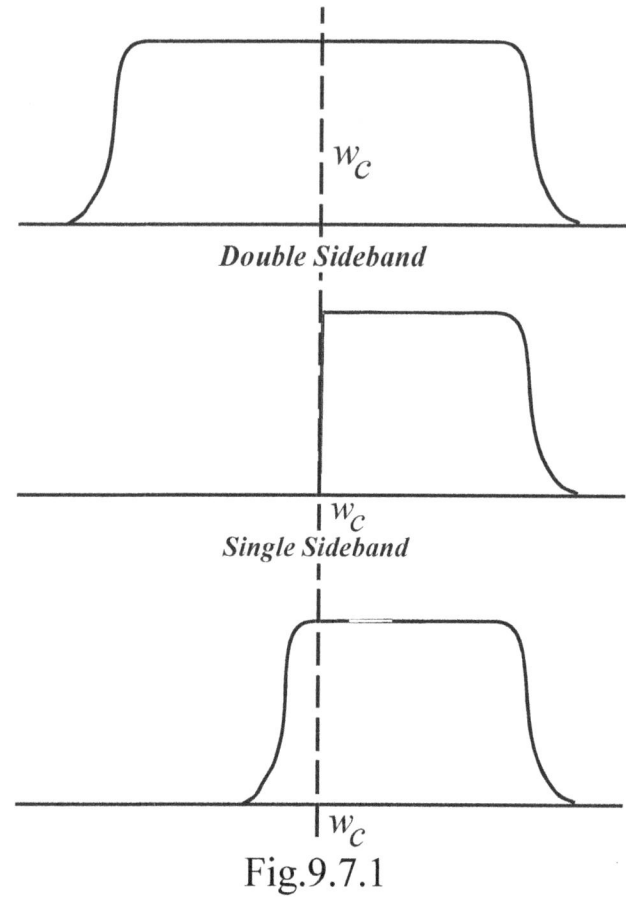

Fig.9.7.1

Such VSB transmission is almost exclusively used in the transmission of T.V signals where information about the d.c level (brightness) of the original signal is required at the receiver. In this case the bandwidth is about 6MHz.

Discussion on generation, reception and *demodulation* of AM waveforms are beyond the scope of this book.

9.8 FREQUENCY MODULATION F.M

As previously stated frequency modulation is the effect of varying the instantaneous frequency of the carrier waveform according to the modulating function. Let the carrier output be sinusoidal of the form

$$y(t) = V_C \, Sin(wt + \phi) \text{ Where the instantaneous phase is}$$
$$\theta = wt + \phi$$
whence $$\frac{d\theta}{dt} = w$$

Thus angular frequency is the differential of instantaneous phase giving a rotating vector (phasor) of constant amplitude. $\hat{V}_C$

If the angular frequency is made to vary according to the modulating function g(t) it may be written as $w = w_C(1 + g(t))$

Giving $$y(t) = \hat{V}_C \, Sin[w_C(1 + g(t))t + \phi]$$

Single Sinusoidal F.M

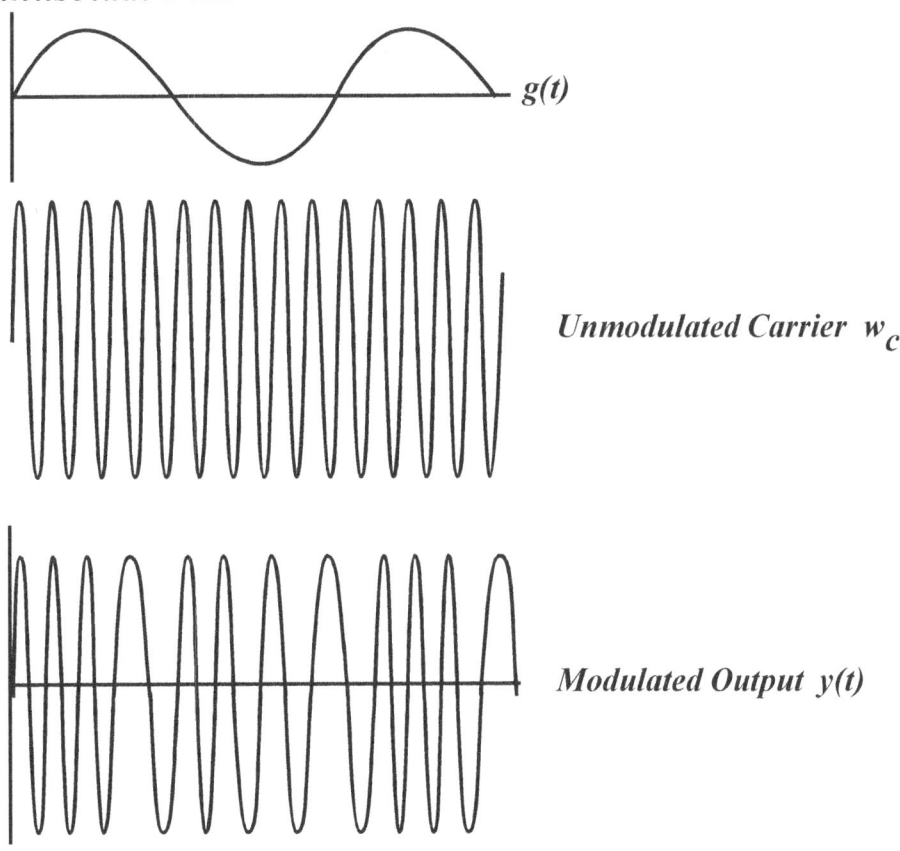

Fig. 9.8.1

Again consider the case where $\phi = 0$ and the modulating function is sinusoidal (fig. 9.8.1) of the form $g(t) = A \cos pt$

Then $w = w_C[1 + A\cos pt]$

giving a maximum value to w of $w_C(1 + A)$

and minimum value of $w_C(1-A)$

Hence $wt = \int_o^t w_C[1 + A\cos pt]dt$

$$= w_c t + \frac{w_C A}{p} \sin pt$$

Substituting in the expression for y(t) gives $y(t) = V_C \sin\left(w_C + \frac{w_C A}{p}\sin pt\right)$

$w_c A$ above is proportional to the modulating signal amplitude which also determines the frequency derivation from the carrier.

The ration $\frac{w_C A}{p}$ is called the modulation factor, modulation index or deviation ratio δ

$$y(t) = \hat{V}_C \sin(w_c t + \delta \sin pt)$$
$$= \hat{V}_C \{\sin w_c t \cos(\delta \sin pt) + \cos w_c t \sin(\delta \sin pt)\}$$

The terms $\cos(\delta \sin pt)$ and $\sin(\delta \sin pt)$ may be expanded by BESSEL functions (see fig. 9.8.2)

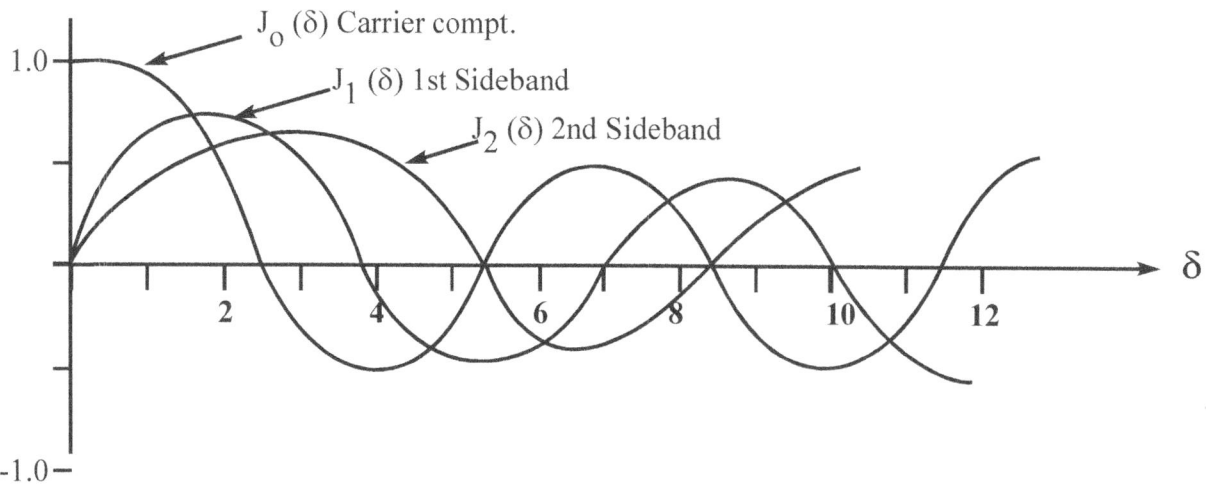

Fig. 9.8.2

where

$$Cos(\delta Sinpt) = J_O(\delta) + 2\{J2(\delta)Cos2pt + J_4(\delta)Cos4pt + J_6(\delta)Cos6pt + ---\}$$
$$Sin(\delta Sinpt) = 2\{J_1(\delta)Sinpt + J_3(\delta)Sin^3 pt + --------\}$$

where J_n are Bessel Coefficients of order n obtained from tables or a plot of such functions

Substitution gives

$$y(t) = \hat{V}_C \begin{bmatrix} Sin w_c t(J_O(\delta) + 2\{J2(\delta)Cos2pt + J_4(\delta)Cos4pt + --\}) \\ + Cos w_c t \cdot 2\{J_1(\delta)Sinpt + J_3(\delta)Sin^3 pt + --------\} \end{bmatrix}$$

$$y(t) = \hat{V}_a \begin{bmatrix} J_O(\delta)Sin w_c t + J_2(\delta)\{Sin(w_C + 2p)t + Sin(\omega_O - 2p)t\} \\ + J_4(\delta)\{Sin(w_C + 4p)t + Sin(w_C t - 4p)t\} + --- \\ + J_1(\delta)\{Sin(w_C + p)t - Sin(w_C - p)t\} \\ + J_3(\delta)\{Sin(w_C + 3p)t - Sin(w_C - 3p)t\} \\ + J_5(\delta)\{---------------\} + -- \end{bmatrix}$$

This equation shows that the spectrum is made up of a fundamental plus an infinite number of sidebands whose amplitudes are Bessell Coefficients.

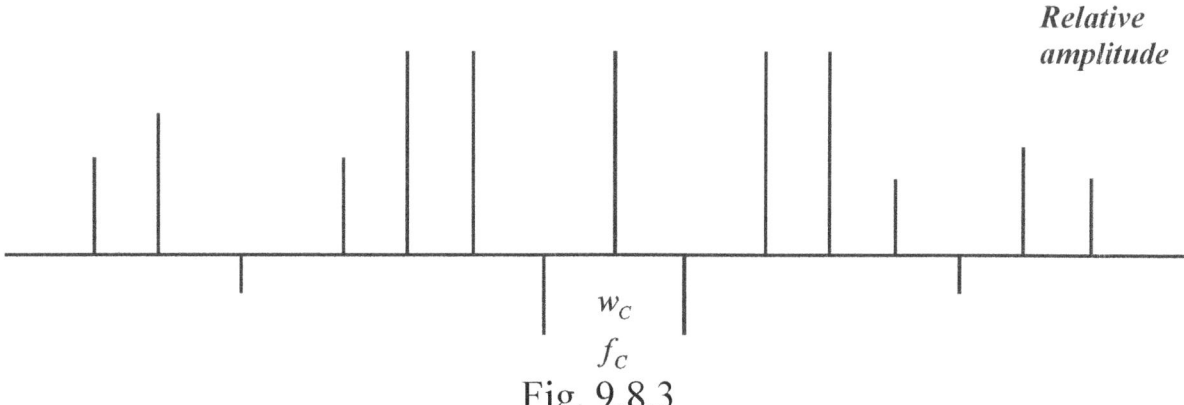

Fig. 9.8.3

Negative coefficients indicates that they are shifted 180^0 from those with positive signs – see fig. 9.8.3

The above analysis implies that the bandwidth required for FM transmission must be much greater than fro AM transmission where the same information is to be conveyed. In practice it is usual to consider a finite number of sidebands where amplitudes are greater than 1% of the largest sidebands or unmodulated carrier.

The frequency deviation is usually limited to the available bandwidth which for a modulating frequency of 15KHz and a typical frequency deviation of ± 75KHz give a deviation ratio of

$$\frac{\Delta f_C}{p} = \frac{75}{15} = 5 = \delta$$

As with the AM case modifications to the basic FM waveform are used namely Narrowband FM and Broadband FM whose operation is self evident. In narrowband FM transmission is usually limited to the carrier and a few (one or two) pairs of sideband frequencies. For broadband FM the bandwidth is limited to approximately $2(\Delta f_C + p)$ which in the above case would be 2(75 + 15)KHz = 180KHz say 200KHz.

The analysis of theses two forms of FM is outside the scope of this course.

9.9 COMPARISONS OF AM AND FM

1. From the foregoing it is evident that AM only produces two sidebands whereas FM produces a large number of sidebands and hence FM requires a wider bandwidth.

2. AM depends on amplitude variations which are prone to fading and alternation. FM is constant amplitude and therefore is less susceptible to interference and has a better signal/ noise ratio.

3. AM is simpler and cheaper to operate than FM

9.10 PHASE MODULATION

In PM the instantaneous phase of the carrier signal is made to vary in sympathy with the modulating function $g(t)$

For the carrier $y(t) = \hat{V}_C \, Sin(w_C t + \phi)$

Where instantaneous phase $\theta = w_C t + \phi$

Let the modulating function $g(t)$ vary the phase angle
$$\phi = \phi_C (1 + kg(t)) \qquad k \text{ factor}$$
Substitution gives $\theta = w_C t + \theta_C (1 + kg(t))$

Consider again sinusoidal modulation with a single tone $g(t) = Sin\, pt$ giving
$$\theta = w_C t + \theta_C (1 + k\, Sin\, pt)$$
and the modulated output in
$$y(t) = \hat{V}_C \, Sin(w_C t + \theta_C (1 + k\, Sin\, pt))$$
$$= \hat{V}_C \, Sin(w_C t + \theta_C + \theta_C k\, Sin\, pt)$$
$$= \hat{V}_C \, [Sin(w_C t + \phi_C) Cos(\phi_C k\, Sin\, pt) + Cos(w_C t + \phi_C) Sin(\phi_C k\, Sin\, pt)]$$

Thus can be expanded using Bessel functions to give

$$y(t) = \hat{V}_a \begin{bmatrix} J_O(\phi_C k) Sin(w_C t + \phi_C) + J_2(\phi_C k) Sin\left(w_C + \frac{\theta_C}{t} + 2p\right)t \\ + Sin\left(w_C + \frac{\theta_C}{t} - 2p\right)t + J_4(\phi_C k)[Sin - - - - - - - - \\ + J_1(\phi_O k)\left\{Sin\left(w - \frac{\theta_C}{t} + p\right)t - Sin\left(w - \frac{\theta_C}{t} - p\right)t\right\} \\ + J_3(\phi_O k)\left[Sin\left(w - \frac{\theta_C}{t} + 3p\right)t - - - - -\right] + - - - - - - - \end{bmatrix}$$

Hence there is a carrier and infinite number of sidebands. Note that the fundamental difference between FM and PM lies in the fact that the term $(\phi_O k)$ is constant and independent of the modulation frequency $\frac{p}{2\pi}$

Whereas in FM the deviation ratio δ is proportional to the frequency deviation and inversely proportional to the modulating frequency $\frac{p}{2\pi}$
As previously stated frequency in the differential of phase and this relationship can be used to practical advantage as shown in fig. 9.9.1

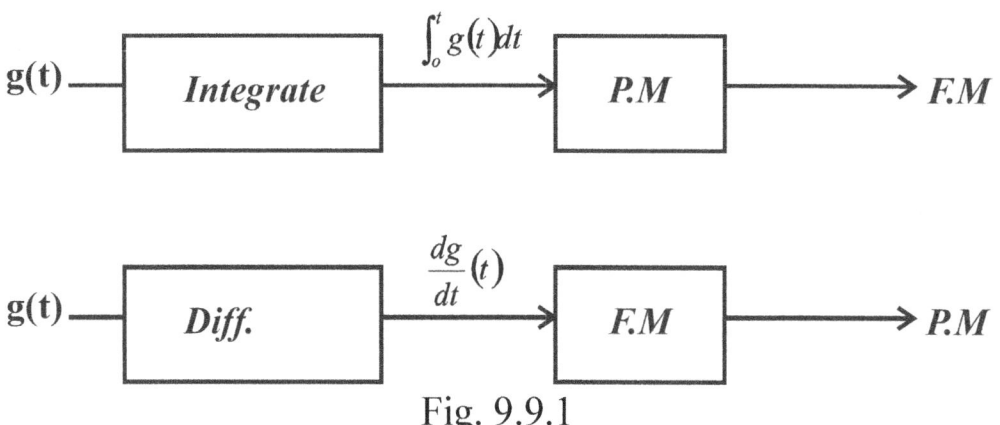

Fig. 9.9.1

The former case is useful in FM generation in which the phase of a very stable crystal oscillator is shifted by a small but constant amount.

DIGITAL ELECTRONICS

CHAPTER TEN

10.0 DIGITAL ELECTRONICS
10.1 BINARY ARITHMETIC & BOOLEAN ALGEBRA

In the principles and practice of Electrical Engineering we have seen how the AND, OR and other Gates can have only two states (0,1) and cannot have any other. This time we shall look into the concept of using the two states 0, 1 for purposes which suit us.

This will be done in two parts;

i) First we shall see how the two states o and 1 behave as Nos. and how to use them for arithmetic calculations.

ii) Next we shall see how these two states 0 and 1are handled by the gates (AND, OR, NAND, NOR etc. and their combinations) this will tell us what the circuits are capable of doing.

Once we know these two concepts ie; i) how to make the Nos. do the arithmetic calculations, and ii) how the circuits handled the Nos. we can go on make combinations of gates to make them do the calculation..

(a) THE ARITHMETIC OF 0 AND I

The word "numbers" brings to mind the ten symbols 1 2 3 4 5 6 7 8 9. Numbers express quantity and these symbols form the code for expressing them. This is called a decimal system as 10 symbols are used. Systems with more symbols or less symbols can be formed. This total number of symbols used in a system is called the radix of the system or the base of the system.

250

In the decimal system after the count reaches NINE the next number is formed by putting a '1' to the left of zero; this is to indicate that the count has reached ten once. Thus every time the count reaches ten a unit is added to hand digit.

Thus -1 5 means 1 ten +5 one's

2 4 ,, 2 ten's +4 one

At 99 the symbols are again exhausted and so we repeat the change that occurred at 10. That is, to put a '1' to the left of the two digits to indicate that the count at the adjacent position has reached 10.

The numbers run as:

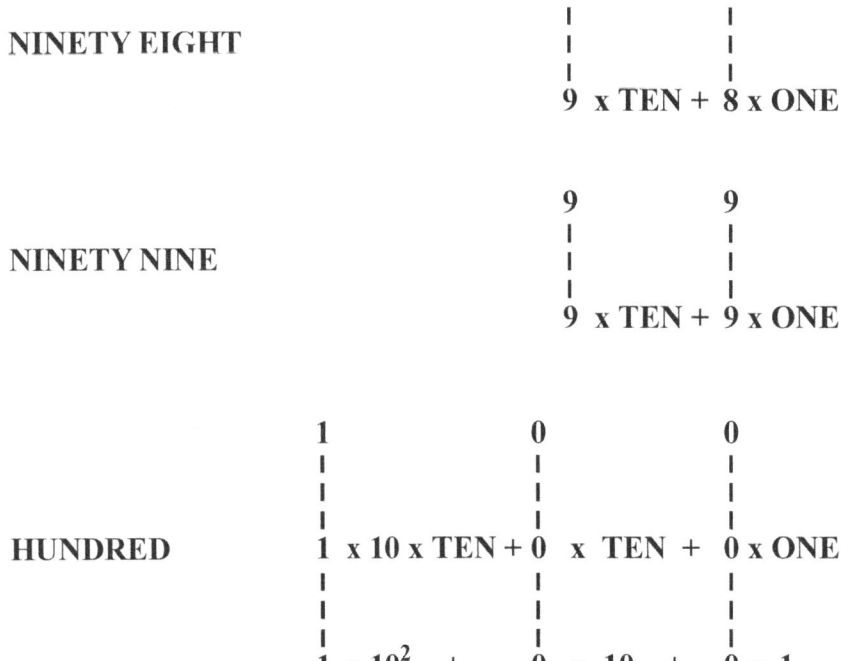

Thus the count can proceed. The valued of the numbers increases by power of ten at every place to the left.

For example, a number like 34,215 means-

$$3 \quad 4 \quad 2 \quad 1 \quad 5$$

$$3 \times 10^4 + 4 \times 10^3 + 2 \times 10^2 + 1 \times 10 + 5 \times 1$$

DIGITAL ELECTRONICS

These principles of forming numbers are true with all systems of numbers even though their base may be different from ten.

Thus for an eight symbol system the first place (from the right) represents the one's, the second place represents the 8's, the third place represents the 8^2, the fourth place the 8^3 and so on.

Thus the number 125 in an eighty system will mean,

        1   2   3

$1 \times (eight)^2 + 2 \times (eight) + 5 \times one$

This system is called the **OCTAL** system since it uses only eight symbols. These are; 0 1 2 3 4 5 6 7 (note that symbols 8 and 9 are not used in this system).

If we now consider a sixteen symbol system called the **HEXADECIMAL** system the number 125 will mean as below;

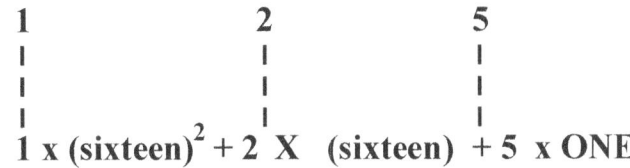

$1 \times (sixteen)^2 + 2 \times (sixteen) + 5 \times ONE$

The sixteen symbols used are 0 1 2 3 4 5 6 7 8 9 A B C D and E. The decimal equivalent of these last five symbols and the progress of further count in the hexadecimal system are given later.

Along the same lines we can consider the 2 symbol (called the **BINARY** system). The two symbols used are 'o' and '1' (and will not contain any other symbol like 2, 3 etc).

The valued of a multi – digit number follows the same principles as for the other systems i.e.; the first digit from right represents the '1'; the next represents 'two'; the third represents two^2, and the fourth represents $(two)^3$ and so on (as two is the "base" of the system.)

For example, number 101101 will mean

1   0   1   1   0   1

$1 \times (two)^5 + 0 \times (two)^4 + 1 \times (two)^3 + 1 \times (two)^2 + 0 \times two + 1 \times one$

$= 2^5 \quad + \quad 2^3 \quad + \quad 2^2 \quad + \quad 0 \quad + \quad 1$

= 45 in decimal system

The count of binary numbers (using the earlier stated principles) runs as below.

Binary number -   0   1   10   11   11   100   101 ....

Decimal number-   0   1   2   3   4   5   ....

From the above we can see that the following relations exist for binary numbers (for addition)

$$\left.\begin{array}{l} 0 + 0 = 0 \\ 0 + 1 = 1 \\ 1 + 0 = 1 \\ 1 + 1 = 10 \end{array}\right\}$$ All are **BINARY** numbers

The least line is worth attention. '10' in binary stands for the decimal number '2'. Thus whenever we add '1' and '1' of binary system there is a '1' which is carried to the left.

The other arithmetic functions also can be similarly derived.

However, addition is the function most easily done by digital circuits. Hence it is desirable if the same circuits can be used for subtraction. This is possible by using the system of complementary numbers. The basic principle used in this is that x −y = x + (-y), thus converting a subtraction into an addition.

DIGITAL ELECTRONICS

In the decimal system the 9's and 10's complements are used. The 9's complement of a number is reached by subtracting each digit from 9. For example 9's complement of 361 is -

$$\begin{array}{r} 999 \\ -\ 361 \\ \hline 638 \end{array}$$

By adding '1' to this we get the 10's complement, i.e.: 638 + 1 = 639 is the 10's complement. To subtract a number A from a number B, the following steps are done if both the numbers belong to decimal system.

i) Find the 9's complement of A (see note 1)

ii) Add this result to B and,

iii) "End around carry". Whenever there is a 1 carry in the last position we remove it and add it on to the remainder.

This is called "end-around carry".

Example 10.1.1 – to subtract 540 from 641

i) 9's complement of 540 is 459

ii) Add this to 641

$$\begin{array}{r} 459 \\ \underline{641} \\ (1)\ \ \ 100 \end{array}$$

iii) End around carry

$$\begin{array}{r} 100 \\ +\ \underline{\ \ 1} \\ 101 \end{array}$$

∴ 641 – 540 = 101

(b) USE OF 10'S COMPLEMENT:

The following steps are involved in subtracting A from B

DIGITAL ELECTRONICS

i) Find the 9's complement of A (see note 1)

ii) Add '1' to this to get 10's complement

iii) Add this to B. if there is a carry at the last place, ignore same (see note 2)

Using the same example as above to subtract 540 from 641

i) 9's complement of 540 is   4   5   9

ii) 10's complement of 540 is   4   5   9 + 1 = 460

iv) Add to 641

$$\begin{array}{r} 460 \\ \underline{641} \\ (1) 101 \end{array}$$

Ignore this
Carry
Therefore, result is 101

A Similar method is applicable for binary numbers. The subtraction is converted into a process of addition by using '1's complement.

The '1's complement is reached by subtracting each digit from one. In other words, the 1's complement can be reached by simply changing all the 1's to '0' and all zero's to 1.

For example, one's complement of

    1 0 0 1 0 1

is   0 1 1 0 1 0

The following are the steps involved in subtracting a binary number A from binary number B.

i) Find the 1's complement of A (see note 2)

ii) Add this to B

iii) End around carry-i.e.: at step (ii) if there is a carry at the most significant digit (i.e. the first digit from the left) take that around to the least significant digit (i.e. the first digit from the right).

Example 10.1.2 - To subtract (A)   10101

                  From   (B)   11101

Step i)   One's complement of (A)   10101

                            is   01010

Step ii)   Add the above to (B)   11101

```
                            11101
                            01010
There is a carry 1          00111
```

Step iii)   End around carry (i) 0 0 1 1 1

                                               1

                                     1 0 0 0

Hence the answer is 1000

An easier method is to use the 2's compliment.

The 2's compliment of a number is reached by first getting the 1's complement and then adding 1 to it.

For example, the 2's complement of 10101 is reached by-

         One's complement is =   01010

                add 1   +   1

∴   2's complement is =   01011

To subtract number A from B-2's complement of A is added to B.

If there is a carry, ignore it

Example 10.1.3

   A   =   10101

   B   =   11101

The 2's complement of A is ….   01011

Therefore, B-A is   11101

                  +   01011

   Carry   1   01000

Ignoring the carry the result is 1000

The 2's complement is more often used in digital circuits as the "end around carry" and is not required in this method.

NOTES

1. it is of important to note that before forming the one's complement it should be ensured that the number A contains as many digits as B. If it is not, then add as many zero's as required to the left of A so that A and B have the same number of digits.

2. If there is no carry, then it indicates that the answer is negative. The result reached is the complement of the actual magnitude,

In handling of logic circuits the systems of circuits most often used are the above three systems, i.e.; Binary, Octal and Hexadecimal. Hence conversion from decimal to any of these systems is of great use.

DECIMAL to OCTAL conversion is done by dividing the decimal number by 8 successively – for example, take the decimal number 236 – to convert to octal –

DIGITAL ELECTRONICS

$$8 \overline{)236}$$
$$8 \overline{)29} \quad) \, 4$$
$$8 \overline{)3} \quad) \, 5$$
$$0 \quad) \, 3$$

Hence $236 = 3 \times 8^2 + 5 \times 8 + 4 \times 1$

     $= 3\ 5\ 4$ (Octal)

Thus $236_{10} = 3\ 5\ 4_8$ The subscript indicates the system to which the number belongs.

To convert **OCTAL** to **BINARY** an easy method exists. Just put the corresponding binary number below each digit and write them together – as shown below;

$$\begin{array}{ccc} 3 & 5 & 4 \\ 011 & 101 & 100 \end{array}$$

Therefore, 354_8 =   011   101   100

Therefore, 11101100_2 = 354_8 = 235_8

     (Binary)   (Octal)   (Decimal)

The subscript indicates the base of the system.

From above it can be seen that one easy way of converting **DECIMAL** to **BINARY** is by first converting **DECIMAL** to **OCTAL** to **BINARY**.

The above are the basic systems of numbers used in digital circuits. Many other systems exist.

10.2 BOOLEAN ALGEBRA

The next part is to see how the states '0' and '1' are handled by the gates. We will also see the methods by which digital circuits can be designed. This will also help us to understand such circuits. These were first introduced by George Boole, a British mathematician and hence called Boolean Algebra.

We know that the function of the AND gate and OR gate

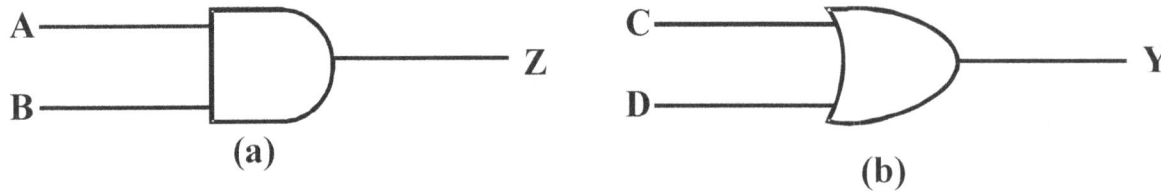

Fig. 10.2.1

The function of an AND gate is represented by a multiplication sign (as A × B or A .B or simply AB)- fig. 10.2.1a

The OR function is represented by the addition sign. Thus in the above diagrams of fig 10 .2.1b

$Z = $ A . B (read as "A and B$''$)

$Y = $ A + B (read as "A or B$'$)

Each of the above variables can have only two valves – 0 or 1

Thus for the **AND** gate

When   A = 0, B = 0, Z = 0   i.e 0 . 0 = 0
            A = 1, b = 0, Z = 0   1. 0 = 0
            A = 0, B = 1, Z = 0   0 . 1 = 0
            A = 1, B = 1, Z = 1   1. 1 = 1

For the **OR** gate

When   A = 0, B = 0, Y = 0   i.e 0 + 0 = 0
            A = 1, B = 0, Y = 1   1 + 0 = 1
            A = 0, B = 1, Y = 1   0 + 1 = 1
            A = 1, B = 1, Y = 1   1 + 1 = 1

Using these relations we can see that

    When A=0, B =0 y =0 i.e.   0+0=0
        A=1, B= 0 Y = 1   1+0=1
        A =0, B = 1 Y = 1   0+1=1
        A = 1, B = 1 Y = 1   1+ 1 =1

Using these relations we can see that

$$\left.\begin{array}{l} A.1 = A \\ A.0 = O \\ A.A = A \\ A.\overline{A} = O \end{array}\right\} \text{Since a can have only 2 values (0 or 1)}$$

$$A + 1 = 1$$
$$A + 0 = A$$
$$A + A = A$$
$$A + \overline{A} = 1$$

$\overline{A}$ is the inverse of A.

So if $A = 1$ $\overline{A} = 0$

If $A = 1$ $\overline{A} = 1$

By using the above relations we can prove the following"

$A + A6$	$= A$	-(1)
$A(A + B)$	$= A$	-(2)
$A(\overline{A} + B)$	$= AB$	-(3)
$A + \overline{A}B$	$= A + B$	-(4)
$(A + B)(A + C)$	$= A + BC$	-(5)
$(A + B)(A + B)$	$= A$	-(6)

There are two other theorems which are of great use.

They are:

$$\overline{A \cdot B \cdot C} = \overline{A} + \overline{B} + \overline{C} \qquad (7)$$
$$\overline{A + B + C} = \overline{A} \cdot \overline{B} \cdot \overline{C} \qquad (8)$$

These are known as De Morgan's Theorems. All the above equations can be used for simplifying Boolean equations. For example;

$$Z = ABC + AB + \overline{A}BC + \overline{A}\overline{B}C$$
$$Z = AB(C + 1) + \overline{A}C(B + \overline{B})$$
$$Z = AB + \overline{A}C \qquad \text{since } B + \overline{B} = 1 \text{ and}$$
$$C + 1 = 1$$

10.3 KARNAUGHS MAPS;

The above method will be unsuitable for more complex equations. Hence a graphical method called Karnaughs maps is used. This basically is a way of recording the normal truth table so that adjacent terms differ by only one digit.

Taking the same example as above;
$$Z = ABC + AB + \overline{A}BC + \overline{A}\,\overline{B}C$$

This has three variables A, B and C. The map for three variables is as shown in fig. 10.3. It provides a square each for all possible combinations of the variables. Then plotting is done by putting '1' in the squares of each of the terms in the equation and a zero in the other places. Then group the terms represented by the 1's. The adjacent terms can be grouped together as shown by the circles. Every pair of adjacent terms can be reduced to the common factor of the pair. Thus the 'a' of the map is AB.

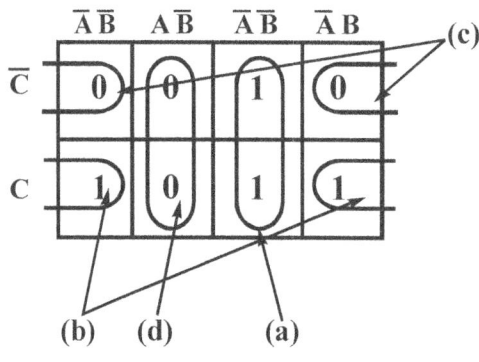

Fig. 10.3.1

The term $\overline{A}\,\overline{B}C$ and $\overline{A}BC$ are adjacent terms as they differ only by one digit although they appear at opposite ends of the map. Thus the term 'b' of the map is $\overline{A}C$. Thus the function has now only these two terms AB, AC. So the equation is

$$Z = AB + \overline{A}C$$

This result is the same as that got earlier by the algebraic method. For equations with more variables maps as shown in fig 10.3.2

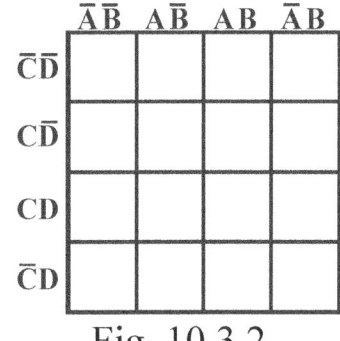

Fig. 10.3.2

The terms appearing at the opposite ends (left and right ends or top and bottom ends) can be grouped together since they differ by one digit.

Thus in the following example term 'a' is $\overline{A}CD$ and term; 'b' is $\overline{C}AB$–Fig 10.3.3

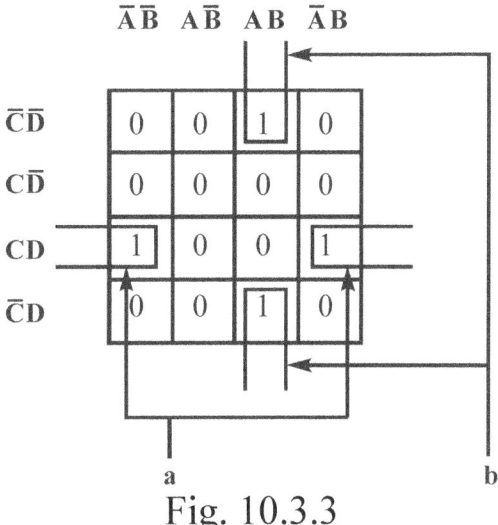

Fig. 10.3.3

This also applies to block of 4 and 8 adjacent; for example –Fig 10.3.4

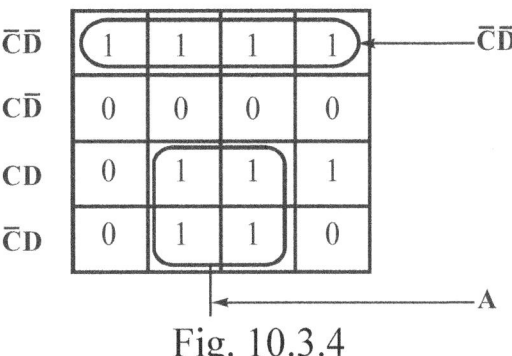

Fig. 10.3.4

For equation with 5 variables the map (Fig 10.3.5 be used;

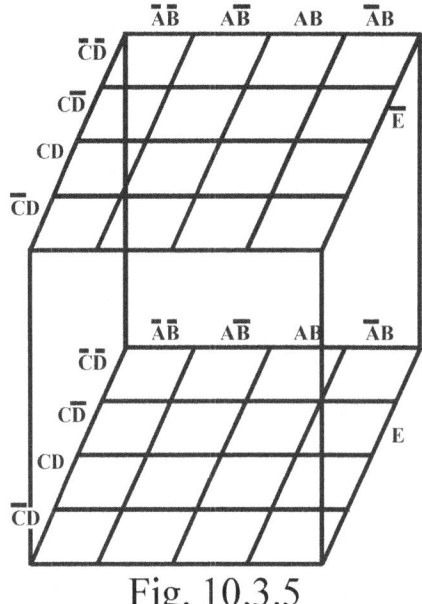

Fig. 10.3.5

These maps can be used for designing circuits.

Example; Fig. 10.3.6. To make a circuit that will compare two inputs and give an output if they are equal – Fig 10.3.1. Since there are only two inputs, the map for 2 variables will be used. Mark 1 in the squares where the inputs are same (i.e. wherever there is an output). Mark 0 at other places. So we have;

	$\overline{A}$	A
$\overline{B}$	1	0
B	0	1

Fig. 10.3.7

On grouping the factors, If the output Z = $\overline{A}\overline{B}$+AB, thus we get the equation for a comparator (handling only 2 inputs and hence called 2 bit comparator).

Now the circuit for the above equation is formed as below;

i) $\overline{A} \cdot \overline{B}$ is an 'AND' gates whose inputs are A and B. $\overline{A}$ and $\overline{B}$ are got by inverting A and B respectively.

ii) A.B is an 'AND' gate whose inputs are A and B.

iii) The outputs of these AND gates are given to an OR gate to get $\overline{A}\,\overline{B}$ +AB. Thus we have as shown in fig 10.3.7

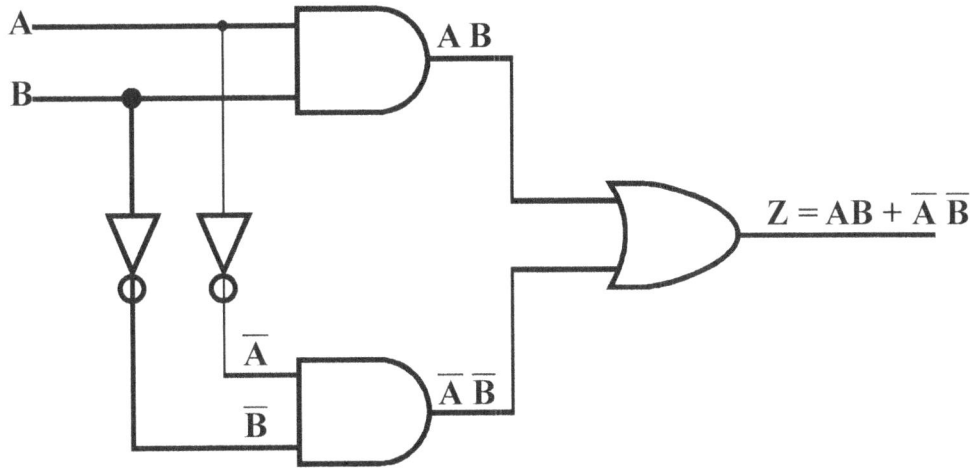

Fig 10.3.7 Two Bit comparator

It should be noted that N AND and NOR gates are more easier to manufacture (see TTL, DTL, ECL etc. gates)

Hence it is better to design circuits using them instead of AND can be re-written as below

Z= AB + AB   Using De Morgan's Law.

$= \overline{\overline{\overline{AB}} + \overline{AB}}$

In the above result $\overline{A}\,\overline{B}$ is a NAND gate

$\overline{\overline{AB}}$ is a NAND gate. Hence the circuit is Fig 10.3.8;

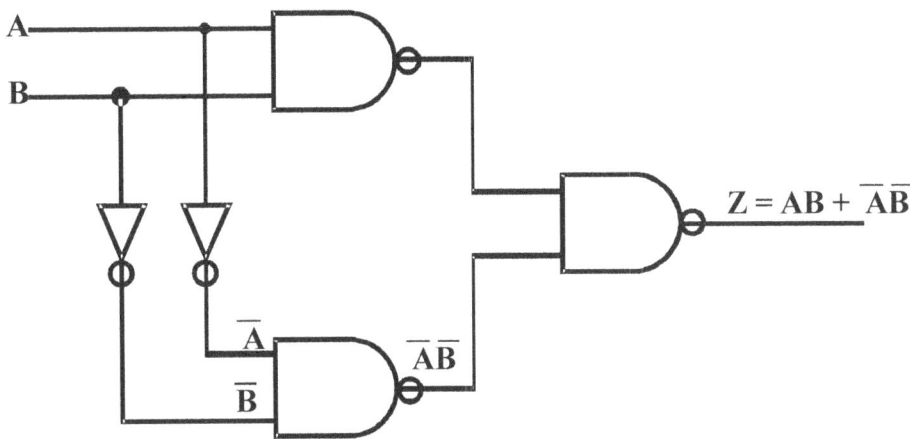

Fig. 10.3.8 NAND gate circuit for a Two Bit comparator

10.4 EXCLUSIVE OR CIRCUIT

To design a circuit which gives output when inputs are not equal. If Z is the output and A and B are the inputs, we have as shown in fig. 10.4.1

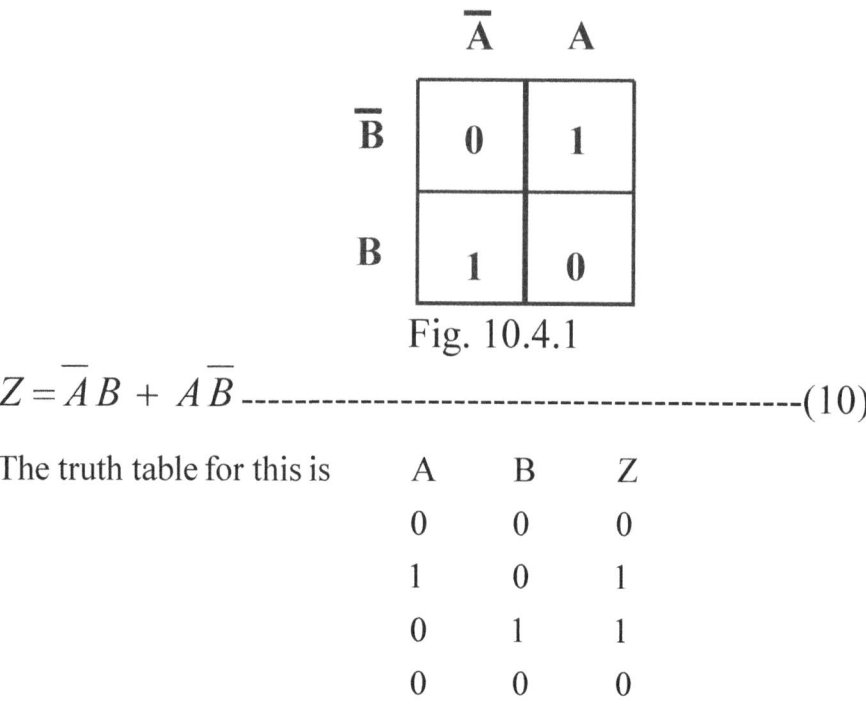

Fig. 10.4.1

$$Z = \overline{A}B + A\overline{B} \quad \text{------------------------------------(10)}$$

The truth table for this is

A	B	Z
0	0	0
1	0	1
0	1	1
0	0	0

The table is very similar to an 'OR' circuit except that the output is present only when any one input is 1, but not when both are 1. This is called an exclusive "OR" since output is present only when A or B is 1. The circuit symbol for such a gate is as shown in fig 10.4.2 and is represented as $Z = A \oplus B$

Fig10.4.2

The circuit can be formed from the equation 10 above;

$Z = \overline{A}B + A\overline{B}$ by De Morgan's Law $= \overline{\overline{\overline{AB}} \cdot \overline{\overline{AB}}}$

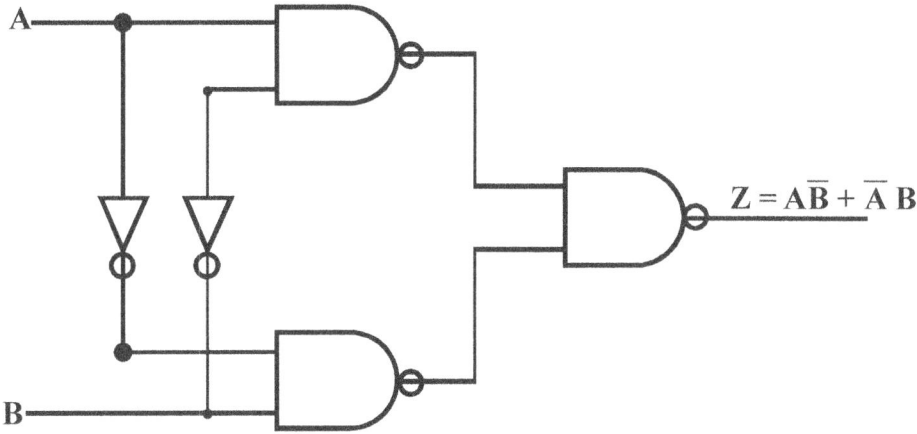

Fig 10.4,3 Exclusive – OR Circuit and Logic Symbols

Thus the karnaughs Maps can be used

    i)   For simplifying the equations

    ii)   For forming the equations

    iii)   For simplifying the circuits.

So far we have seen how the gates can be combined and their inputs and outputs are related by Boolean algebraic equations

We have also seen in the earlier portion of this chapter how to do calculations using the two numbers 0and1.

The fundamental arithmetic function is addition. There are four possibilities which are;

$$\begin{array}{cccc} 0 & 0 & 1 & 1 \\ +0 & +1 & +0 & +1 \\ \hline 0 & 1 & 1 & 1\,0 \end{array}$$

In the first three cases there is a sum (which is zero or 1). No digit is carried to the left. In the fourth case the sum is zero, but a digit is carried to the left. Thus when 2 binary digits are added there are two outputs. The sum and carry. The carry output for the first three cases is 0 and the fourth case is 1.

Now sum is 1 if one of the two digits is '1'. The sum is 0 if both the digits are equal (if both are equal to zero or if both are equal to 1). This is the same as of the condition discussed earlier for "exclusive OR (see equation 10). Hence the karnaughs map for this is the same as for the "exclusive OR".

Therefore, the sum of two digits A and B is given by sum

= AB+AB   (see Fig 10.4.4)

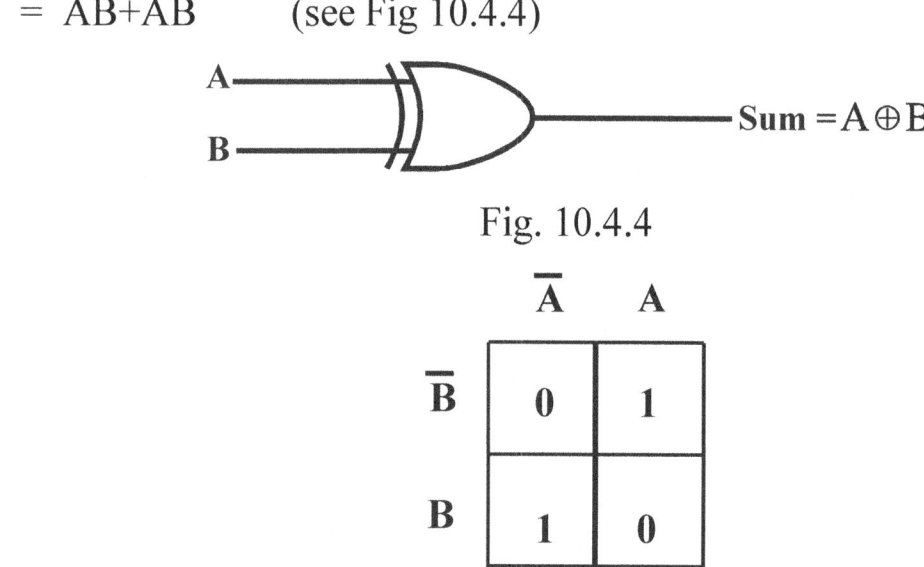

Fig. 10.4.4

The "carry" output is present only when two inputs are 1. The "carry" is zero in other cases. So if A and B are the inputs digits, the karnaughs map for "carry" output is as below.

This is the same as an AND gate hence carry = A.B as represented in fig 10.4.5

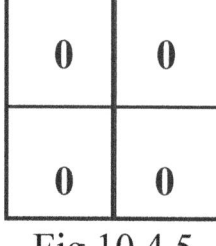

Fig 10.4.5

10.5 HALF ADDER CIRCUIT.

Combining the above two results the circuit for adding two binary digits is as shown in fig 10.5.1.

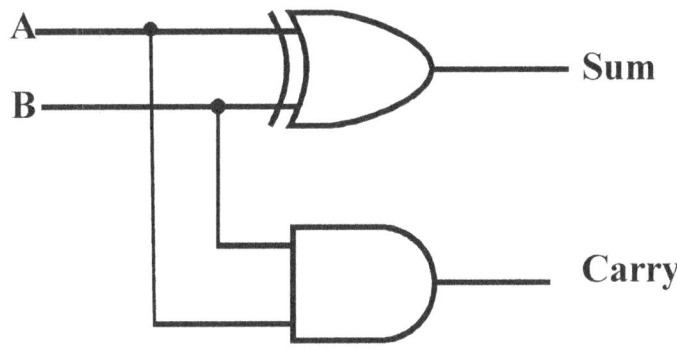

Fig 10.5.1 Half Adder

This is called a half adder.

The result of a half adder is summarised in the truth table of Fig 10.5.2

A	B	Carry	Sum
0	0	0	0
0	1	0	1
1	0	0	1
1	1	1	0

Fig. 10.5.2

In a normal situation of adding the binary numbers, sometimes we have to 'carry' 1 from one column to the other. For example;

```
    1   1   1
+   1   0   1
-----------
1   1   0   0
```

In the least significant column (i.e. the first right hand side column,) we added the two digits;-

$1 + 1 = $ sum $= 0$, carry 1 to the Second column.

In adding the digits in the second column, we have to add the existing two digits (1 and 0) and also the 'carry' from the first column.

Thus we have $1 + 0 + 1 = 0$, carry 1 to the third column

In adding the last column, because of the carry from the previous column, there are three digits to add.

1 + 1 + 1 = 1, carry 1

10.6 FULL ADDER CIRCUIT.

Thus in adding binary numbers electronically, the circuits have to be capable of adding 3 digits at a time (as in needed in columns 2 and 3 of the above example). But the half adder is capable of adding only 2 digits. A full adder capable of handling 3 digits can be formed using 2 half adders;

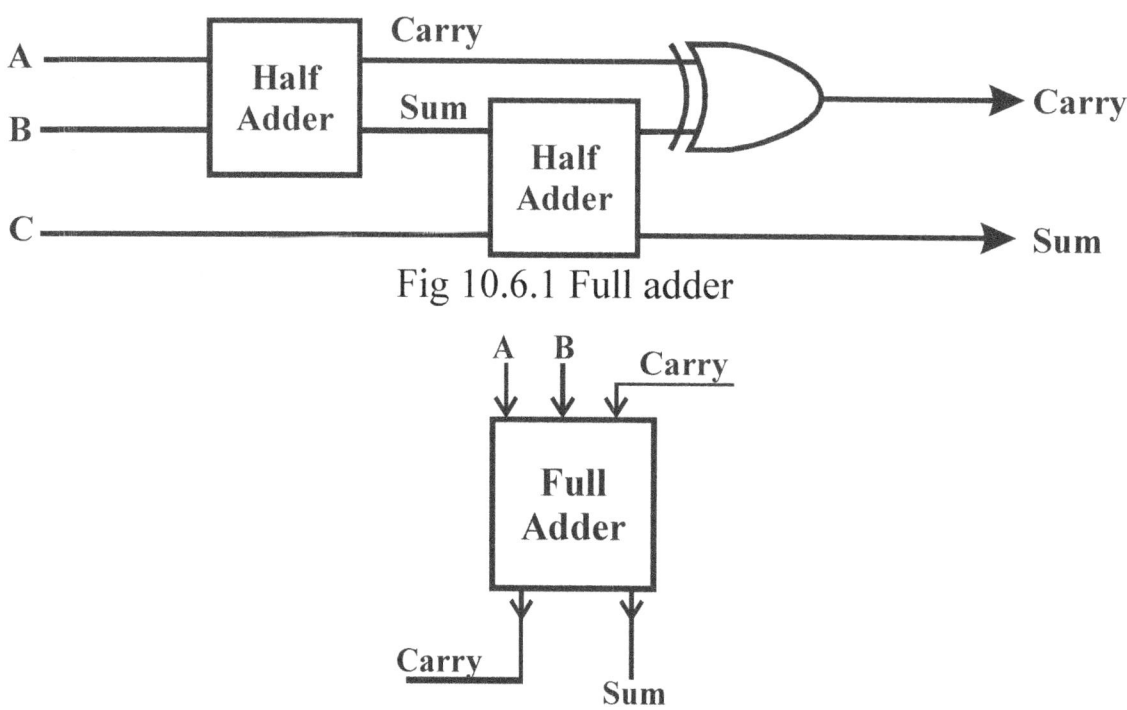

Fig 10.6.1 Full adder

Fig 10.6.2.is a five digit parallel full adder.

DIGITAL ELECTRONICS

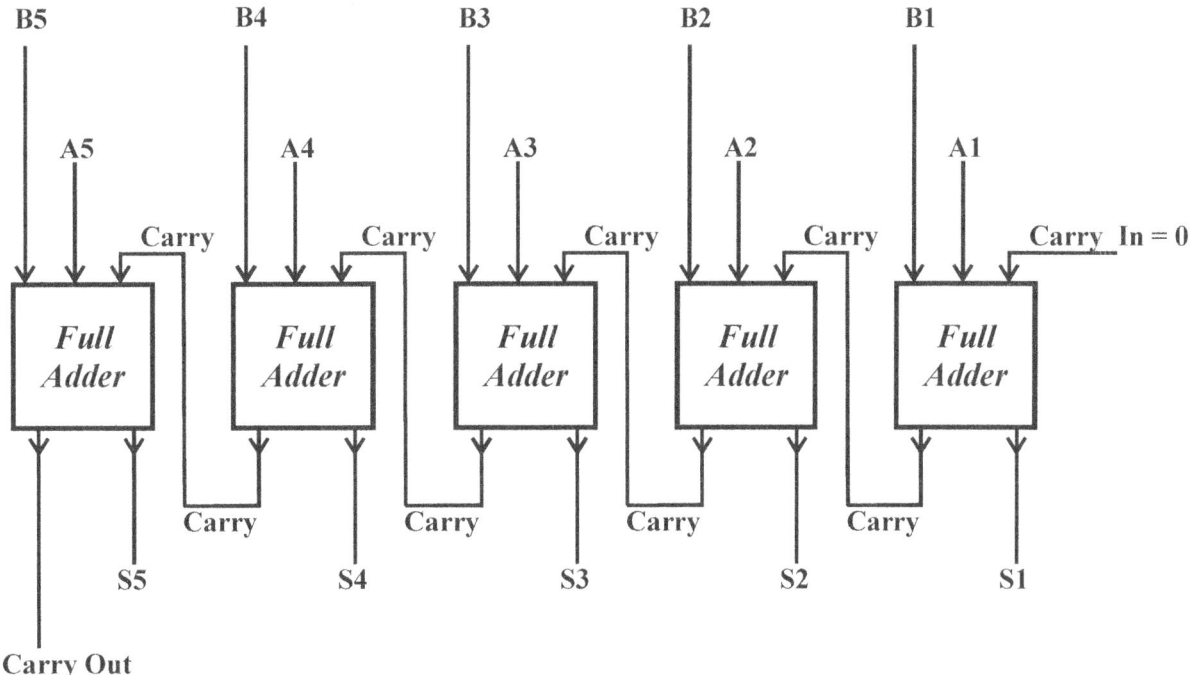

Fig 10.6.2; Five digit parallel full adder

The full adder has three inputs (2 are for the digit from the numbers and one is for the carry from the previous column). It has two outputs, the sum the carry.

A number of full adders can be cascaded to form a multibit adder. It is called a parallel added as all the digits of the numbers are fed at the same time.

A five bit parallel adder (i.e an adder capable of adding 2 numbers each having 5 bits) is given below.

Let the numbers be

$$\begin{array}{ccccc} A_5 & A_4 & A_3 & A_2 & A_1 \\ B_5 & B_4 & B_3 & B_2 & B_1 \\ \hline S_5 & S_4 & S_3 & S_2 & S_1 \end{array}$$

A subtractor can be formed using the same circuit elements as explained earlier, the subtraction is achieved by first forming the 1,s complement (this is formed by the NAND gates in the circuit fig 10.6.3) and then adding this to the other number (done by the full adders), and lastly doing the end around carry (by the full adders), and lastly doing the end around carry (by the feedback loop from the carry output of the last full to the last full adder to carry inputs of the first full adder).

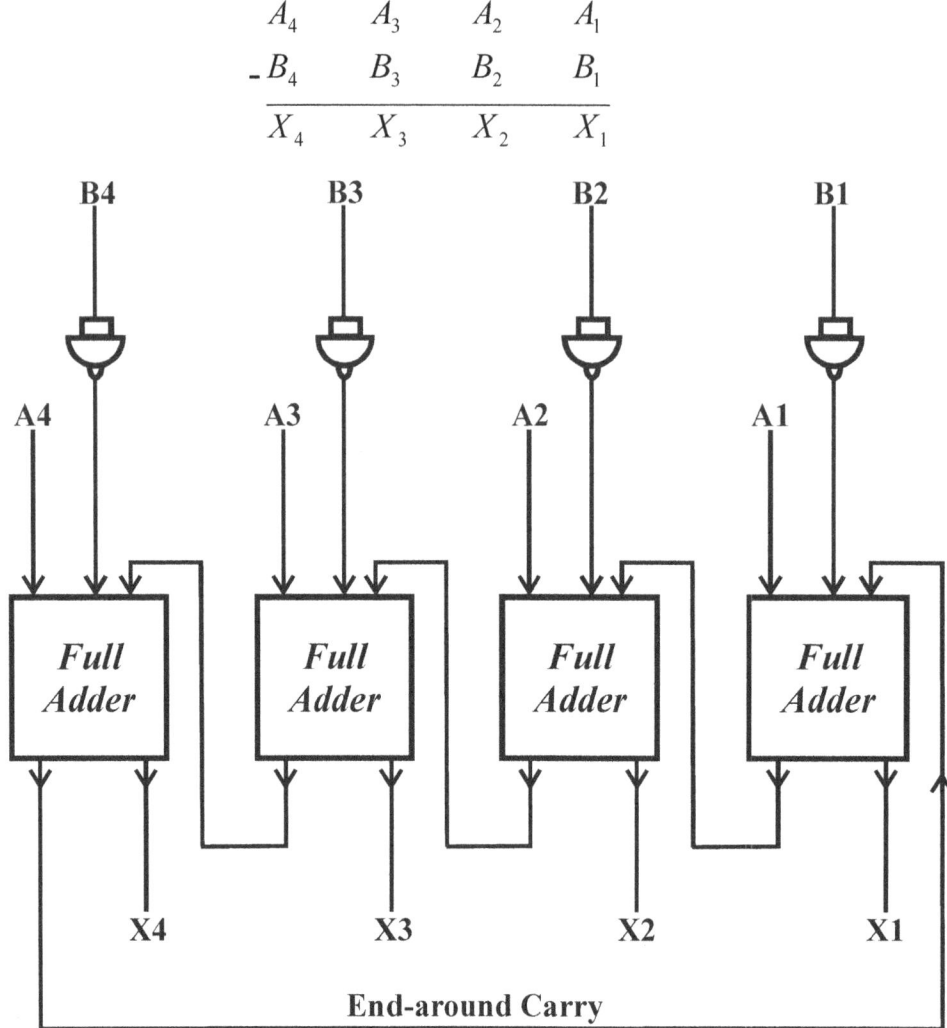

Fig 10.6.3 – This is a 4 bit parallel subtractor.

At the stage we shall have a look into how a full adder looks when the actual gates are inserted. Fig 10.6.4 shows the circuit for full adder from two half adders. A more efficient and fast arrangement for multibit adding using a "look-ahead" carry system will be discussed later.

A slower but error free multiple bit adder can be built out of a single full adder and shift registers to contain the numbers. This will also be discussed later.

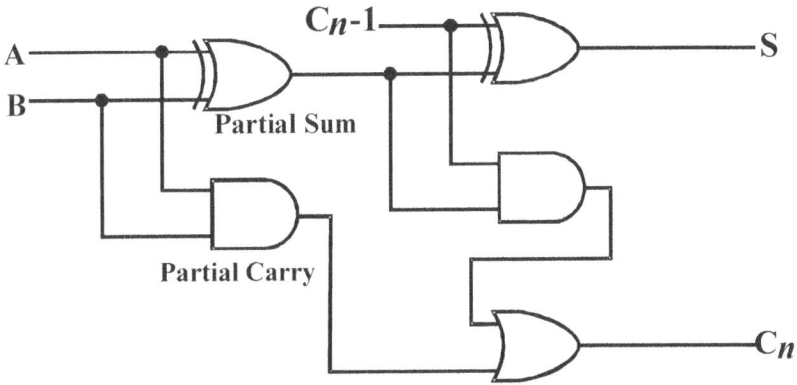

Fig 10.6.2 – Full Adder From two Half Adders.

10.7 BINARY CODES

1. The ordinarily used systems of codes are the decimal, octal, hexadecimal and binary. Of these the decimal suits human beings, while binary suits the logic circuits and computers. There are many binary codes which lie in between these two which either are for some specific purpose, or used as an intermediate step between man and machine.

ii. The BCD (Binary coded decimal): This is a compromise between the decimal and binary numbers. The most common type of Code is the 8421 Code. This expresses each decimal digit by its 4 bit binary equivalent.

For instance take the decimal number 249. Each decimal digit is changed to the binary numbers as below;

```
   2      4      9
 0010   0100   1001
```

Hence in BCD8421 code, 0010 0100 1001 stands for the decimal number 249.

Since decimal digits go up to 9 only, the code make use of the simple binary 4 digit numbers only up to 1001 (i.e. 1010, 1011 etc. up to 1111 is not used).

The 8421 code is so called because the value of the binary digit from the right increase in that order. In the above example;

$$
\begin{array}{rcllll}
 & & 8 & 4 & 2 & 1 \\
2 = & & 0 & 0 & 1 & 0 \\
 = & & \multicolumn{4}{l}{0 \times 8 + 0 \times 4 + 1 \times 2 + 0 \times 1 = 2} \\
4 = & & 0 & 1 & 0 & 0 \\
 = & & \multicolumn{4}{l}{0 \times 8 + 1 \times 4 + 0 \times 2 + 0 \times 1 = 4} \\
9 = & & 1 & 0 & 0 & 1 \\
 = & & \multicolumn{4}{l}{1 \times 8 + 0 \times 4 + 0 \times 2 + 1 \times 1 = 9}
\end{array}
$$

How it is not necessary for the values of the binary digits positions to be 8,4,2 and 1.

For instance it can be 2, 4, 2, 1.

This means that the four positions will have values

2, 4, 2, 1 instead of 8, 4, 2, 1.

Thus the above example of decimal number 249 in this code becomes the following;

```
     2       4       9    Decimal
   0010    0100    1111   2421 code
```

In a similar way these can be 5 digit weighted codes. They help to detect errors made readily. Some example of these types of codes are given below;

<u>BCD 5 bit codes</u>

<u>Decimal</u>	<u>86421</u>	<u>Shift counter</u>
0	00000	00000
1	00001	00001
2	00010	00010
3	00011	00011

4	00100	00100
5	00101	00101
6	01000	01000
7	01001	01001
8	10000	10000
9	10001	10001

iii. **The Gray Code:** The main characteristic of the Gray code is that each number of this code differs from the preceding and succeeding number by a single it. This is of great use in certain situation while handling digital date.

10.8 PARITY ERROR DETECTION IN NUMBERS

The different numerical codes use different methods of representing numbers. But in all case it is possible that while these numbers pass through different stages of manipulation, one of more digits may get replaced by an error. One of the ways of locating this error is by using an additional digit at the end of the number and uses this digit to indicate whether there is an error. One way is to count the total number of '1's existing in a number and if it is odd, add an '0' at the end. If the total number of 1's is even then add '1' at the end. So that now the number of 1's in the number is odd. At the receiving end of the data, it is checked whether the total number of 1's in the numbers received is odd or even.

If it is even it indicates an error.

The above is called ODD parity.

In a similar way "even parity" can be used, by making the total number of digits even, using the parity digit.

Example of odd parity;

Number	Odd parity bit	Number with parity
10011011	0	100110111
10101010	1	101010101
00000001	0	000000010

The above method has the draw back the error can be detected but cannot be corrected as we do not know which digit of number has gone wrong. Further it two errors occur in the number it will not affect the parity check and hence will not be detected. To overcome this, other methods can be used. One method is to use a vertical column parity digit and also a horizontal (row) parity digit.

Let us take the following six numbers, each six digits.

	Number	Even parity digit	Number with parity bit
1)	101001	1	1010011
2)	000110	0	0001100
3)	111011	1	1110111
4)	101010	1	1010101
5)	110100	1	1101001
6)	101000	0	1010000

To this we add an even digit as explained earlier. Then we get as in the last column above. Now let us consider each column of the block of numbers and check whether they are odd or even. Then add an even parity digit at the bottom of each column. Thus we ultimately get six numbers, the seventh being the parity number.

Parity vertical row

1)   101001   1
2)   000110   0

3)	111011	1
4)	101010	1
5)	110100	1
6)	101000	1
	100010	0 Parity horizontal row

This method helps us to locate an error, because the error will show up both in the row and column. For example let us assume that an error occurs in the fourth number and it reads as 101110. Then the parity bit on the 4th row and fourth column will indicate an error. This shows the position of error. This method is used with magnetic tapes.

10.9 MULTIPLE BIT SERIAL ADDER

The arrangement for multiple bit binary adder using full address has been discussed. A lower, but error free multiple bit adder can be built using a single full adder and shift registers. Fig 10.9.1 shows the circuit for addition using 3 shift registers, one full adder. The theory and working of shift registers need to be understood by students. The shift register A and B store the two binary numbers which are to be added. This is fed serially to the shift registers. The least significant bit (LSB) comes out first and the most significant digit (MSB) last. Thus the numbers are fed to the full adder one digit at a time. The carry output goes to a 'D' type bistable. The bistable and the shift registers are all clocked by a common clock pulse.

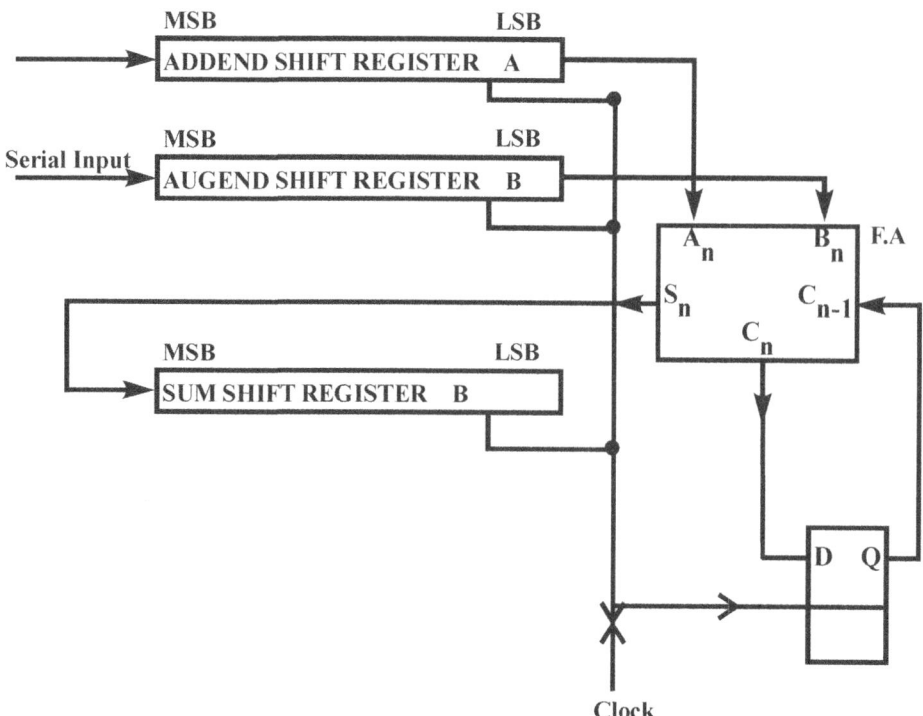

Fig 10.9.1 – multiple bit serial adder

In other words, while the nth digits are being added, the bistable output is the "carry" from "$n-1^{th}$" addition. The input to the bistable is the carry from the nth addition.

To take an example, consider adding 1 0 1 0 to 1 1 so the shift register A gets 1 0 1 0 – shift register B gets 0 0 1 1

When the digits are being shifted out the following sequence of events take place.

No. of Clock Pulses	input to F A from A	input to F A from B	Carry out From F A (i.e; Bistable Input)	Bistable output ("carry In" to FA)	Sum output from F A to C
1	0	1	0	0	1
2	1	1	1	0	0
3	0	0	0	1	1
4	1	0	0	0	1

Thus the shift register will have the number 1 0 1 1 after the 4th clock pulse is over. In other words, the addend and augend are processed one bit at a time. The sum can be stored either in a separate sum register or in the addend or augend register. The carry out is temporarily stored in the bistable so that it can be used as the "carry in" for the next stage of addition. The size of the numbers can then be added is limited only by the storing capacity (no. of bits) of the shift registers.

CHAPTER ELEVEN

11.1 ELECTRONIC SWITCHING

In some applications, it is often required to make or break an electrical circuit. It is sometimes desirable and necessary that this making and breaking should be very quick and without sparking. Mechanical switches cannot be used for the purpose for two main reasons. Firstly, a mechanical switch has high inertia which limits its speed of operation.

Secondly, there is always sparking at the contact during the breaking operation, which results in the burning of the contacts.

Investigations have shown that tubes and transistors can serve as switching devices. The can turn ON and OFF power in an electrical circuit at a very high speed without any sparking. Such switched are known as electronic switches. Electronic switches are being used extensively to produce non-sinusoidal wave's e.g square, rectangular, triangular or sawtooth waves. These non-sinusoidal waves are extensively used in solid-state switching circuit. We shall here confine ourselves to transistor as a switch.

11.2 ELECTRONIC SWITCH

It is an electronic device which can turn ON or OFF current in an electric circuit e.g transistor or tubes.

Electronic switches have become very popular because of their high speed of operation and absence of sparking. A transistor can be used as a switch by driving it back and forth between saturation and cut-off. This is illustrated in the following discussions.

ELECTRONIC SWITCHING

1. When the base input voltage is enough negative, the transistor is cut off and no current flows in the collector load (see fig. 11.2-1a)

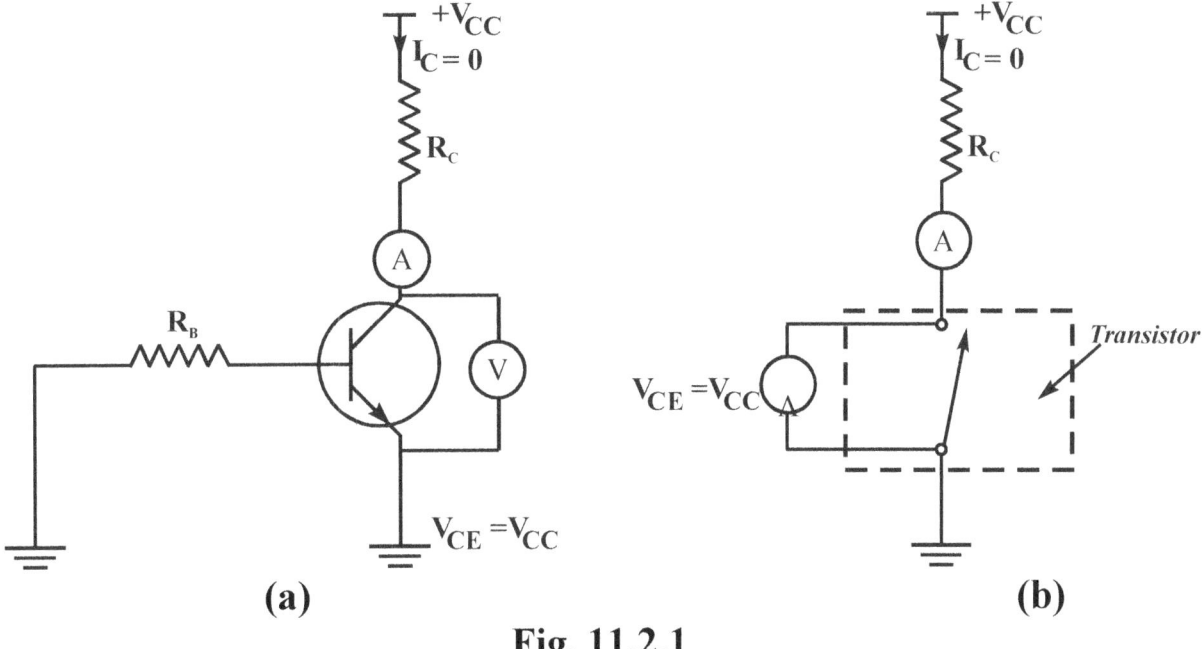

Fig. 11.2.1

As a result, there is no voltage drop across Rc and the output voltage is ideally Vcc ie.

Ic = 0 and Vcc = V_{CE}.

This condition is similar to that of an open switch (ie. OFF state) as shown in fig. 11.2-1b

2. When the input base voltage is positive enough that the transistor saturates, then Ic(sat) will flow through Rc. Under such conditions, the entire Vcc will drop across collector load Rc and output voltage is ideally zero ie.

Ic = Ic(sat) = $\dfrac{V_{cc}}{R_c}$ and $V_{CE} = 0$.

This condition is similar to that of a closed switch (ie. On state) as shown in fig. 11.2-2b.

ELECTRONIC SWITCHING

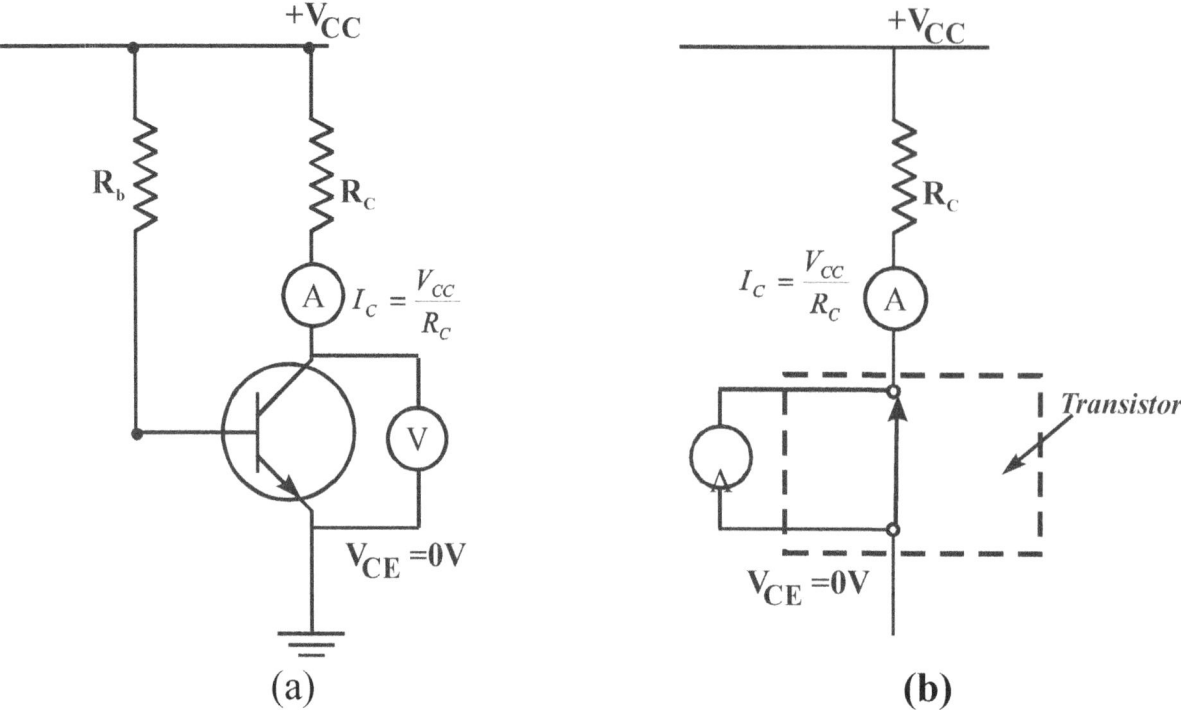

(a)   (b)

Fig. 11.2.2: Switching action of a transistor in terms of d.c load line

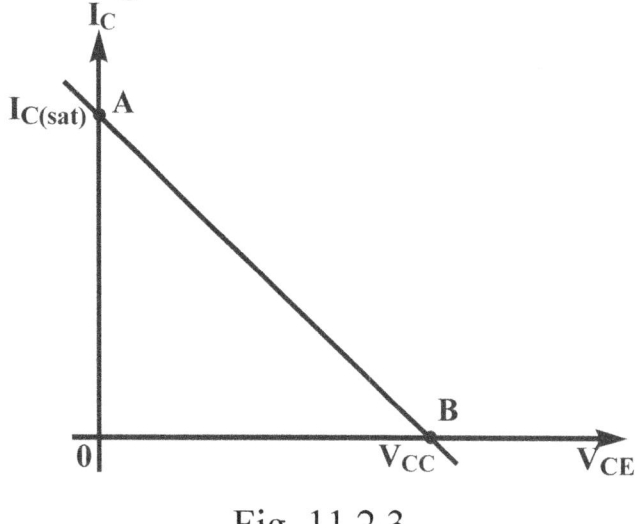

Fig. 11.2.3

Therefore it can be concluded that a transistor can behave as a switch under proper conditions. In order words, if the input base voltages are enough negative and positive, the transistor will be driven between cut off and saturation. These conditions can easily be fulfilled in a transistor circuit, thus a transistor can act as a switch. Fig. 11.2-2 shows the switching action of a transistor in terms of d.c loadline. The point A of the load line represents the on condition while point B represents the OFF condition.

Example 11.2.1

Determine the minimum high input voltage (+v) required to saturate the transistor switch shown below.

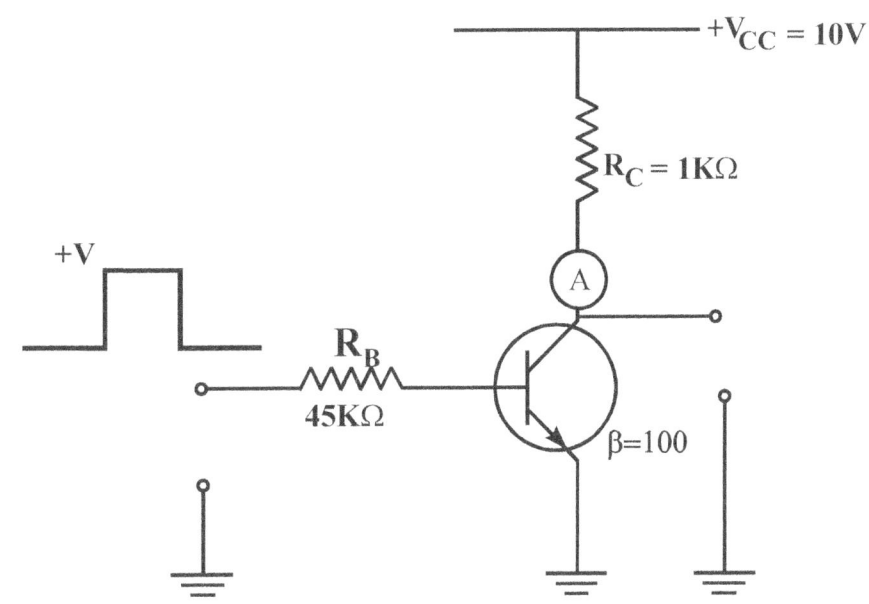

Solution: Assume transistor to be ideal

$$I_{c(sat)} = \frac{V_{cc}}{R_C} = \frac{10V}{1K\Omega} = 10mA$$

$$I_B = \frac{I_{C(sat)}}{\beta} = \frac{10mA}{1000} = 0.1mA$$

Now, $+V = I_B R_B + V_{BE}$

$= (0.1mA)(45) + 0.7$

$= 4.5 + 0.7 \qquad = \underline{5.2V}$

Hence in order to saturate the transistor, we require +5.2V

11.3 ADVANTAGES OF ELECTRONIC SWITCHES

The advantages of transistor switch over other types of switches are

i. It has no moving part and hence there is little wear and tear. Therefore it give noiseless operation.

ii. It has smaller size and weight.

iii. It gives trouble-free service because of solid state.

iv. It is cheaper than other switches and required little maintenance.

v. It has a very fast speed of operation say up to 10^9 operations per second. On the other hand the mechanical switches have a small speed of operation e.g. less than 5 operative in a second.

Important Terms: The transistors we have so far considered is ideal. An ideal transistor has $V_{CE} = V_{CC}$ (or $I_C = 0$) in the OFF state and $V_{CE} = 0$ (or $I_C = I_{C(sat)}$) in the ON state. However, such conditions are not realised in practice. In practical transistor, the output voltage is neither V_{CC} in the OFF state no is it zero in the ON state. In the design of transistor switching circuit; the following points must be considered.

a. **Collector Leakage Current**: When the input circuit is reverse biased or input voltage is zero a small current (a few µA) flows in the collector. This is known as collector leakage current and it is due to the minority carriers. The value of this leakage current is quite large in *Ge* transistors, but in modern silicon transistors, the value of leakage current is low enough to be ignored.

b. **Saturation Collector Current**: It is the maximum collector current for a particular load in a transistor.

Consider an NPN transistor having a load R_C in its collector circuit as shown in fig. 11.3.1.

ELECTRONIC SWITCHING

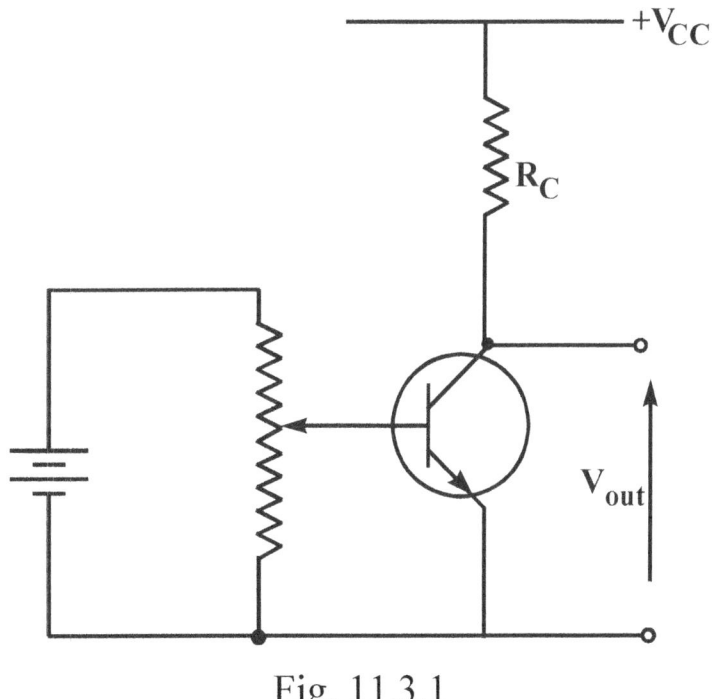

Fig. 11.3.1

As the input forward bias is increased, the collector current I_C also increases because $I_C = \beta I_B$. However, with the increase in I_C, the voltage drop across R_C increases. This results in the decrease of V_{CE}. When V_{CE} drops to knee voltage (V_{Knee}) any further increase in collector current is not possible since β decreases sharply when V_{CE} falls below knee voltage. This maximum current is known as saturation collector current.

$\therefore$ Saturation Collector Current, $I_{C(sat)} = \dfrac{V_{CC} - V_{base}}{R_C}$.

11.4 SWITCHING ACTION OF A TRANSISTOR

The output characteristics of a transistor help us to explain the switching action of a transistor. Fig. 11.4.1b shows the output characteristics of a typical transistor biased in a common emitter mode. The characteristics are arranged in three regions: OFF, ON or saturation and active regions.

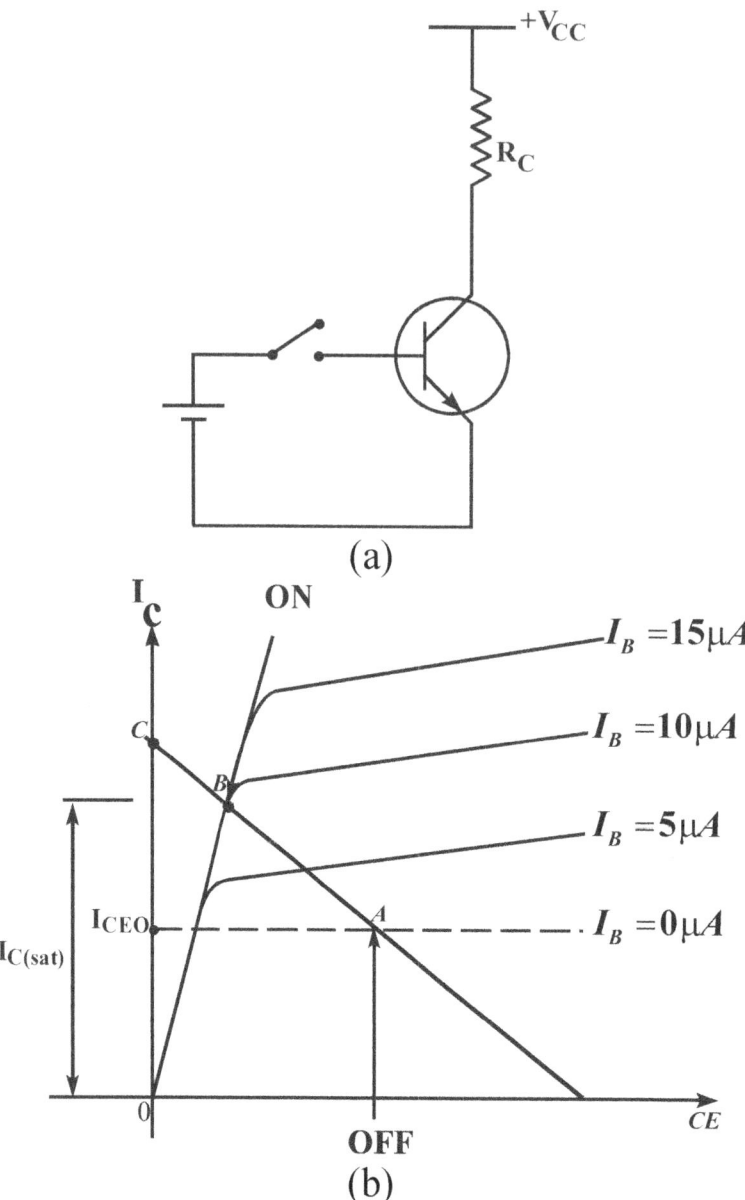

Fig. 11.4.1: Output characteristics of a typical transistor biased in a common emitter mode

a. OFF Region:

When the input base voltage is zero or negative, the transistor is said to be in the OFF condition. In this condition, $I_B = 0$ and the collector current is equal to the collector leakage current I_{CEO}. The value of I_{CEO} can be obtained from the characteristics if we know V_{CE}.

Power loss = Output voltage × Output current.

As already noted, in the OFF condition, the output voltage = V_{CC} since voltage drop in the load due to I_{CEO} is negligible.

$$\text{Power loss} = V_{CC} \times I_{CEO}$$

Since I_{CEO} is very small as compared to full-load current that flows in the ON condition, power loss in the transistor is quite small in the OFF condition. It means that the transistor has a high efficiency as a switch in the OFF condition.

b. ON or Saturation Region:

when the input voltage is made so much positive that saturation collector current flows, the transistor is said to be in the ON state. In this condition the saturation collector current is given by;

$$I_{C(sat)} = \frac{V_{CC} - V_{knee}}{R_C}$$

Power loss = Output voltage × Output current

The output voltage in the ON condition is equal to V_{knee} and output current is $I_{C(sat)}$.

$$\therefore \text{Power loss} = V_{knee} \times I_{C(sat)}$$

Again the efficiency of the transistor as a switch in the ON condition is high. It is because the power loss in this condition is quite low due to small value of V_{knee}.

c. Active Region:

It is the region that lies between OFF and ON conditions. The OFF and ON regions are the stable regions of operation. The active region is the constable (or transient) region through which the operation of the transistor passes while changing from OFF state to the ON state. The path AB in fig. 11.4.1b is the active region. The collector current increases from I_{CEO} to $I_{C(sat)}$ along the path AB as the transistor is switched ON. However, when the transistor is switched OFF, the collector current decreases from $I_{C(sat)}$ to I_{CEO} along BA.

11.5 MULTIVIBRATORS

An electronic circuit that generates square wave (or other non-sinusoidal such as rectangular sawtooth waves) is known as a multivibrator. A multivibrator is a switching circuit which depends on positive feedback for operation. It is basically a two-stage amplifier with the output of one fedback to the input of the other as shown in fig. 11.5.1.

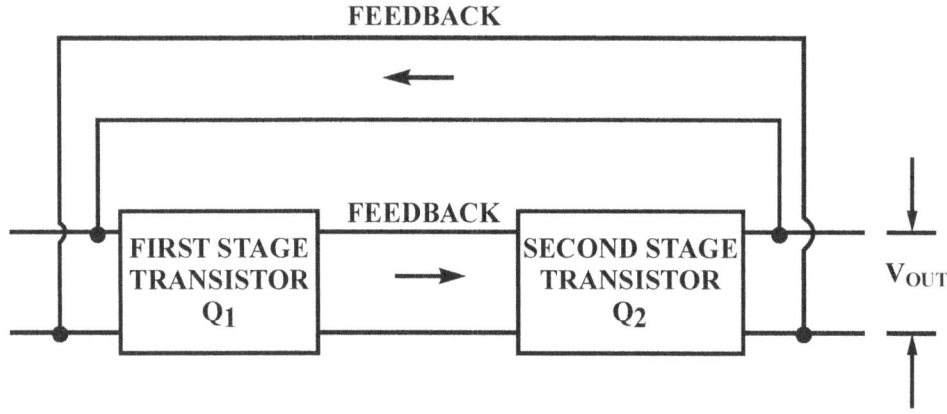

Fig. 11.5.1

The circuit operates in two states (Viz ON and OFF) controlled by circuit conditions. Each amplifier stage supplies feedback to the other in such a manner that will drive the transistor of one stage to saturation (ON state) and the other to cut off (OFF state).

After a certain time controlled by circuit conditions, the action is reversed i.e. saturated stage is driven to cut off and the cut off stage is driven to saturation. The output can be taken across either stage and may be rectangular or square wave depending upon the circuit conditions.

11.6 TYPES OF MULTIVIBRATORS

Depending upon the manner in which the stages interchange their states, the multivibrators are classified as:

i. Astable or free running multivibrator.

ii. Monostable or one-shot multivibrator

iii. Bistable or flip-flop multivibrator

Fig. 11.6.1 shows the input/output relations for the three types of multivibrators.

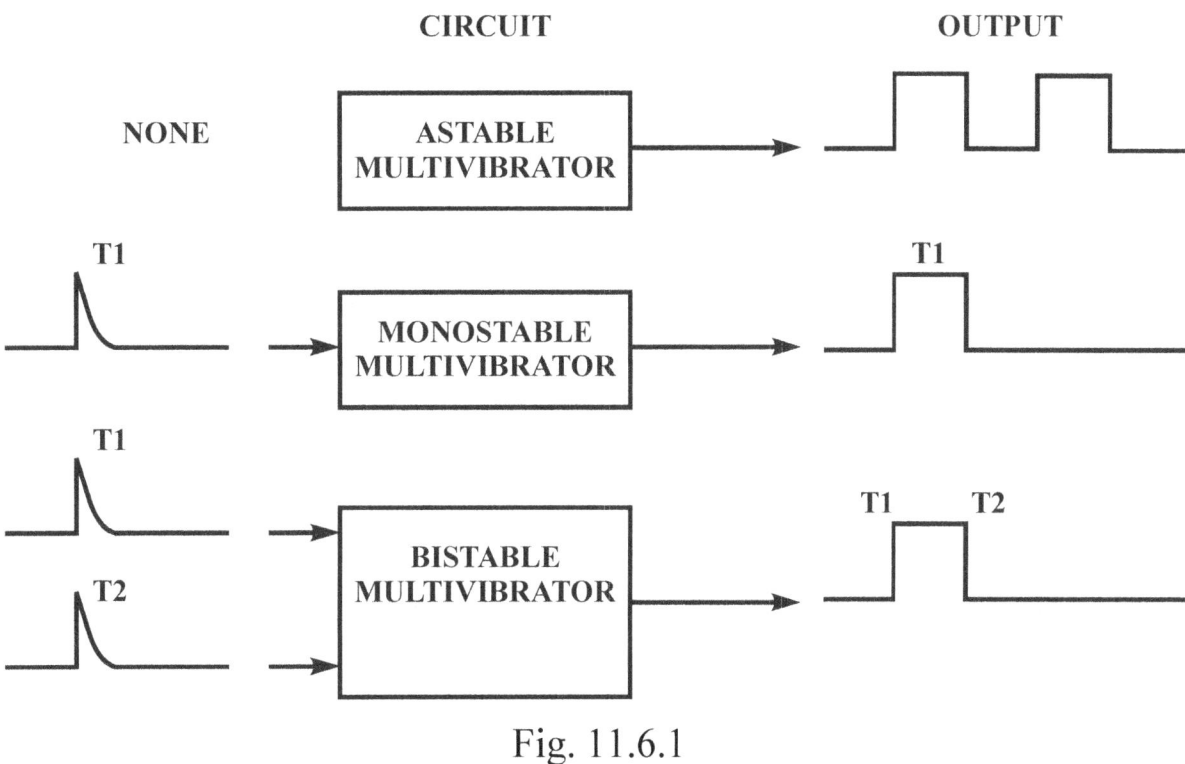

Fig. 11.6.1

1. The Astable or Free running Multivibrator alternates automatically between the two states and remains in each for a time dependent upon the circuit constants. Thus it is just an oscillator since it requires no external pulse for its operation but d.c power. Because it continuously produces the square-wave output it is often referred to as a free running multivibrator.

2. The Monostable or one-shot multivibrator has one state stable and one quasi-state (i.e, half stable) state. The application of input pulse triggers the circuit into its quari-stable state in which it remains for a period determined by circuit constant. After this period of time, the circuit returns to its initial stable state. The process is repeated upon the application of each trigger pulse. Since the monostable multivibrator produces a single output pulse for each input trigger pulse it is generally called one-shot multivibrator.

3. The bistable multivibrator has both the two states stable. It required the application of an external triggering pulse to change the operation from either one state to the other. Thus, one pulse is used to generate half-cycle of square wave and another pulse to generate the next half-cycle of square wave. It is also referred to as a flip-flop multivibrator because of the two possible states it can assume.

11.7 ASTABLE MULTIVIBRATOR CIRCUIT:

We have already noted that an astable multivibrator generates square waves of its own (i.e. without any external triggering pulse). It has no stable state. It switches back and forth from one state to the other, remaining in each state for a time determined by circuit constants. The circuit detail for a transistor astable multivibrator is shown in fig. 11.7.1. The circuit uses two transistors Q_1 and Q_2. The circuit essentially consists of two symmetrical CE amplifiers stages each providing a feedback to the other. The collector loads of the two stages are equal i.e. $R_1 = R_4$ and the biasing resistors are also equal i.e. $R_2 = R_4$. The output of transistor Q_1 is coupled to the input of Q_2 though C_1 while the output of Q_2 is fed to the input of Q_1 through C_2. The square wave output can be taken from Q_1 or Q_2.

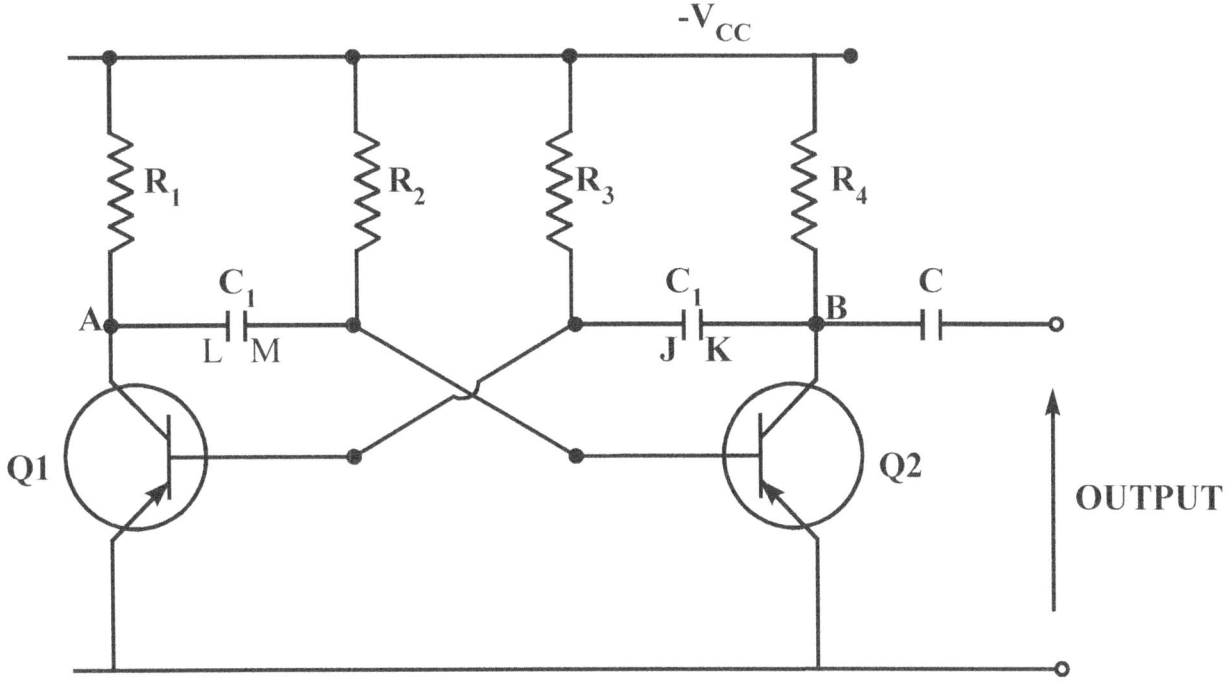

Fig. 11.7.1

At the application of V_{CC}, the collector current start to flow in Q_1 and Q_2. In addition the coupling capacitors C_1 and C_2 also start charging up. As the characteristics of two transistors (i.e. β, V_{BE}) are not exactly alike, therefore one transistor say Q_1, will conduct more rapidly than the other. The rising collector current in Q_1 drives its collector more and more positive. The increasing positive output at point A is applied to the base of transistor Q_2 and its collector current starts decreasing. As the collector of Q_2 is connected to the base of Q_1 through C_2 therefore, base of Q_1 becomes more negative i.e. Q_1 is more forward biased.

This further increases the collector current in Q_1 and causes a further decrease of collector current in Q_2. The series of actions are repeated until the circuit drives Q_1 to saturation and Q_2 to cut-off. These actions occur very rapidly and may be considered practically instantaneous. The output of Q_1 (ON state) is approximately zero and that of Q_2 (OFF state) is approximately V_{CC}. This is shown by **ab** in fig. 11.7.2.

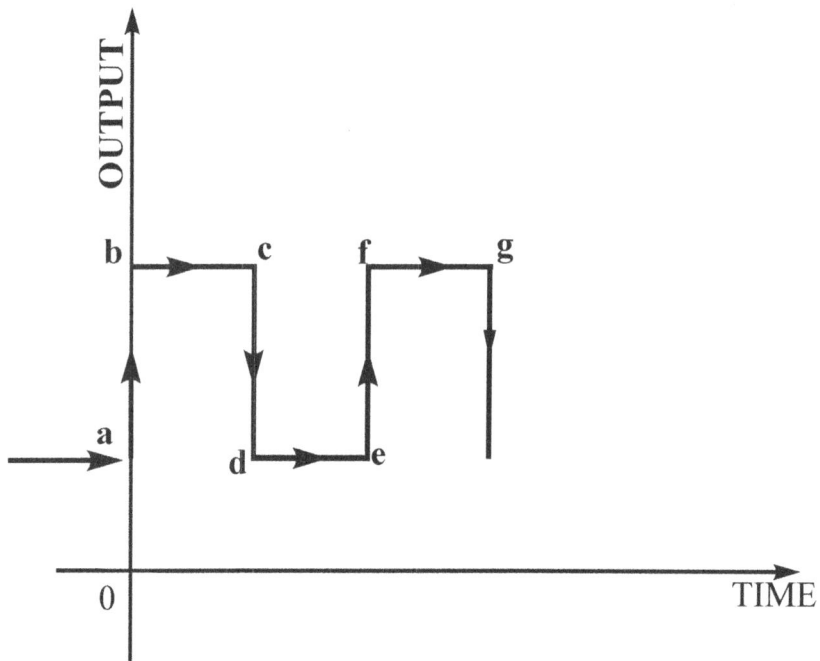

Fig. 11.7.2

When Q_1 is at saturation and Q_2 is cut off, the full voltage V_{CC} appears across R_1 and voltage across R_4 will be zero. The changes developed across C_1 and C_2 are sufficient to maintain the saturation and cut off conditions at Q_1 and Q_2 respectively. This condition is repeated by time interval **bc** in fig. 11.7.3a. However, the capacitors will not retain the charges indefinitely but will discharge through their respective resistors. The discharge path for C_1 with plate L negative and Q_1 conducting is $LAQ_1 V_{CC}R_2M$ as shown fig. 11.7.3a. The discharge path for C_2, with plate K negative and Q_2 cut off, is KBR_4R_3J as shown in fig. 11.7.3b. As the resistance of the discharge path for C_1 is lower than that of C_2, therefore C_1 will discharge more rapidly. As C_1 discharges, the base bias at Q_2 becomes less positive and at a time determined by R_2 and C_1, forward bias is re-established at Q_2. This causes the collector current to start in Q_2. The increasing positive potential at collector of Q_2 is applied to the base of Q_1 through the capacitor C_2. Hence the base of Q_1 will become more positive i.e. Q_1 is reverse biased. The decrease in collector current in Q_1 sends a negative voltage to the base of Q_2 through C_1 thereby causing further increase in the collector current of Q_2.

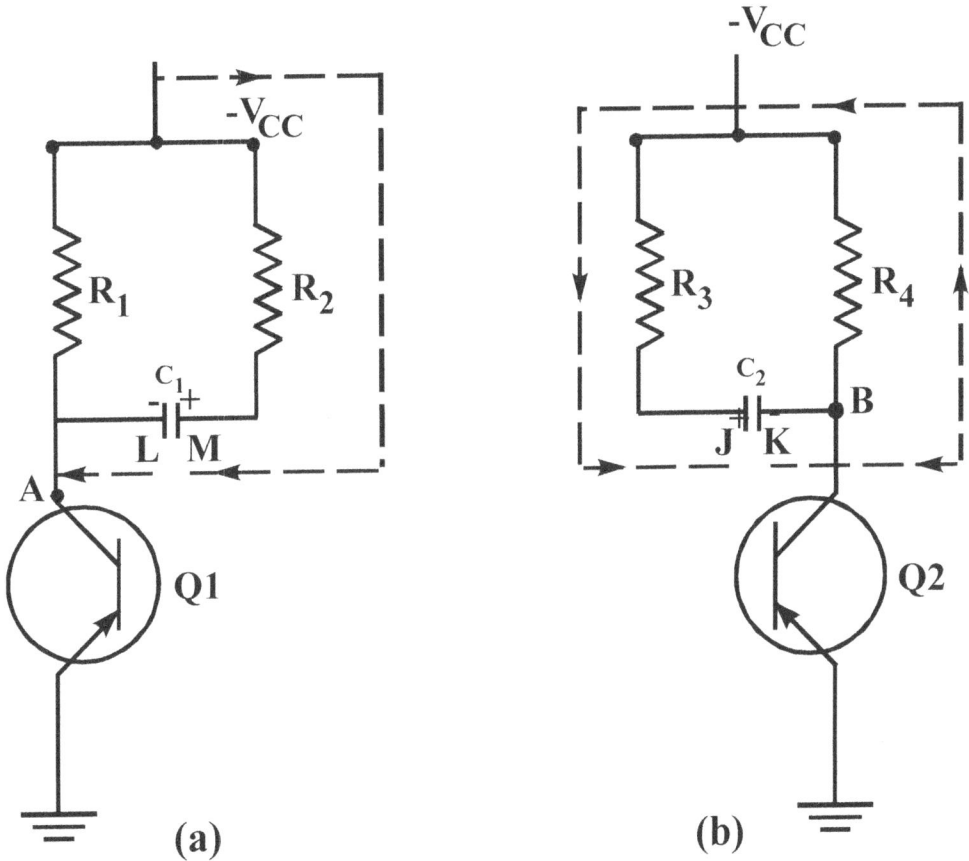

Fig. 11.7.3

With this set of actions taking place, Q_2 is quickly driven to saturation and Q_1 to cut off. This condition is represented by **cd** in fig. 11.7.2. The period of time during which Q_2 remains at saturation and Q_1 at cut off is determined by C_2 and R_3.

The time for which either transistor remains ON or OFF is given by.

ON time for Q_1 (or OFF time for Q_2) is

$$T_1 = 0.694\ R_2 C_1$$

OFF time for Q_1 (or ON time for Q_2),

$$T_2 = 0.694\ R_3C_2$$

Total time period of the square wave is

$$T = T_1 + T_2 = 0.694\ (R_2C_1 + R_3C_2)$$

As $R_2 = R_3 = R$ and $C_1 = C_2 = C$.

$$T = 0.694\ (RC + RC) = 1.4RC \text{ seconds.}$$

Frequency of the square wave is

$$f = \frac{1}{T} = \frac{0.7}{RC} Hz.$$

It may be noted that in these expressions, R is in ohms and C in farad.

11.8 MONOSTABLE MULTIVIBRATOR CIRCUIT

A multivibrator in which one transistor is always conducting (i.e. in the ON state) and the other is non-conducting (i.e. in the OFF state) is called a monostable multivibrator. By definition, a monostable multivibrator has only one state stable in order words, if one transistor is conducting and the other is not conducting, the circuit will remain in this position. It is only with the application of external pulse that the circuit will interchange the states. However, after a certain time the circuit will automatically switch back to the original stable state and remains there until another pulse is applied. Thus a monostable multivibrator cannot generate square waves of its own like an astable multivibrator. Only a pulse can cause it to generate the square wave.

Fig. 11.8.1 shows a monostable multivibrator circuit it consist of two similar transistors Q_1 and Q_2 with equal collector loads i.e $R_1 = R_4$. The values of V_{BB} and R_5 are such as to reverse bias Q_1 and keep it at cut off. The collector supply V_{CC} and R_2 forward bias Q_2 and keep it at saturation. The input pulse is given through C_2 to obtain the square wave. Output can be taken from Q_1 or Q_2.

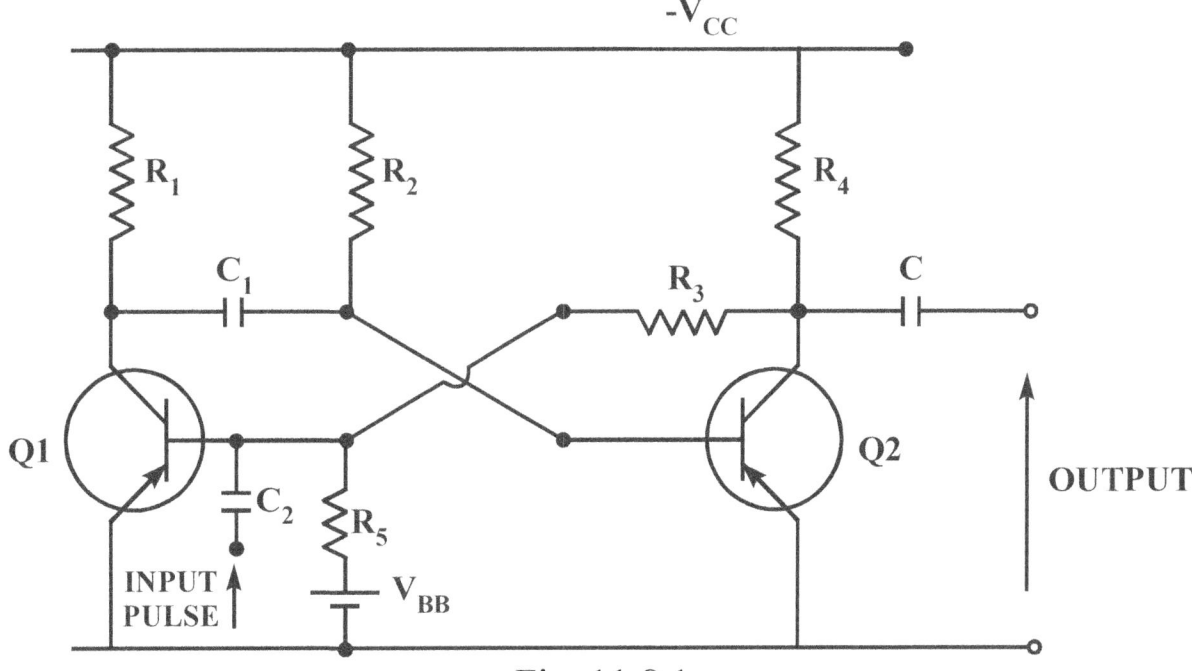

Fig. 11.8.1

With the circuit arrangement shown, Q_1 is at cut off and Q_2 at saturation. This is the stable state for the circuit and it will continue to remain in this state until a triggering pulse is applied at C_2. When a negative pulse of short duration and sufficient magnitude is applied to the base of Q_1 through C_2, the transistor Q_1 starts conducting and positive potential is established at its collector.

The positive potential at the collector of Q_1 is coupled to the base of Q_2 through capacitor C_1. This decreases the forward bias on Q_2 and also the collector current. The increasing negative potential on the collector of Q_2 is applied to the base of Q_1 through R_3. This further increases the forward bias on Q_1 and hence its collector current. Thus, Q_1 is quickly driven to saturation and Q_2 to cut off. The circuit will remain in that state for some time and will change to the original state after some time has elapsed. The circuit is now in its stable state and will remain in that state until another trigger pulse causes it to switch over the state.

11.9 BISTABLE MULTIVIBRATOR CIRCUIT

This type of multivibrator has both the states stable. It remains in which ever state it happens to be until a trigger pulse causes it to switch to the other state. If left to itself without triggering the bistable multivibrator will stay in this position forever. However, when an external pulse is applied to the circuit then changes the state. Another trigger pulse is required to switch the circuit back to the original state.

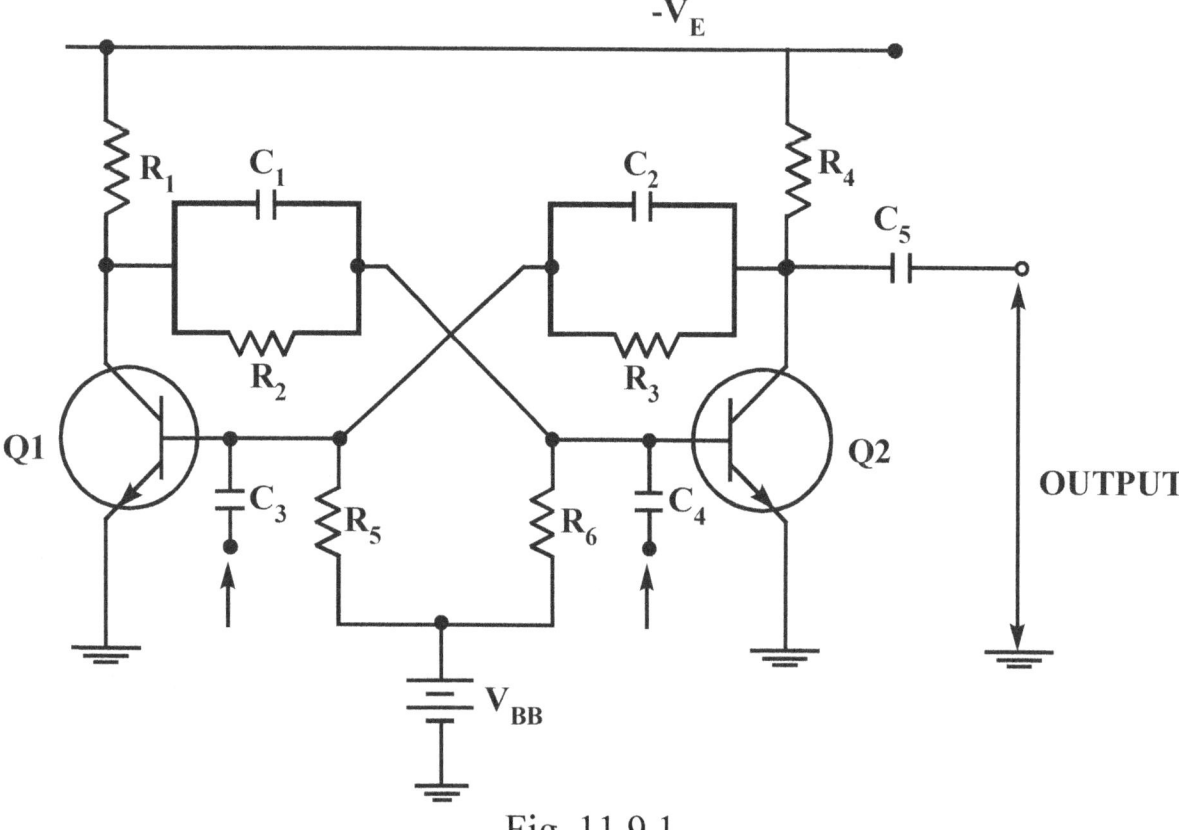

Fig. 11.9.1

Fig. 11.9.1 shows the circuit for a typical transistor bistable multivibrator. It consists of two identical CE amplifiers stages with output of one fed to the input of the other. The feedback is coupled through resistors (R_2, R_3) shunted by capacitors C_1 and C_2. The main purpose of capacitors C_1 and

C_2 is to improve the switching characteristics of the circuit by passing the high frequency components of the square wave. This allows fast rise and fall times and hence distortionless square wave output. The output can be taken across either output.

When V_{CC} is applied, one transistor will start conducting slightly ahead of the other due to some differences in characteristics of the transistor. This will drive one transistor to saturation and the other to cut off. Assuming that Q_1 is turned ON and Q_2 is cut off. If left to itself as stated already, the circuit will remain in this condition. In order to switch the multivibrator to its other state, a trigger pulse must be applied. A negative pulse applied to the base of Q_1 through C_3 will cut if off or a positive pulse applied to the base of Q_2 through C_4 will cause it to conduct.

Assuming a negative pulse of sufficient magnitude is applied to the base of Q_1 through C_3. This reduces the forward bias on Q_1 and cause a decrease in its collector current and an increase in collector voltage.

The rising collector voltage is coupled to the base of Q_2 where it forward biases the base-emitter junction of Q_2. This will cause an increase in its collector current and decrease in collector voltage. The decreasing collector voltage is applied to the base of Q_1 where it further reverse biases the base-emitter junction of Q_1 to decrease its collector current. With these actions taking place, Q_2 is quickly driven to saturation and Q_1 to cut off. The circuit will now remain stable in this state until a negative trigger pulse at Q_2 (or a positive trigger pulse at Q_1) changes that state.

Example 11.9.1

In the astable multivibrator shown in fig. 11.9.1 $R_2 = R_3 = 12k\Omega$ and $C_1 = C_2 = 0.01\mu f$. Determine the time period and frequency of the square wave.

Solution

Here $R = 12k\Omega = 12 \times 10^3 \Omega$; $C = 0.01\mu F = 0.01 \times 10^{-6}$

Time period of square wave is,
$$T = 1.4RC = 1.4 \times 12 \times 10^3 \times 0.01 \times 10^{-6} \text{ second}$$
$$= 0.17 mS.$$

$$\text{Frequency} = \frac{1}{T} = \frac{1}{0.17 \times 10^{-3}} = \underline{5.88 kHz}$$

11.10 DIFFERENTIATING CIRCUIT

A differentiating circuit is one in which the output voltage is directly proportional to the derivative of input.

i.e., Output voltage $\propto \frac{d}{dt}(input\,voltage)$.

A differentiating circuit is a simple RC series circuit with output taken across the resistor R. The circuit is suitably designed in such a way that the output is proportional to the derivative of the input. Thus if a d.c or constant input is applied to such a circuit, the output will be zero. This is because the derivative of a constant is zero.

Fig. 11.10.1 shows a typical differentiating circuit. The output across R will be the derivative of the input. It is **important** to note that merely using voltage across R does not make the circuit a differentiator; it is also necessary to set the proper circuit values.

ELECTRONIC SWITCHING

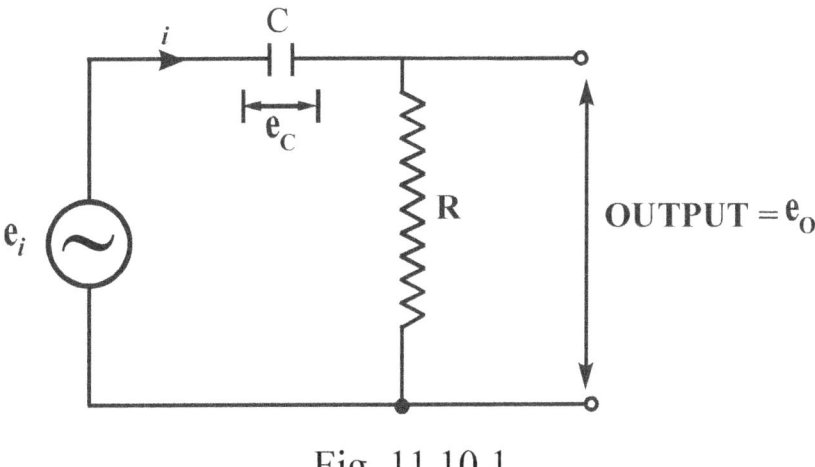

Fig. 11.10.1

The two conditions necessary to achieve a good differentiating are;

a. The time constant of the circuit should be much smaller than the time period of the waveform.

b. The value of X_C should be 10 or more time larger than R at the operating frequency. When these conditions are fulfilled the output across R in fig. 11.10.1 will be the derivative of the input.

CIRCUIT ANALYSIS:

Let e_i be the input alternating voltage and let i be the resulting alternating current. The change q on the capacitor C is given by;

$$Q = Ce_c$$

Now
$$i = \frac{dq}{dt} = \frac{d}{dt}(Ce_c)$$

$$= C\frac{d}{dt}(e_c)$$

Since by design, the capacitive reactance is very much larger than R, the input voltage can be considered equal to the capacitor voltage with negligible error i.e. $e_c = e_i$

$$i = C\frac{d}{dt}(e_i)$$

Output voltage, $e_o = iR$

$$= RC\frac{d}{dt}(e_i) \qquad (RC \text{ is constant})$$

$$\propto \frac{d}{dt}(e_i)$$

Output voltage $\propto \dfrac{d}{dt}$ (input voltage)

OUTPUT WAVEFORM

The output waveform from a differentiator depends upon the time constant and shape of the waveform. Here we are going to consider three types of inputs to a differentiating circuit.

a. Square Wave Input

When the input fed to a differentiating circuit is a square wave, the output will consist of sharp narrow pulses as shown in fig. 11.10.2c.

During part OC of the input waveform, its amplitude changes abruptly and hence the differentiated wave will be a sharp narrow pulse as shown in fig. 11.10.2c. But during the constant part CB of the input, the output will be zero because the derivative of a constant is zero. In explaining the behaviour of the circuit, we can see that since the time constant RC of the circuit is very small with respect to time period of input wave and $X_C \gg R$, the capacitor will become fully charged during the early part of each half cycle of the input wave. During the remaining part of the half-cycle, the output of the circuit will be zero because the capacitor voltage e_c neutralises the input voltage and there can be no current flow through R. Thus the output waveform will have sharp pulse during the start of each half cycle of input waveform while for the remaining part of the half cycle the output will be zero.

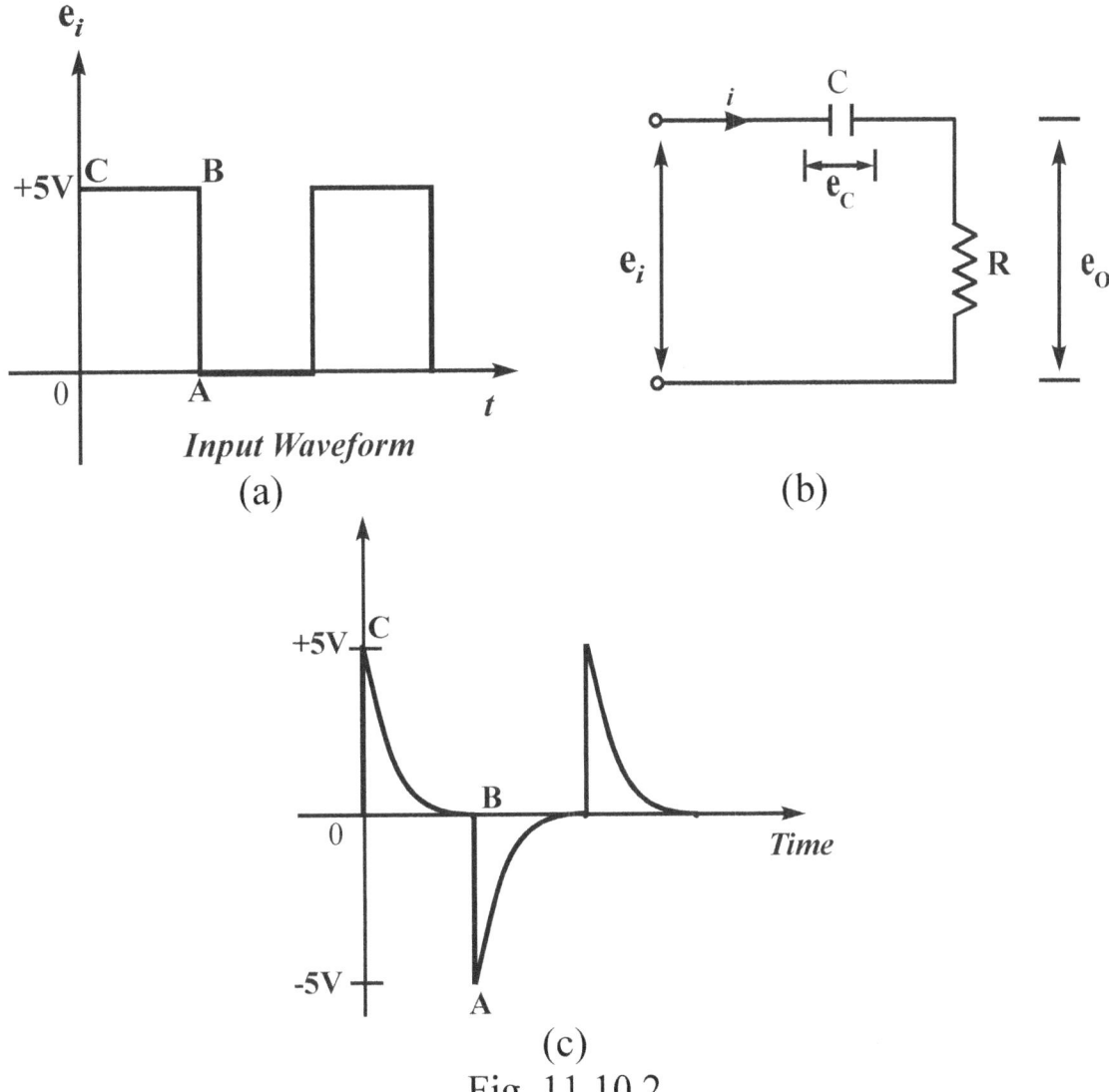

Fig. 11.10.2

We therefore have a symmetrical output waveform with sharp positive and negative peaks. These pulses are used extensively in electronic circuits e.g. in television transmitters and receivers, in multivibrators to initiate action etc.

b. Triangular Wave Input

When the input to a differentiating circuit is a triangular wave, the output is a rectangular wave as shown in fig. 11.10.3. During the period OA of the input wave, its amplitude changes at a constant rate and therefore, the differentiated wave has a constant value for each constant rate of change.

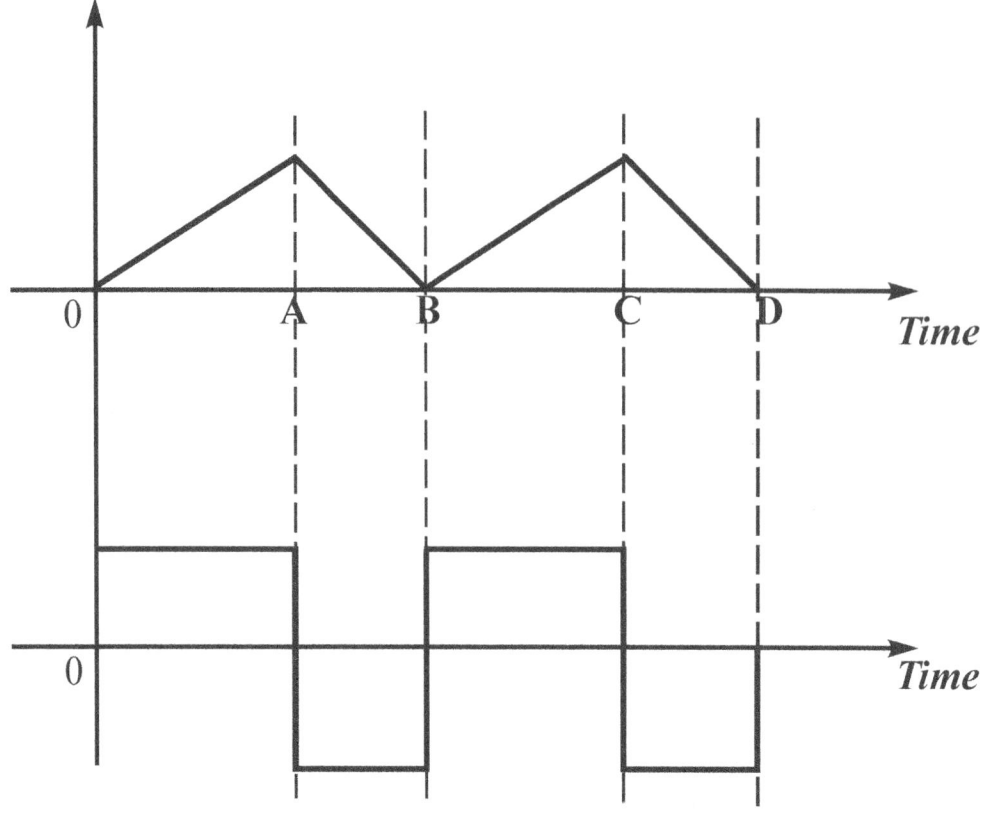

Fig. 11.10.3

During the period AB of the input wave, the change is less abrupt so that the output will be a very narrow pulse of rectangular form. Therefore when a triangular wave is fed to a differentiating circuit, the output waveform consists of a succession of rectangular waves of equal or unequal duration depending upon the shape of the input waveform.

c. Sine Wave Input

A sine wave input becomes a cosine wave and cosine wave input becomes an inverted sine wave at the output.

11.11 INTEGRATING CIRCUIT:

This is a circuit in which the output voltage is directly proportional to the integral of the input voltage, i.e.

$$\text{Output voltage} \propto \int \text{Input voltage}$$

As represented in fig. 11.11.1 an integrating circuit is a simple RC series circuit with output taken across the capacitor C.

It will be observed that R and C of the differentiating circuit have changed positions.

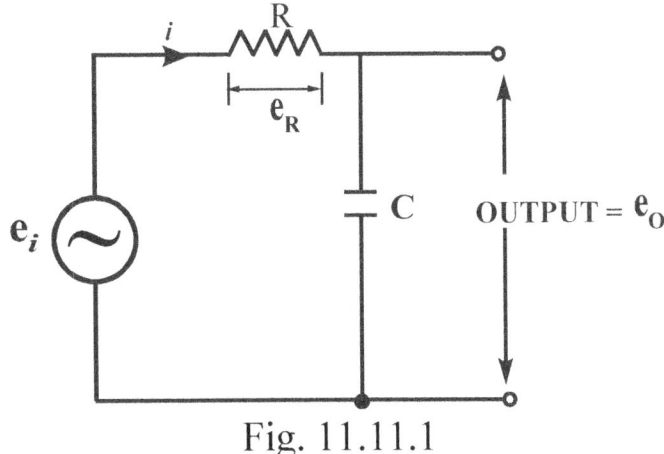

Fig. 11.11.1

For the circuit to render good integration, the following conditions should be fulfilled.

a. The value of R should be 10 or more times larger than X_c.

b. The time constant RC of the circuit should be very large as compare to the time period of the input wave.

If e_i is the input alternating voltage, i the resulting alternating current and R is very large compared to capacitive reactance X_C of the capacitor, it will be reasonable than to assume that $e_i = e_R$

Now, $\quad i = \dfrac{e_R}{R} = \dfrac{e_i}{R}$

The charge q on the capacitor at any instant is,

$$q = \int i \, dt$$

Output voltage, $e_o = \dfrac{q}{C} = \dfrac{1}{C}\int i \, dt$

$$= \dfrac{1}{C}\int \left(\dfrac{e_i}{R}\right) dt$$

$$= \frac{1}{RC}\int e_i\, dt$$

$$\text{Output voltage} \propto \int e_i\, dt$$

$$\propto \int Input$$

Output Waveform

The output waveform from an integrating circuit depends upon time constant and shape of input waveform. Two important waveforms will be considered.

a. Square wave input.

When the input fed to an integrating circuit is a square wave, the output will be a triangular wave as shown in fig. 11.11.2(a) since integrator means summation, the output from an integrating circuit will be a sum of all the input waves at any instant. This sum is zero at A and goes increasing till it becomes maximum at C. After this, the summation goes on decreasing for input CD.

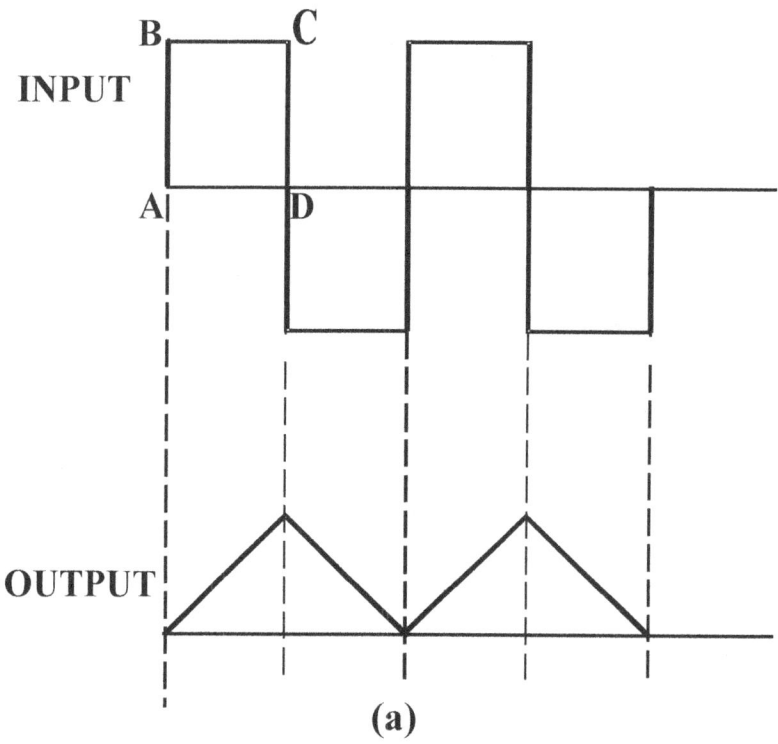

(a)

ELECTRONIC SWITCHING

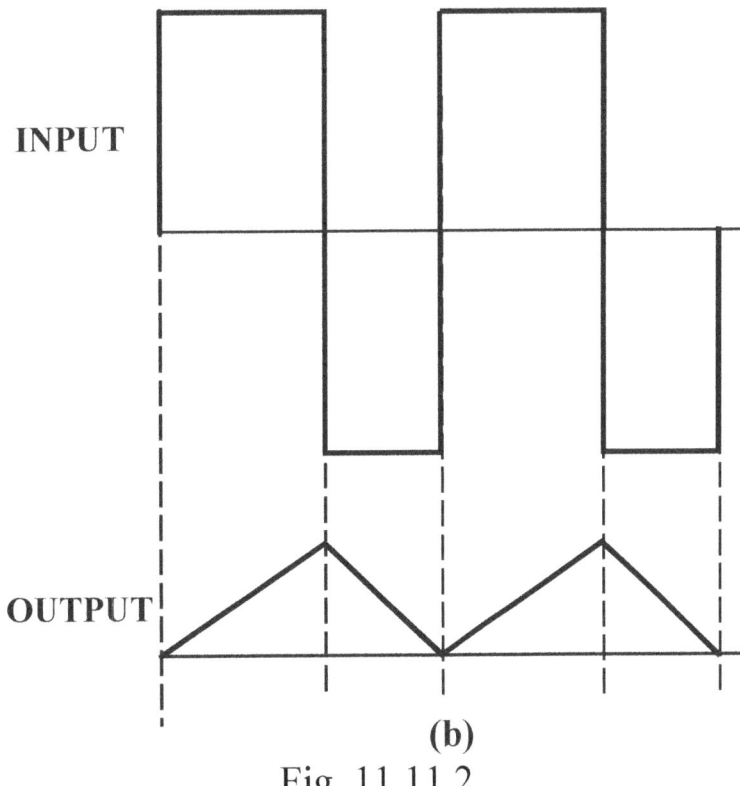

Fig. 11.11.2

b. Rectangular Wave Input

When the input is a rectangular wave, the output will be a triangular wave as shown in fig. 11.11.2(b)

Example 11.11.1

(i) What is the effect of time constant of RC circuit on the differentiated wave?

(ii) Sketch the output waveform from the differentiating circuit when input is square wave for T = 100RC, T = 10RC, T = RC.

Solution:

i. The output of a differentiating circuit depends on the time constant of the circuit. The circuit will function as a differentiator if the product RC is many times smaller than the time period of the input wave.

ii. Square wave input-Fig. 11.11.3 shows the input square wave fed to the differentiating circuit. Fig. 11.11.4 shows the output waveforms for different values of time period of the input wave.

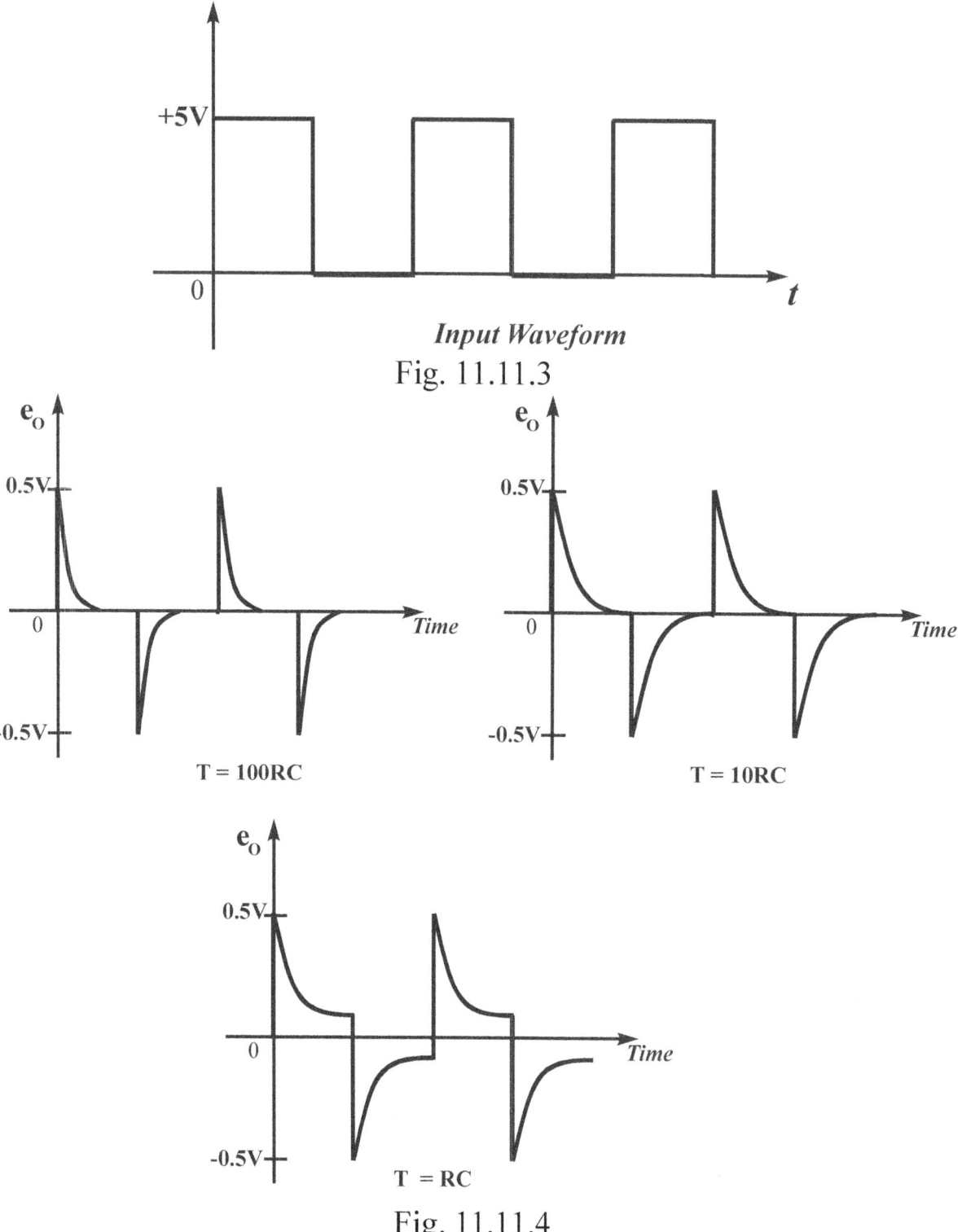

Input Waveform
Fig. 11.11.3

Fig. 11.11.4

Example 11.11.2

In a differentiating circuit, R = 12kΩ and C = 2.0μF. if the input voltage goes from 0 to 10v at a constant rate in 0.5s, determine the output voltage.

Solution:

$$e_o = RC\frac{d}{dt}(e_i) = RC\frac{de_i}{dt}$$

Here R = 12kΩ, C = 2μF; $\frac{de_i}{dt} = \frac{10-0}{0.5} = 20s.$

$\therefore e_o = (12 \times 10^3) \times 2.0 \times 10^{-6} \times 20 = 480 \times 10^{-3}$ V

$\qquad\qquad\qquad\qquad\qquad = \underline{0.48\text{V}}$

CHAPTER TWELVE

12.0 INTEGRATED CIRCUITS

12.1 TECHNOLOGY REVIEW

Both bipolar transistors and F.E.T.'s may be combined in a chip and we may consider two different rypes of intergrated circuit, viz; the complementary symmentry MOS F.E.T in fig. 12.1.1 and the linear micro circuit using NPN. Bipolar transistors in fig. 12.1.2

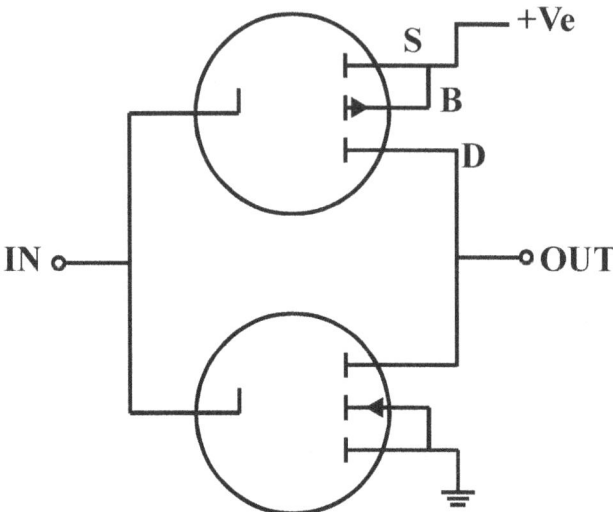

Fig. 12.1.1: Complementary Symmetry MOS FET.

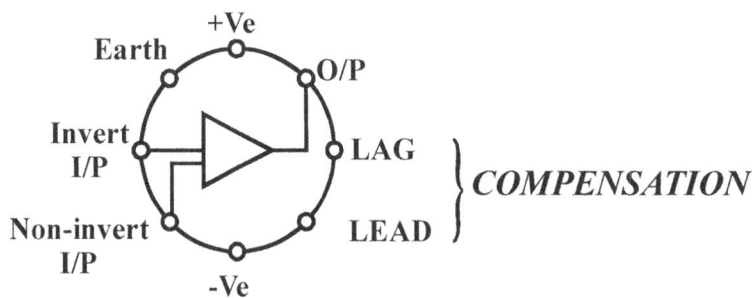

Fig. 12.1.2 Linear Micro- Circuit

The COS-MOS technology is particularly suited to digital applications, the basic circuit of which will be given in later section, therefore this section be primarily devoted to linear micro-circuits. The design of complete circuit using monolithic linear i.c.s is restricted to some extent by the limitations imposed by the manufacturing technique. It is simpler in manufacture to diffuse in only N.P.N. devices, to use D.C coupling throughout to avoid using capacitors (which would increase the cost) and to have resistors which are not guaranteed to a high stability (20% or more of their absolute value). However, on the plus side, doubling the number of junctions used will not increase the proportional cost and tolerance of better than 5% may be achieved in the ratio of resistors of the same value by having the semiconductor resistors in close proximity with identical layouts. A differential input stage will minimise offset voltage since D.C coupling is used.

In order to avoid using bypass capacitors a different method of biasing must be used than would be used in a circuit using discrete components. In the I.C resistors and transistors can easily be made identical.

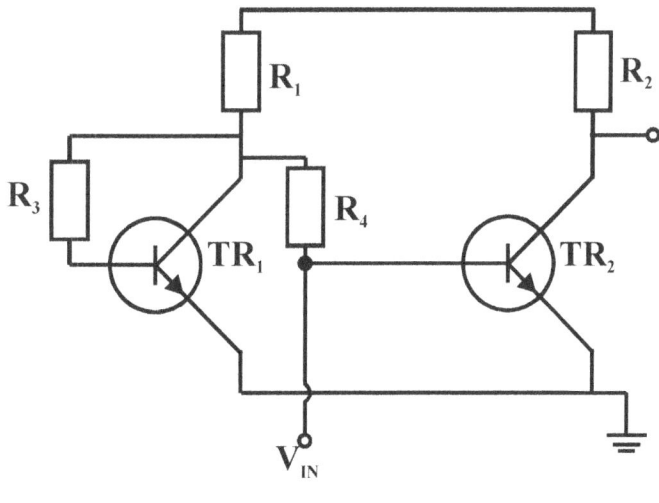

Fig. 12.1.3

If R_3 and R_4 and TR_1 and TR_2 (fig. 12.1.3) are identical, the collector currents of the two transistors will be equal, since their bases are driven from a common voltage point through equal resistance.

Thus the amplifier biased at its optimum operating point at one half the supply voltage independent of the supply voltages as well as temperature and dependent only on how well the parts within the integrated circuit match.

12.2 LINEAR INTEGRATED CIRCUIT

A typical linear I.C. would require low offset, high input impedance, high gain and low power consumption. If a Darlington pair is used to increase input impedance it has the disadvantage with a D.C. amplifier of considerably higher offset and thermal sensitivity than a non-Darlington differential pair. Thus the input stage would be a differential stage as in basic I.C. shown in Fig. 12.2.1.

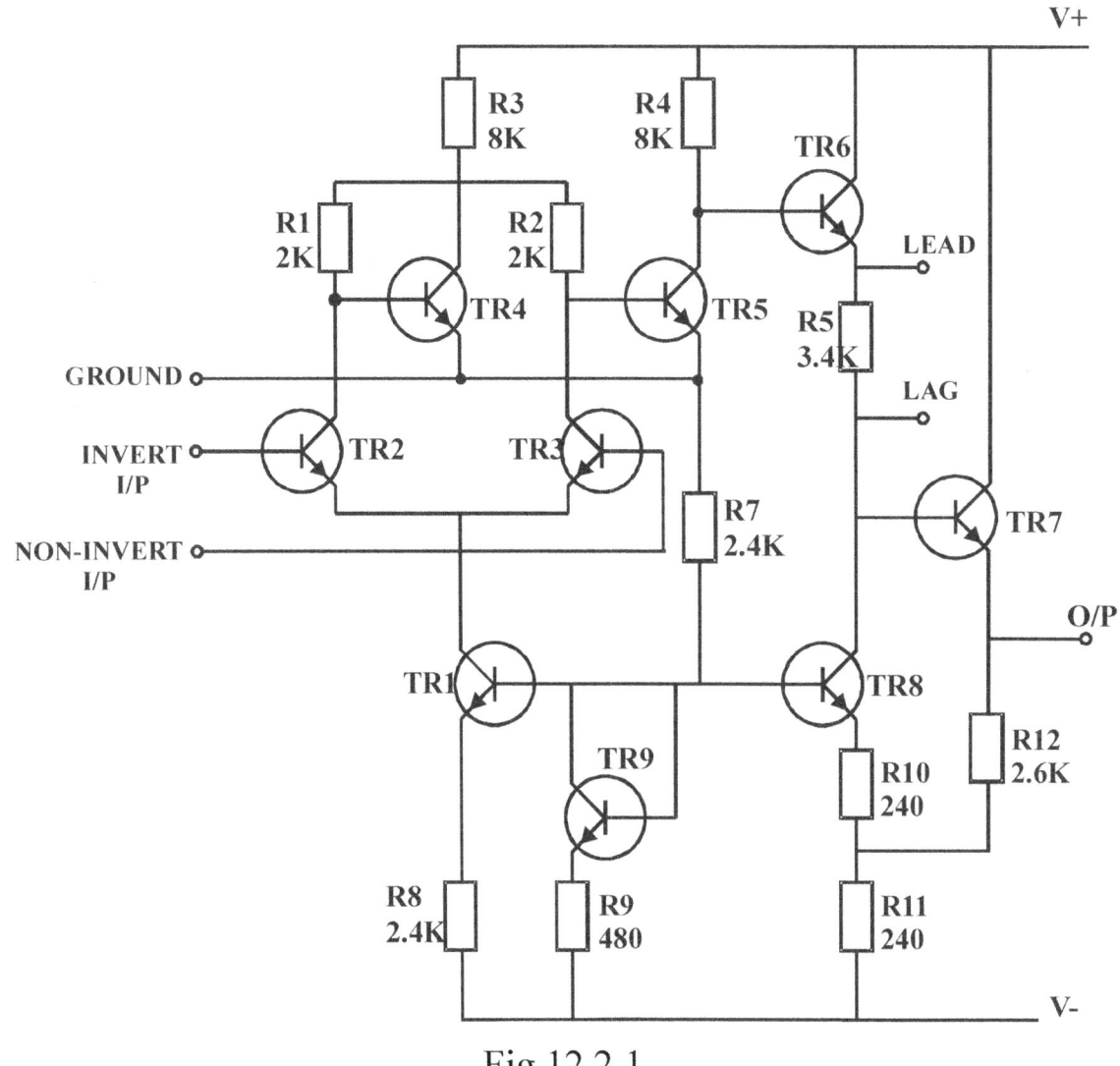

Fig.12.2.1

INTERGRATED CIRCUITS

In fig. 12.2.1, TR2 and TR3 from the differential input stage with TR1 providing a constant current source with biasing isolation between TR2 and TR3. the second stage transistors TR4 and TR5 are identical transistors placed close together with their bases fed by identical resistors thus when TR2 and TR3 collector currents are equal, the collector currents of TR4 and TR5 are also equal and the second stage is balanced. The first two stage provide almost all the voltage gain for the amplifier and the output stage provides level shifting by the voltage drop across R5 and a low output impedance with the emitter follower TR7. the load resistor of TR7 is fed to the emitter of TR8 giving some positive feedback thus the output stage does have some gain.

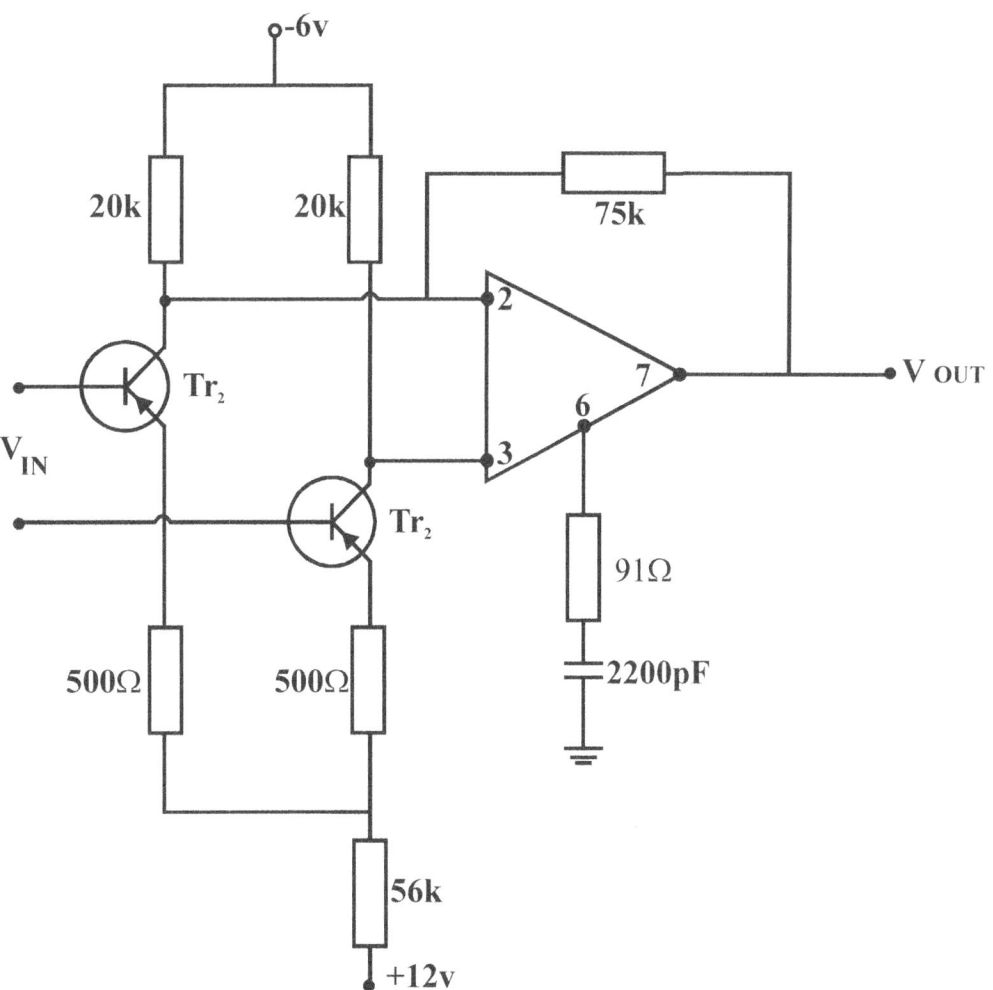

Fig. 12.2.2. A Low Noise, High Input Impedance Audio Pre- Amplifier

A linear I.C. may be used in a variety of ways, as a high gain A.C or D.C. amplifier, as an operational amplifier or as an active filter network. A low noise, high input impedance audio pre-amp is shown in fig. 12.2.2. TR1 and TR2 are a matched pair of low noise transistors and series applied feedback increases the input impedance.

To improve the frequency response negative feedback must be used.

12.3 NEGATIVE FEEDBACK

The behaviour of an amplifier may be rectified by applying part of the output signal to the input. For example an amplifier may be used as an oscillator if positive feedback is used. That is if the loop gain is sufficient and the feedback signal is in phase with the input signal, then the device will maintain oscillations.

Negative feedback is used because of several advantages, viz:-

1. The gain of the amplifier is largely independent of variation due to circuit component changes, change of active devices and supply voltage variations.
2. Reduces the effect of noise and distortion.
3. Allows the input and output impedance to be modified.
4. Improves the temperature stability of the amplifier.
5. Provides for control of the frequency response of the amplifier.

The last three advantages obviously are of paramount importance for video amplifiers etc. All these advantages are at the expense of reduction in gain.

Fig. 12.3.1 shows a general schematic diagram for any feedback amplifier.

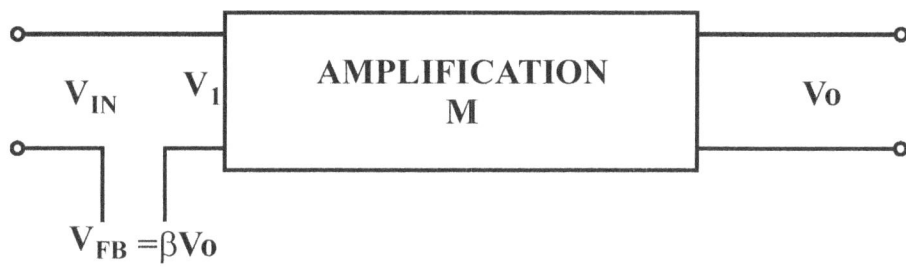

Fig. 12.3.1

$V_O = mV_1$, where m is the gain of the amplifier without feedback.

But $V_1 = V_{in} - \beta V_O$ where β is the feedback factor.

$\therefore V_O = m(V_{in} - \beta V_O)$

$V_O(1 + m\beta) = m V_{in}$

$\therefore m^1 = \dfrac{V_O}{V_{IN}} = \dfrac{m}{1+\beta m}$ \hfill (1)

Thus the modified gain is given by the formula (i) for almost all feedback amplifiers.

The term m is the loop gain of the amplifier and it is important to ensure that the feedback voltage does not become positive until its magnitude is sufficiently low.

This is to stop the circuit going into oscillation since

$m^1 = \dfrac{m}{1+\beta m} = \dfrac{m}{0}$ if loop gain changes phase by 180^0 and is unity.

i.e. the gain $m^1 = \dfrac{V_O}{V_{IN}} = \infty$

the amplifier acts as an oscillator.

If the amplification without N.F.B is M the amplification with N.F.B is $\dfrac{m}{1+\beta m}$. Therefore the change of gain is $20 \log \dfrac{m(1+\beta m)}{m}$. i.e, change of gain due to N.F.B. is $20 \log (1 + \beta m) = $ NdBs.

This gives the meaning of the term Ndb's of negative feedback.

INTERGRATED CIRCUITS

The amplification may be made constant if $\beta m \gg 1$ as is usually the case then $m^1 = \dfrac{m}{\beta m} = \dfrac{1}{\beta}$

Thus the amplification m^1 is independent of variations which would normally affect the factor m such as change of the device, value or transistor changes in component values, resistance and radiance and changes in H.T. If it becomes necessary to alter the gain of an N.F.B amplifier then it is necessary either to alter or to attenuate the input or output signals since altering the m would have no effect since $\beta m \gg 1$.

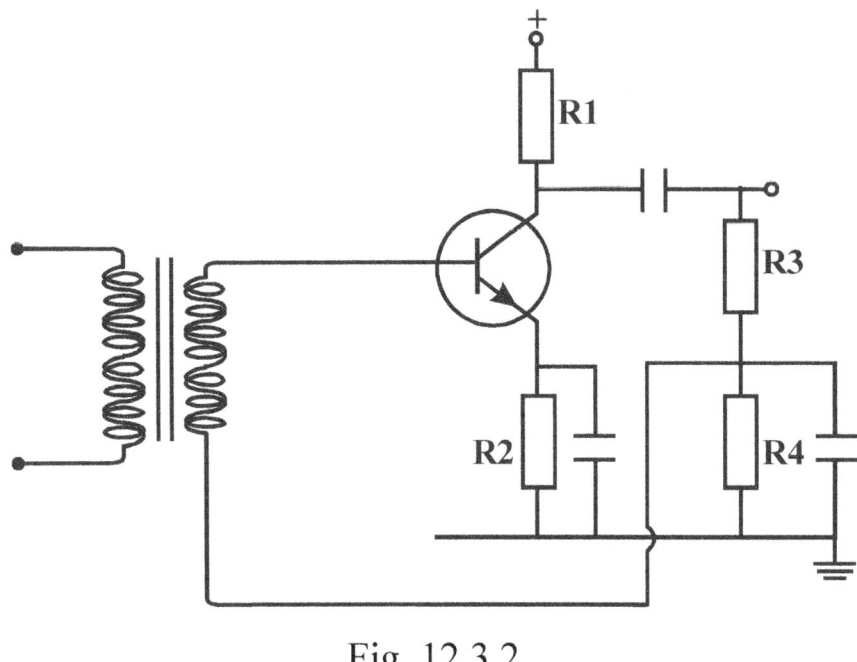

Fig. 12.3.2

Altering m however would affect the terminal impedances of the amplifier so it is preferable to attenuate the input or output signals. A simple circuit for this is show in fig. 12.3.2. At high frequencies the feedback is less due to the presence of the capacitor **C**. The output impedance of the N.F.B amplifier is modified according to whether the N.F.B is voltage derived or current derived.

12.4 VOLTAGE DERIVED N.F.B

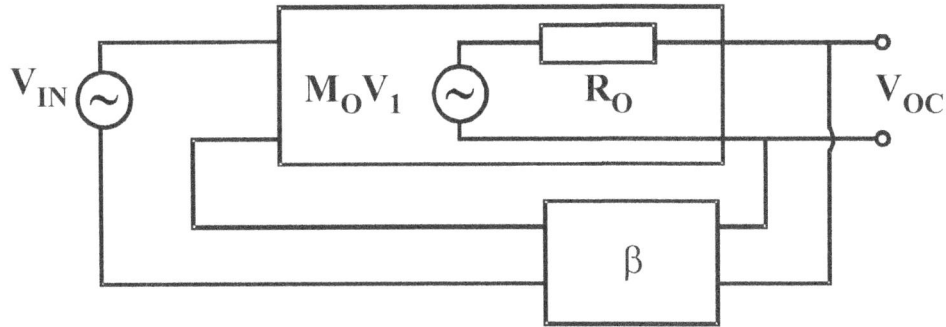

Fig. 12.4.1

$$V_{OC} = M_{O1} \quad ; \quad V_1 = V_{IN} - \beta V_{OC}$$
$$V_{OC} = m_O(V_{IN} - \beta V_{OC})$$
$$V_{OC} = \frac{m_O V_{IN}}{1 + \beta M_O}$$

Note: $Z\,out = \dfrac{V_{OC}}{I_{SC}} \quad I_{SC} = \dfrac{m_O V_{IN}}{R_O}$

$$Zout = \frac{R_O}{1 + \beta m_O}$$

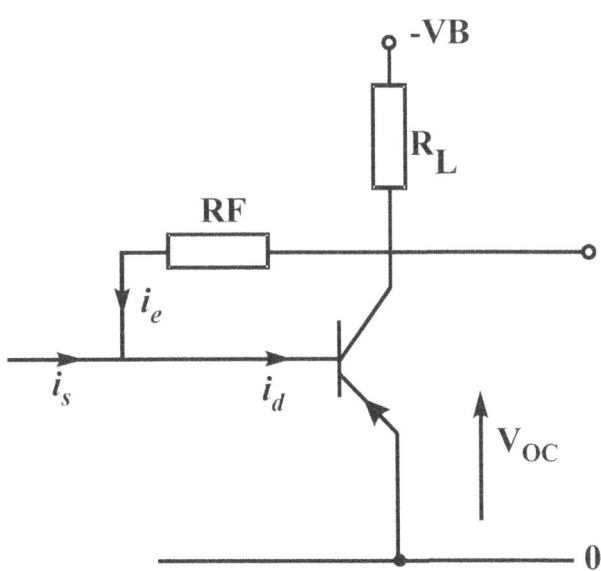

Fig. 12.4.2 Voltage derived N.F.B

A practical circuit is shown in fig. 12.4.2

Where $Zo^1 = \dfrac{R_L}{1 + hfe\dfrac{R_L}{R_F}}$

12.5 CURRENT DERIVED N.F.B

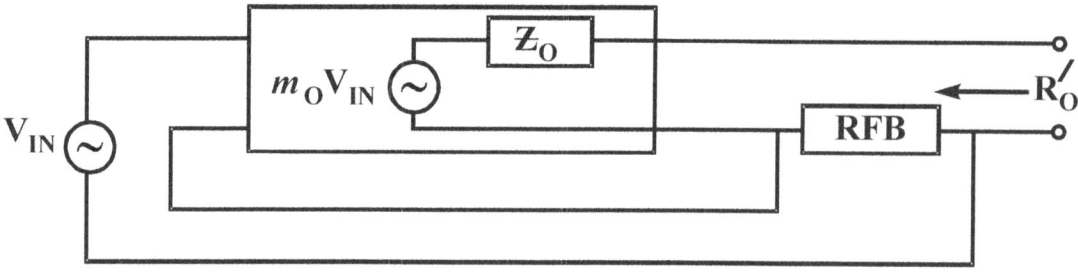

Fig. 12.5.1 Current derived N.F.B

This is current derived N.F.B. and we can ascertain the output impedance by using Thevenin theorem.

The way in which N.F.B is applied to the input terminals determines the input impedance. If we consider fig. 12.5.2 where the feedback voltage could be either voltage or current derived.

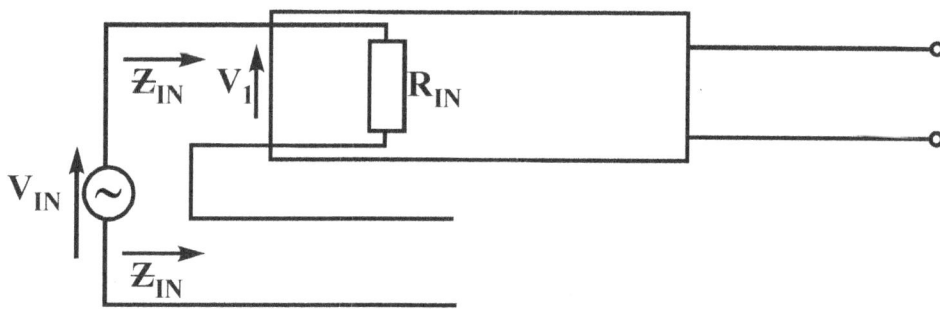

Fig. 12.5.2

Thus, series applied N.F.B. increases the input impedance by the factor $(1 + \beta m)$. The input impedance is again modified if the N.F.B is applied in parallel at the input terminals, as in fig. 12.5.3.

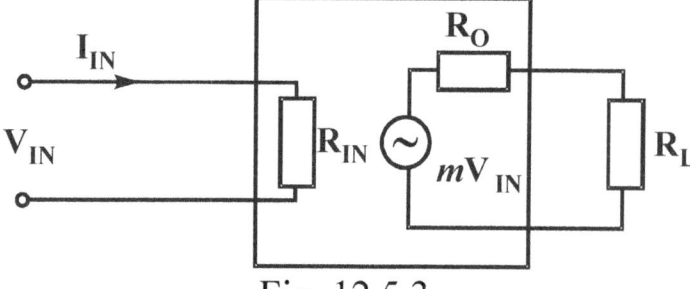

Fig. 12.5.3

Application of N.F.B to a high gain amplifier requires the open loop gain and phase response to be controlled far outside the band of interest.

INTERGRATED CIRCUITS

Linear I.C.'s provide connections such that external feedback compensation can be applied to improve the frequency response or part of the I.C. itself may be so connected to provide frequency compensation with of course, a subsequent loss in gain. It is useful in television to be able to change from one waveform to another and this is of course, performed by differentiation and integration, for example if a square wave output of a multivibrator is integrated it results in a triangular waveform. If an input voltage E is applied to the circuit of fig. 12.5.4

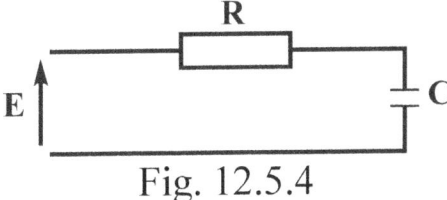

Fig. 12.5.4

Then $Eo = \dfrac{1}{RC}\int (E - Eo)dt$

When E is much greater than Eo then

$Eo = \dfrac{1}{RC}\int E dt$

Thus the output waveform is a true integral if it is small compared to the input. If the circuit of fig. 12.5.5 is also considered.

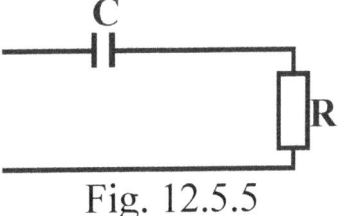

Fig. 12.5.5

$Eo = RC \cdot \dfrac{d}{dt}(E - Eo)$

When Eo is small compared to E then $Eo = \dfrac{RC\, dE}{dt}$, that is the output voltage is a differential of the input. An operational amplifier may be used in order to obtain waveform of reasonable amplifier. The linear I.C. is particularly useful in this respect.

A particular differentiator is shown in fig. 12.5.5.

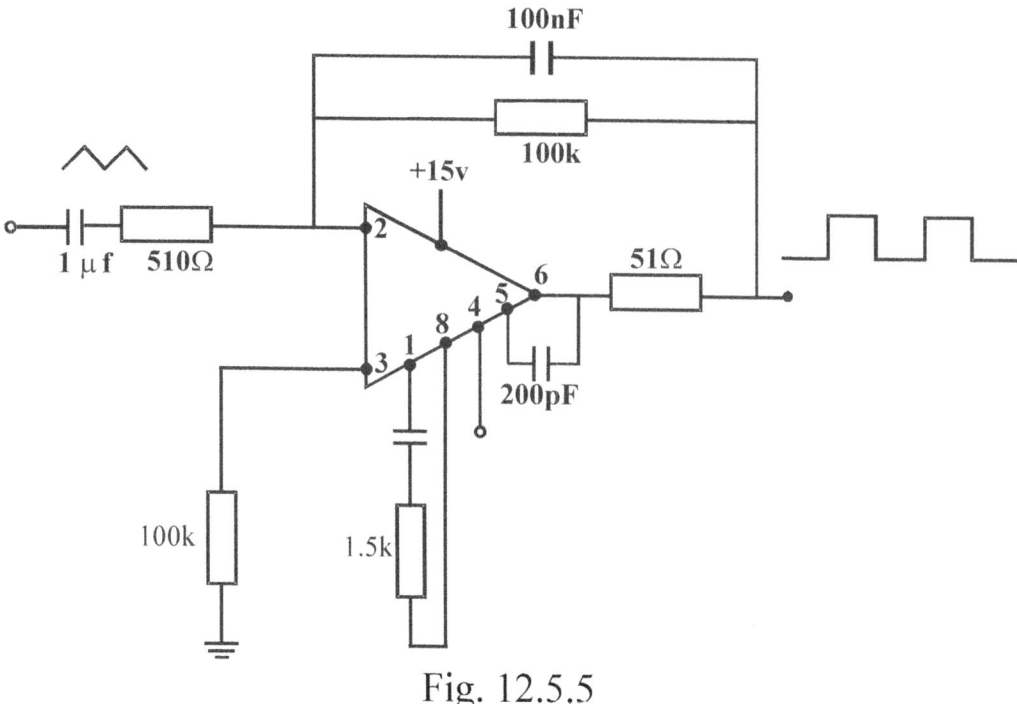

Fig. 12.5.5

A practical integrator is shown below in fig. 12.5.6

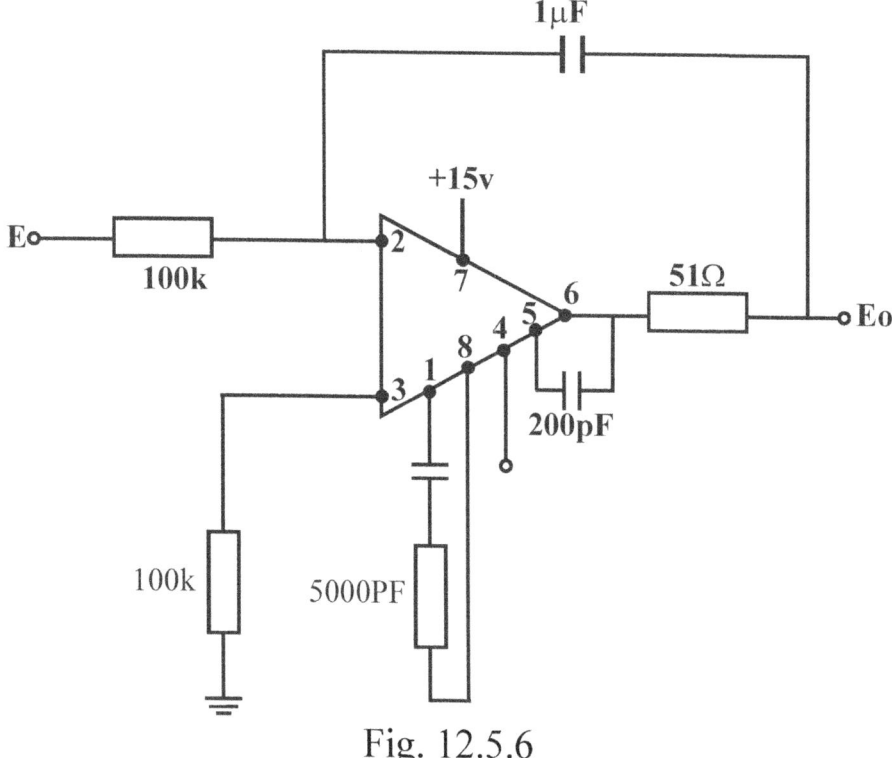

Fig. 12.5.6

The output voltage is given by $Eo = -RdCd \cdot \dfrac{dE}{dt}$

CR and R1 are added to give the necessary frequency response stability to the differentiator circuit.

$$Eo = -\frac{1}{R}\int E\,dt$$

Since the current $\frac{E}{R}$ flowing across R the input resistor must pass through the feedback capacitor. There are many other applications of linear I.C.S such as logarithmic amplifiers, multivibrators, comparators, synchronous detectors, modulators etc. which will be considered in future.

12.6 PINOUT DIAGRAM FOR TTL INTEGRATED CIRCUITS

Fig. 12.6.1 shows the pinout diagram for TTL integrated circuit.

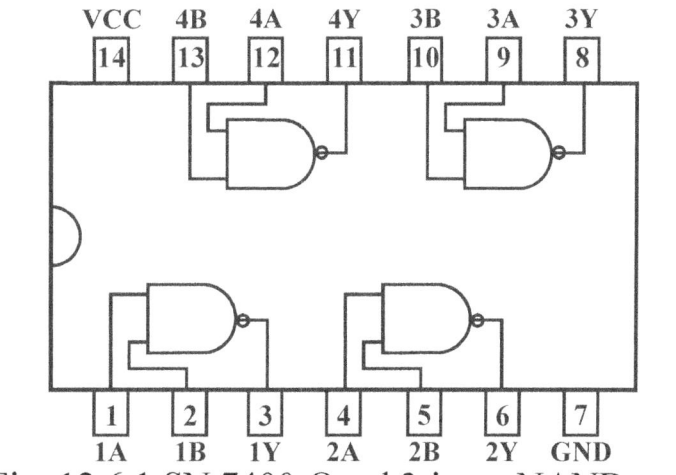

Fig. 12.6.1 SN 7400-Quad 2-input NAND gate

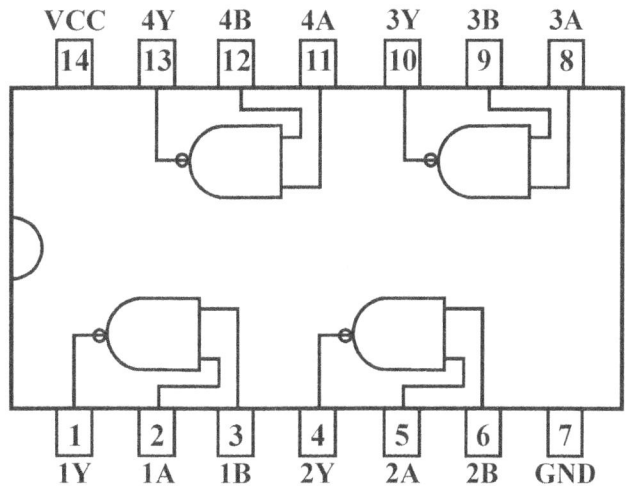

Fig. 12.6.2 SN 7401-Quad 2-input NAND gate with open-collector outputs.

INTERGRATED CIRCUITS

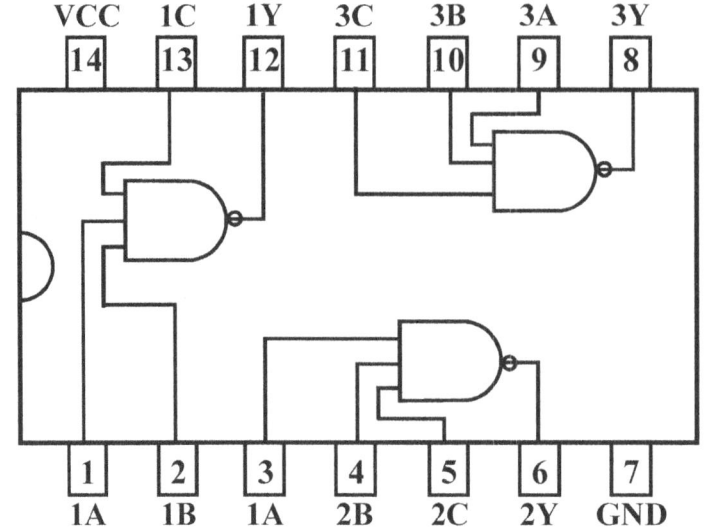
Fig. 12.6.3 SN 7410- Triple 3-Input NAND gate

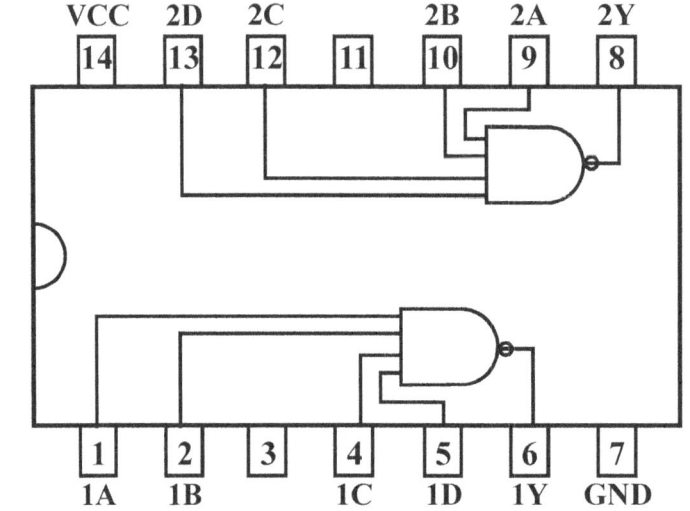
Fig. 12.6.4 SN 7420 -Dual 4-Input NAND gate

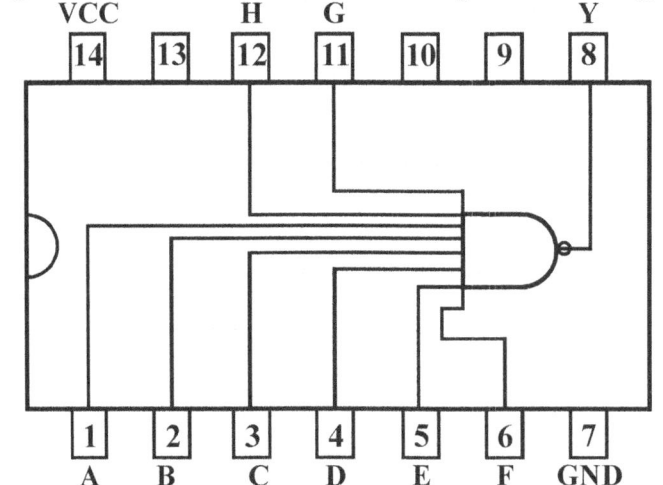
Fig. 12.6.5 SN 7430-Single 8-Input NAND gate

319

INTERGRATED CIRCUITS

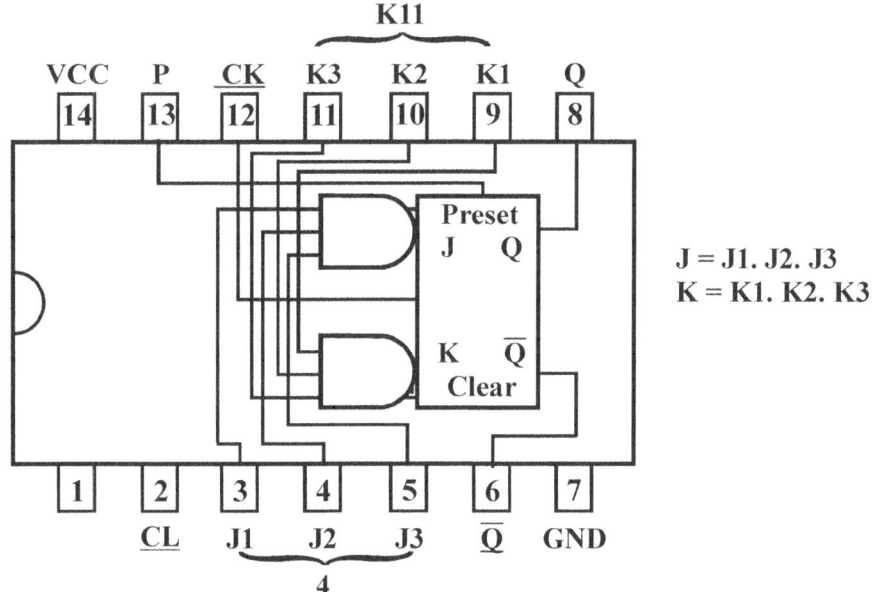

Fig. 12.6.6 SN 7472-Single J-K master-slave flip-flop

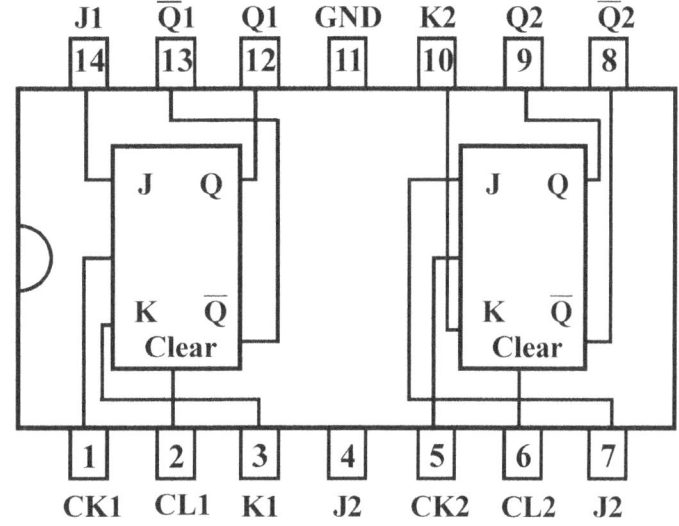

Fig. 12.6.7 SN 7473-Dual J-K master-slave flip-flop

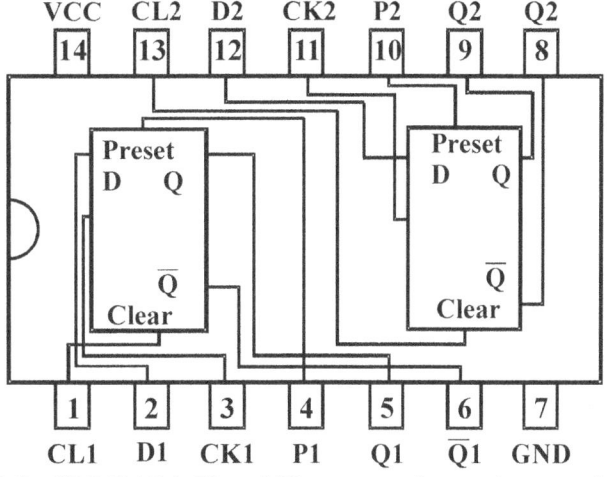

Fig. 12.6.8 SN 7474-Dual D-type edge triggered flip-flop

INTERGRATED CIRCUITS

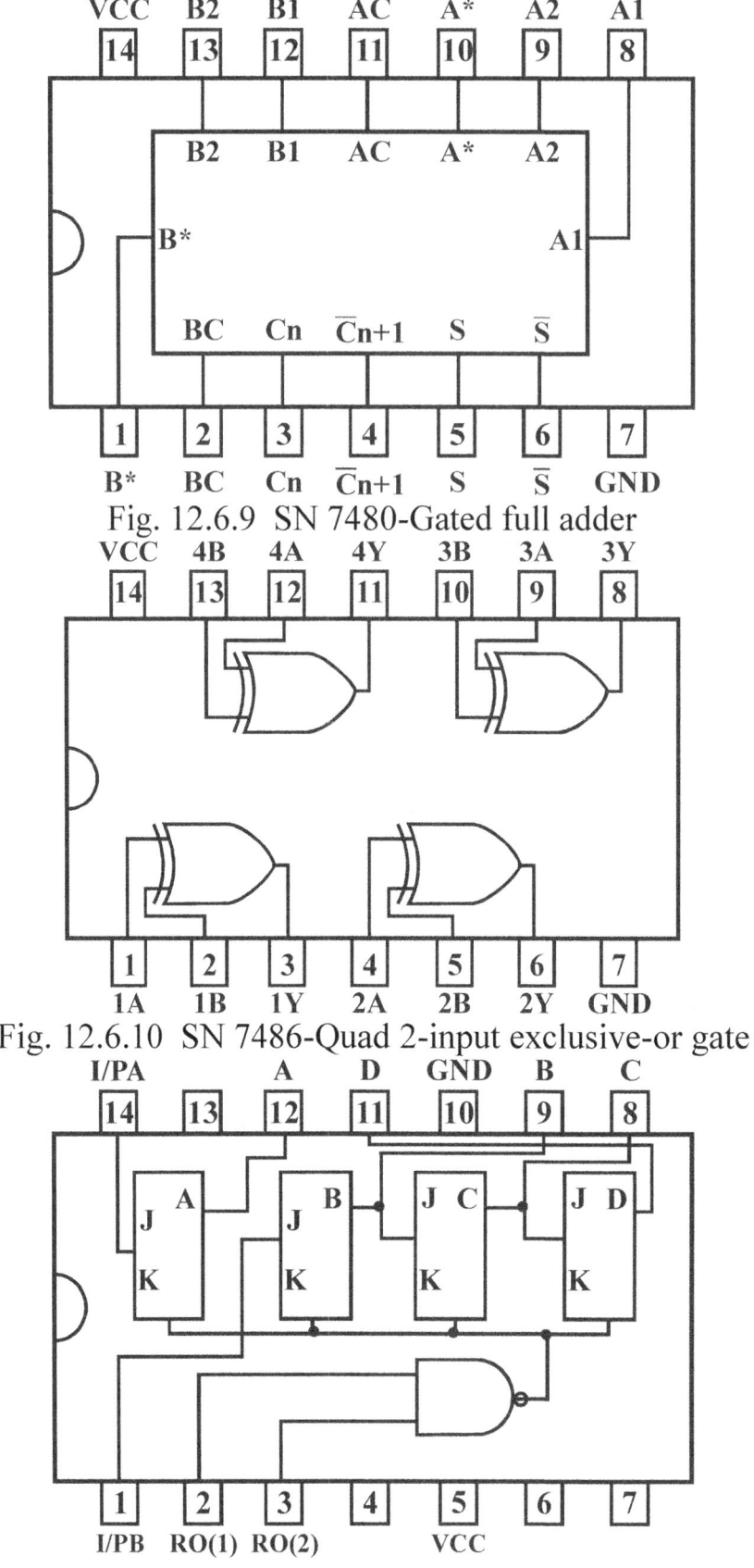

Fig. 12.6.9 SN 7480-Gated full adder

Fig. 12.6.10 SN 7486-Quad 2-input exclusive-or gate

Fig. 12.6.11 SN 7493-4-bit binary counter

INTERGRATED CIRCUITS

Type No.	Technology	Manufacture	No. of Pins	Fan Out No.	Fan Out mA	Prop. Delay Nsec/gate	Comments	Dissipation mW/gate
7400N	Saturated	Many	14 or 16	10	16	10	Commercial spec.	10
7400 J	TTL	Independent	,,	,,	,,	,,	5% supply, 0^0, $+70\ ^0C$ temp.	,,
5400J	,,	Sources	,,	,,	,,	,,	Military spec.	,,
5400W	,,	e.g T.I	,, f.p	,,	,,	,,	10% supply, -55^0 to $+125\ ^0C$ temp.	,,
74L00N	,,	,,	,,	20	4	33	Low power TTL	1
74H00N	,,	,,	,,	10	20	6	High speed TTL	22
74500N	Schottley	,,	,,	10	20	3	Very fast, compatible with normal TTL	20
74L500N	TTL	,,	,,	20	8	10	Speed traded to conserve power	2
830	DTL	,,	14	8	12	30	Super ceded by TTL	11
HC10102	ECL	Motorola	16	10-70	19	2	Very fast with excellent characteristics	25
HC 1662L	,,	,,	16 f.p	,,	25	1	Ultra-fast logic for CPUs	55
H102	HLL	S.G.S	14	25	9	100	High a.c & d.c noise immunity	43
ZNA100	CDI	Ferrant	24-40	20	5	35	187 uncommitted gates (U.L.A)	2.5
CD400IA	Cos/Mos	R.C.A	14	50	1	35	Extremely low power consumption.	0.01
8080	NNOS	Intel	40	-	1.6	50	8 bit micro-processor C.P.U	0.5

Fig. 12.6.1 Typical performance characteristics of digital integrated circuit

Except for those marked **f.p** (flat-pack) packages are dual-in-line configuration. Device encapsation is in plastic or ceramic. Prices are based on the unit price when purchasing 1000 up quantities.

12.7 DIGITAL INTEGRATED CIRCUITS- BASIC GATES

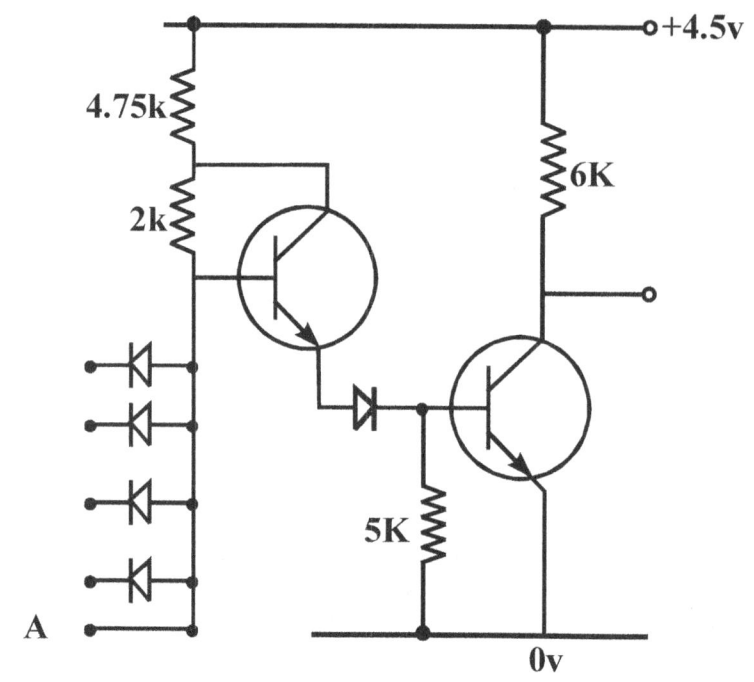

(i) DTL 830

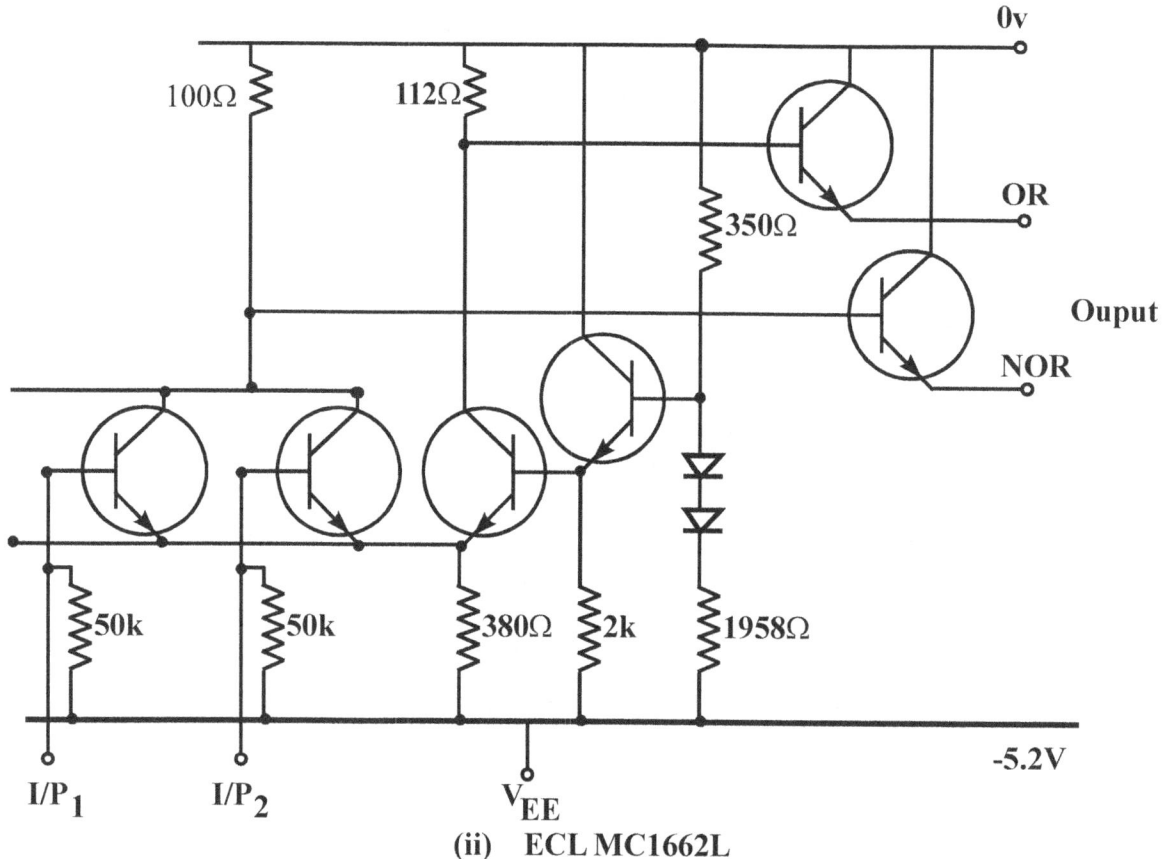

(ii) ECL MC1662L

INTERGRATED CIRCUITS

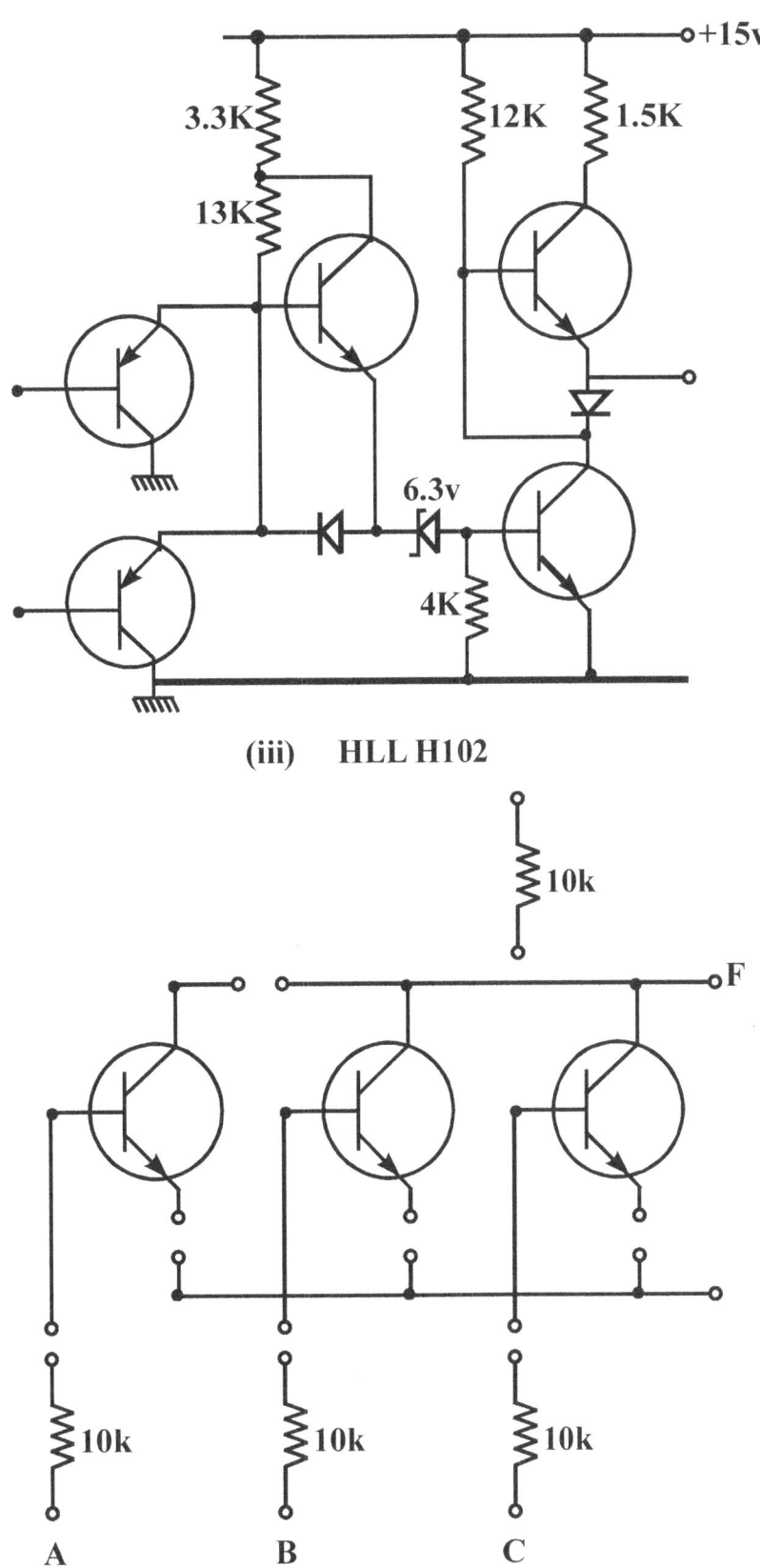

(iii) HLL H102

(Iv) CDI RTL *(Uncommited cell)*

INTERGRATED CIRCUITS

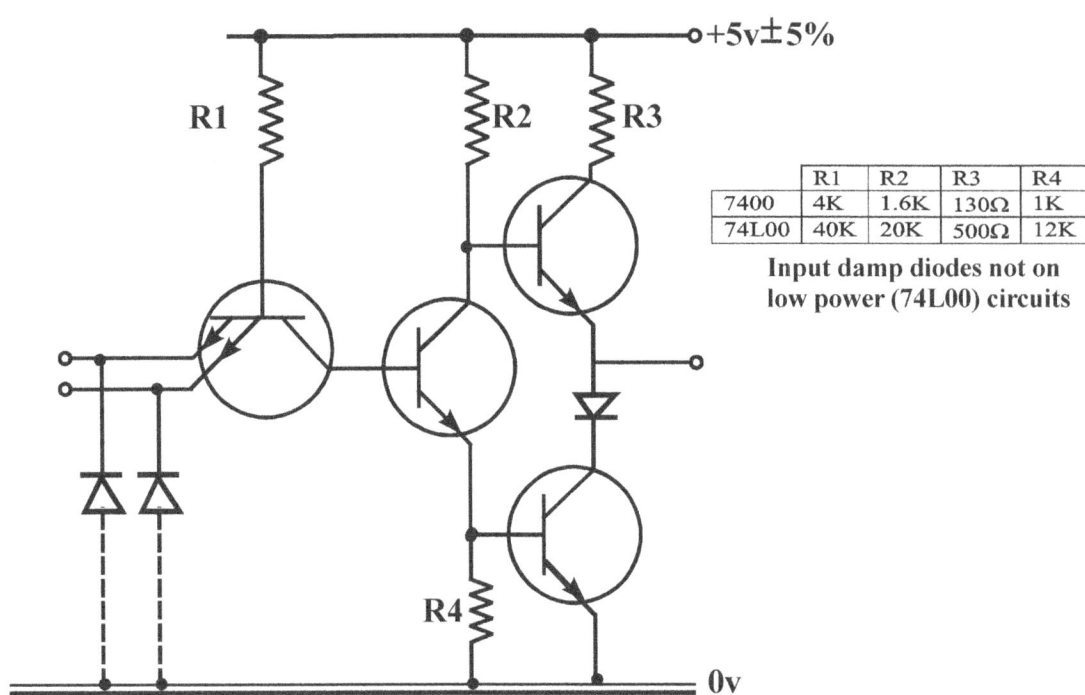

	R1	R2	R3	R4
7400	4K	1.6K	130Ω	1K
74L00	40K	20K	500Ω	12K

Input damp diodes not on low power (74L00) circuits

(v) (7400/74L00) TTL GATES

Complimentary MOS (CD 4001 A- NOR)

(vi) (74H00) TTL GATES

NMOS MEMORY CELL using change pump transistor

325

INTERGRATED CIRCUITS

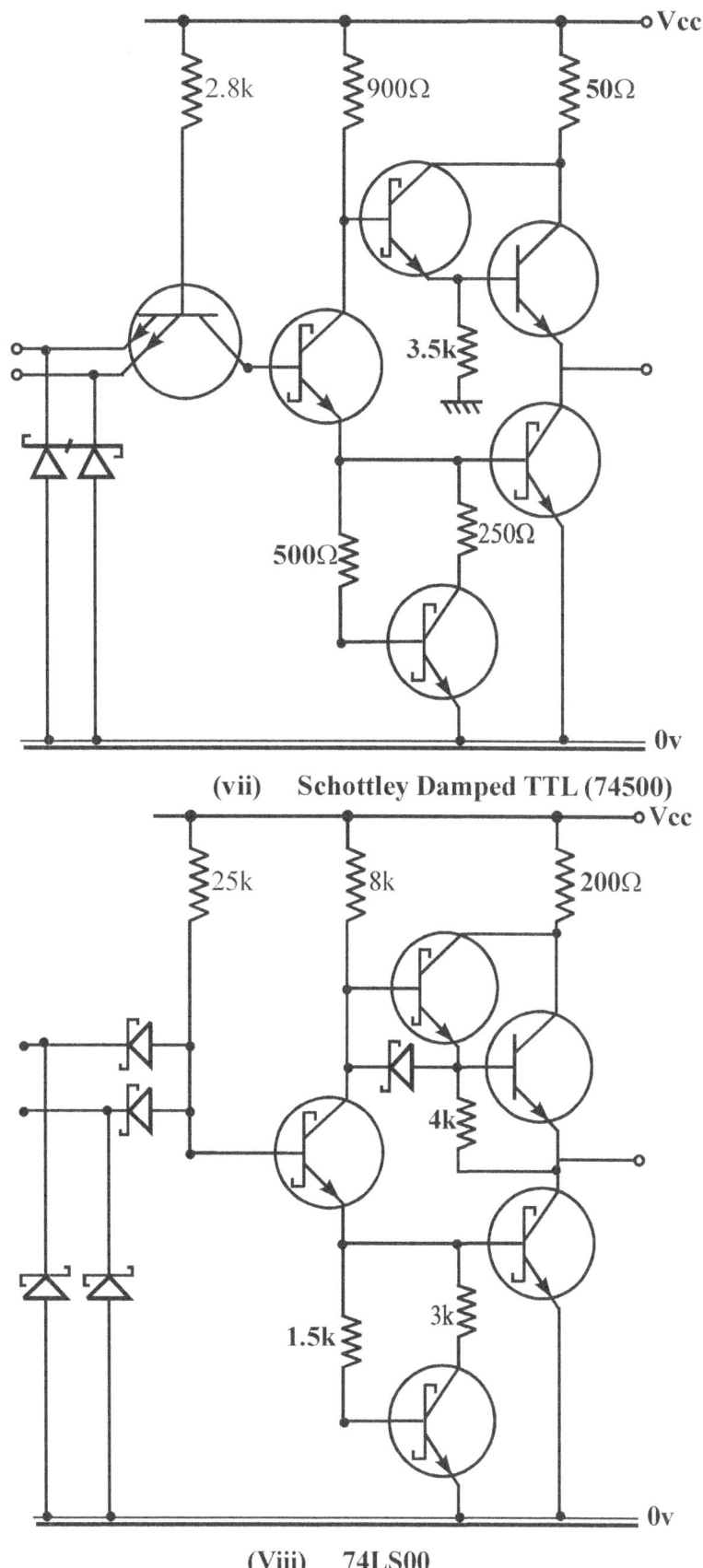

(vii) Schottley Damped TTL (74500)

(viii) 74LS00

Fig. 12.5-1 (i-vii) shows different types of Digital Integrated Circuits

CHAPTER THIRTEEN

13.1 BASIC CONCEPT OF MICROPROCESSORS

Over the past two decades increasing miniaturisation, and in particular the advent of integrated circuits has revolutionised the electronics industry. Nowhere is this truer than in the field of digital electronics, especially in the area of data processing and calculating machines where large scale integration (LSI) has made possible low prices units of a small size and high performance previously unheard of. One of the best examples of this trend is the advent of the Microprocessor.

The basic concept of a microprocessor is shown in fig. 13.1.1. The heart of the system is the unknown gate which we call a 74XX. This gate can perform **AND, OR, NOR, NAND** functions etc but it must first be primed as to what function it should perform. The information or data on which this gate will operate is entered via two latches, latch A and latch B. The instruction of what to do with this data is entered via the control input latch C and the result is fed to the output latch. All the data to and from these latches is entered on a common data bus and a timing unit enables each latch at the appropriate time.

If normal TTL gates were used on such a system as that shown in fig. 13.1.1 then the output of the output latch would always be on the data bus and so would interface with any data being routed to the input or control latches.

BASIC CONCEPT OF MICROPROCESSORS

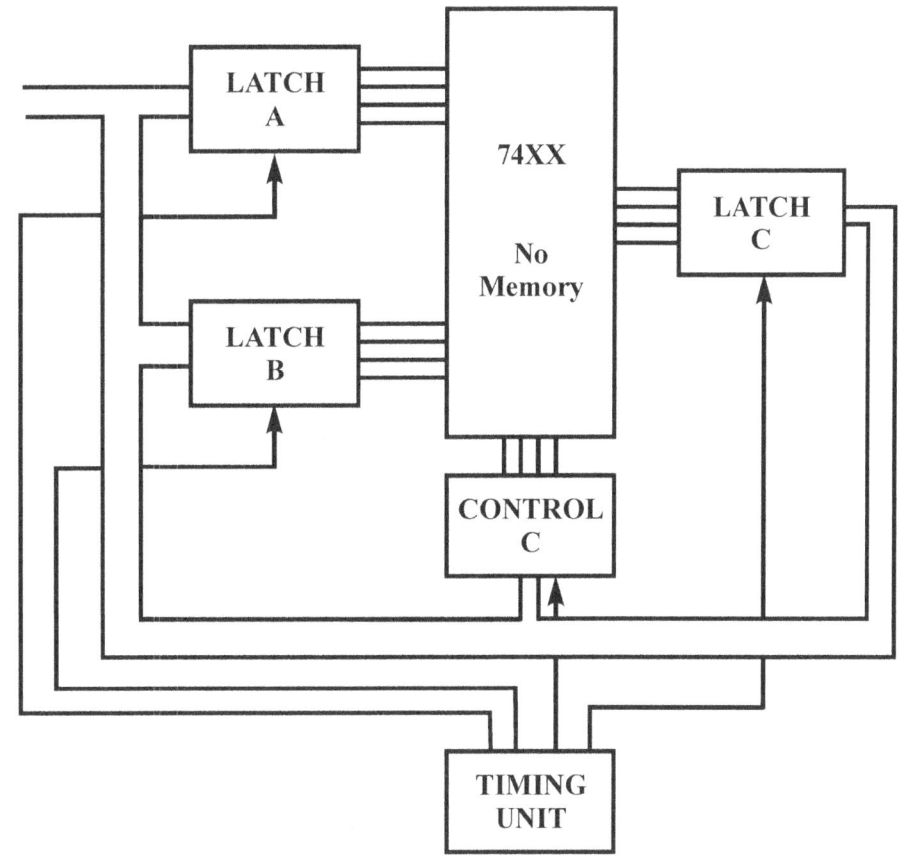

Fig. 13.1-1: Basic Concept of Microprocessors

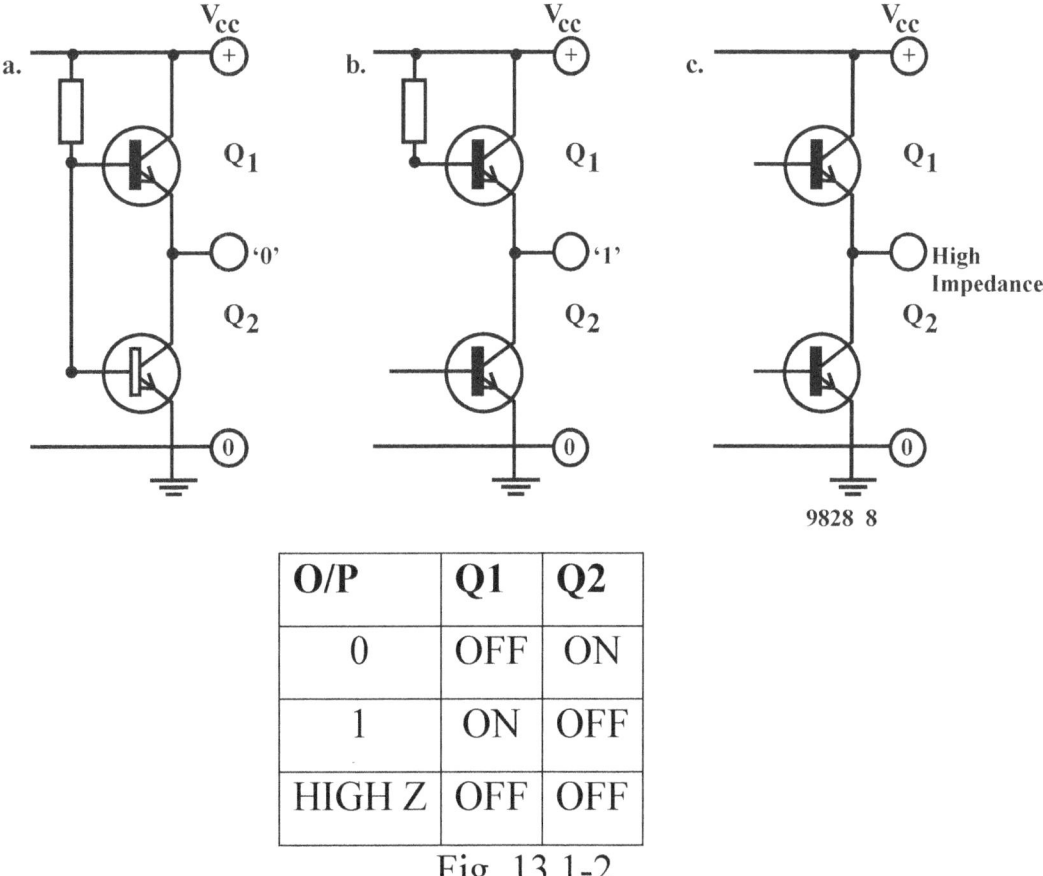

Fig. 13.1-2

For this reason TTL outputs cannot be used and we must go to a system which has three states instead of the normal two, i.e LOGIC "1", LOGIC "0" and an "OFF" state. Such a system is shown in fig. 13.1.2 with the corresponding truth table and is known as "TRI-STATE" logic.

13.2 HEXADECIMAL CODE

Microprocessors work mostly in binary. Long chains of "1"s and "0"s are difficult to handle and mistakes are easily made, so HEXAGONAL convention has been adopted. The binary word is split into groups of 4. Each group of 4 has 16 possible values from zero (all "0"s) to 15 (all "1"s). To avoid confusion the values of 10 to 15 are replaced by the letter A to F as shown in table 13.2.1

Table 13.2-1

BINARY	**HEXADECIMAL**	**DECIMAL**
0000	0	0
0001	1	1
0010	2	2
0011	3	3
0100	4	4
0101	5	5
0110	6	6
0111	7	7
1000	8	8
1001	9	9
1010	A	10
1011	B	11
1100	C	12
1101	D	13
1110	E	14
1111	F	15

BASIC CONCEPT OF MICROPROCESSORS

So in Hex   C4 = 1100 0100
In decimal $593 = 5 \times 10^2 + 9 \times 10^1 + 3 \times 10^0$
In Hex $593 = 5 \times 16^2 + 9 \times 16^1 + 3 \times 16^0$

All computers from the largest main frame installation to the smallest all operate on a system of stored instruction and data, and this also applies to a microprocessor. The instructions and data to be used for a specific task are laid down in a "**PROGRAMME**" and this programme must be stored in some form of memory, so that it can be recalled and used by the computer when it is required. This memory can use semi conductor technology such as in **RAM** and **ROM**, or can use ferrite core stores.

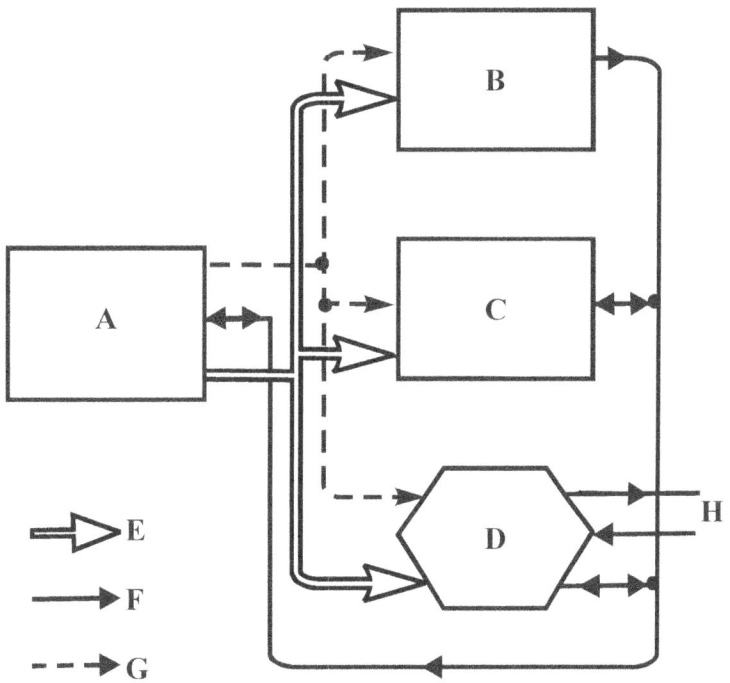

Fig. 13.2-1 A Simple Microcomputer System.

As far as a microprocessor is concerned this "**SYSTEM MEMORY**" is not normally contained within the processor chip but is a separate IC package or packages depending on the amount of memory required. However the system memory is necessary to the operation of the processor and so must be considered an integral part of it although physically not housed in the same package. See fig 13.2-1

A- Microprocessor unit;   B-ROM;   C-RAM;   D-Input/output unit

E- Address Bus (16 lines); F- Data Bus (8 lines); and G – Control lines

13.3 REGISTERS

It is necessary that the processor chip does have some inbuilt storage facilities in order that is may store instructions and data that are in current use. These memories are known collectively as Registers and are normally either 8 or 16 bits wide. Different microprocessors have different arrangement of numbers and names of these registers. The programming model of the Motorola 6800 MPU is shown in Fig. 13.3.1 and shows the registers used.

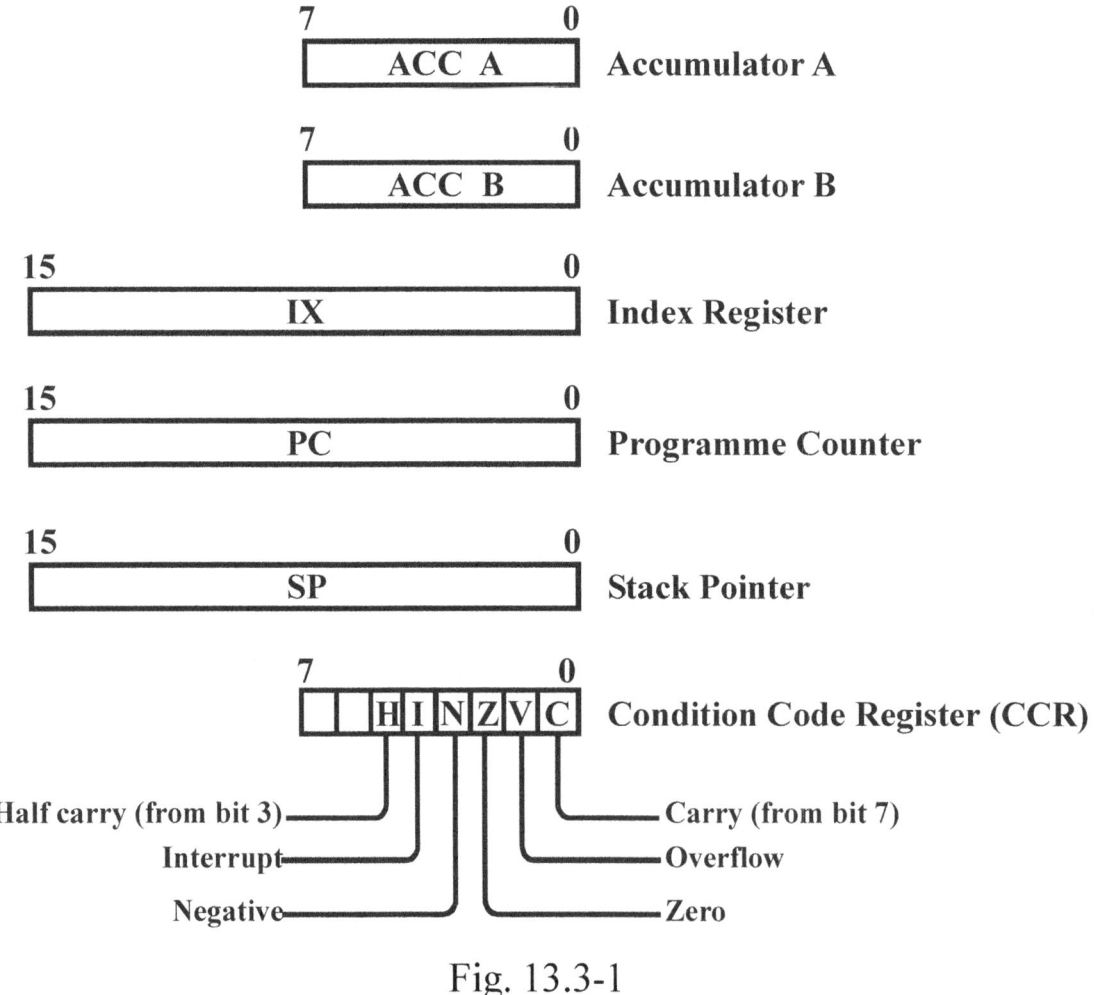

Fig. 13.3-1

Accumulator:- The main working register used to store data and the results of any arithmetic operation.

Programme Counter:- This register simply counts the steps of the programme and acts as a bookmark.

Index Register:- This acts as a memory pointer and can hold a specific memory address which could be used as a reference for addressing other memory locations.

Stack Pointer:- This also acts as a memory pointer and usually contains the address in system memory where the CPU can temporarily store information in a "least in first out" basis. This facility is required when servicing an interrupt routine.

Condition Code Register:- This register consists of six independent flip flops which can be set so as to signal or "flag" certain information about to last operation performed, e.g. result negative, overflow, etc. So much for the generalities. Let us now consider the system in more detail. Fig. 13.2.1 shows the basic layout of any microprocessor system. The CPU can send or receive information normally eight bits wide on the Data Bus and this can be from the system memory or it may be from a previously programmed ROM or even some sort of Input/Output devices.

The CPU also puts out on the Address Bus an address to what part of the system the data should be stored in, or received from. Normally in an eight bit microprocessor system 16 address lines are provided giving a total of (2^{16-1}) memory locations i.e 65,536.

The CPU also issues some control signals to other units to tell the memory for instance, whether it will be giving out (READ) or storing data (WRITE) in the specified memory location. *Note* that any 8 bit byte of information could either be an instruction which the CPU will execute, or date in the form of a numerical value on which the CPU will operate. There are also some other inputs to the CPU which can tell it whether some peripheral device requires attention, such as updating the display on a VDU. Note that the data flowing between the RAM and the CPU is bi-directional while that between the ROM and the CPU only flows into the processor. The system shown in fig. 13.2.1 although very basic is typical of any microprocessor system, independent of manufacturer or complexity. Let us now look at the Central Processing Unit itself in more detail.

Fig.13.3.2 shows the internal structure of a typical microprocessor chip. The Address Register will output onto the Address Bus any required address, e.g that specified by the Programme Counter.

The Arithmetic Logic Unit is the heart of the microprocessor and here it performs all the logic operations such as addition, AND, OR, comparisons etc. The Data Register will send or receive data from outside the CPU and if receiving data, can then pass it on to the appropriate section for processing. As has already been mentioned, not every byte is numerical data but could be an instruction. One area which the data register must therefore serve is the Instruction Decoder. This, as its name would imply, looks at the byte of information and decides what the instruction is. It then sets all the necessary internal logic to perform that function, e.g . LOAD.

BASIC CONCEPT OF MICROPROCESSORS

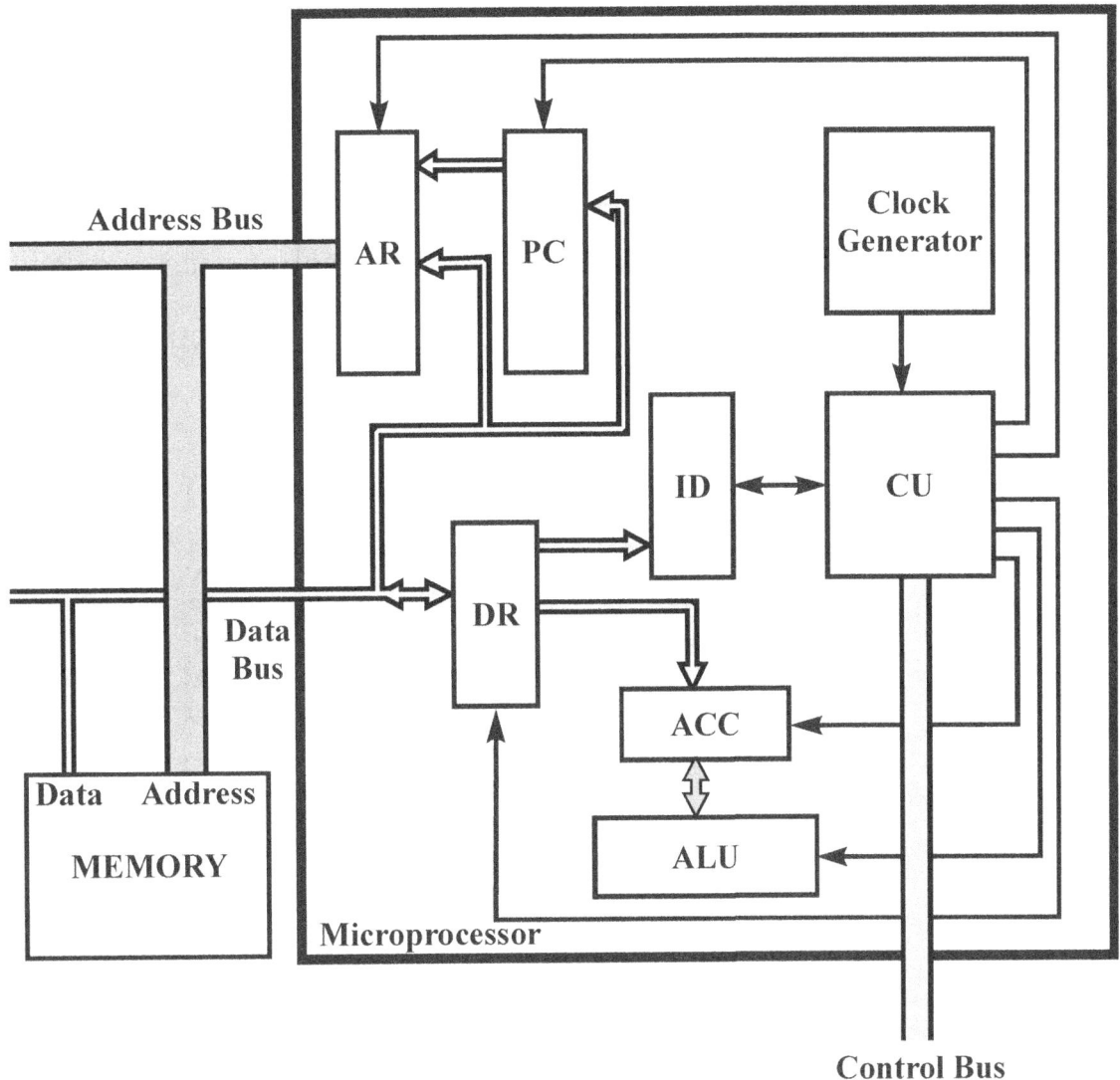

AR = Address Register
PC = Programme Counter
DR = Data Register
ID = Instruction Decoder
ACC = Accumulator
ALU = Arithmetic Logic Unit

Fig. 13.3-2 An Internal Structure of a Typical Microprocessor Chip.

The Accumulator stores the result of anything performed in the ALU and can notify the outside world via the Status Register of certain conditions which may have occurred such as the generation of any borrow or carry bits or an overload. To keep track of where we are in the programme we have a Programme Counter. The programme counter will always contain the address of the next instruction in memory, but the Address register will contain the address of the next memory location to be accessed whether it be data or instruction.

13.4 INTERRUPT ROUTINES

An interrupt routine occurs when some external device causes the microprocessor to suspend the main programme. This outside request typically comes from some peripheral device such as a VDN display which requires updating.

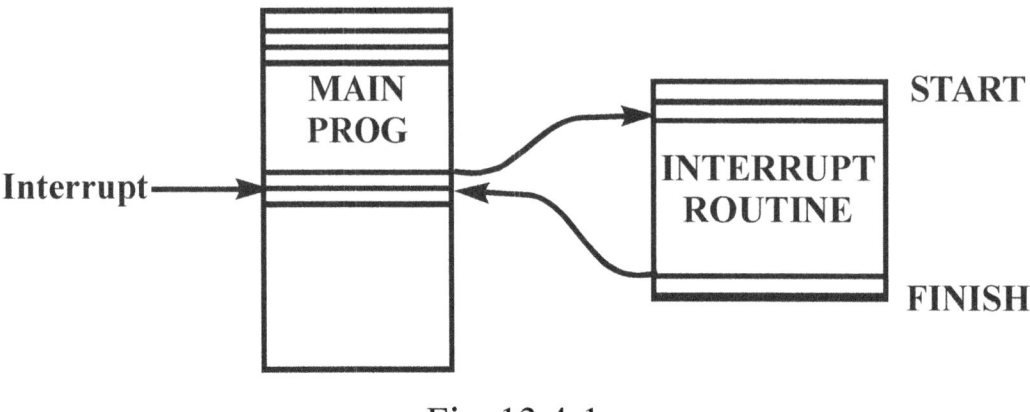

Fig. 13.4-1

When the CPU acknowledges this request it will suspend the main programme and store the contents of the registers somewhere in RAM. Any such store of data from the CPU which is likely to be recalled is known as the STACK, and within the CPU there will be some form of Stack Pointer register which will contain the address of the next available location in this section of memory. The CPU will then call up the appropriate programme from RAM and ROM and will complete this programme before resuming where it left off in the main programme. See fig.13.4.1

In some applications a number of different peripherals may be connected to the same microprocessor. In this case a priority list must be made up so that the CPU will know, if presented with two or more simultaneous request, what one should be serviced first, and in what order the remainder should be services.

13.5 MICROPROCESSOR PROGRAMMING

Every microprocessor has a unique repertoire or list of instructions which the manufacture has designed to understand and obey. This list is known as the "Instruction Set". The size of the instruction set of an MPU is not necessarily a valid guide as the versatility of the device, as some instructions, particularly address instructions, can be written using different modes of addressing, and this can tend to inflate the size of the instruction set without actually increasing the number of commands on which it can operate.

Before proceeding any further it would be useful to consider some of the different methods of memory addressing used. Remember that being able to address the memory in order to fetch or store data is fundamental to the operation of the microprocessor and as programmes increase in complexity so there is a need for different ways of addressing the memory for convenience and flexibility.

Let us now look at 4 of the different mode of addressing used by the Motorola 6800 microprocessor.

a. Inherent
b. Immediate
c. Relative
d. Indexed

a)　INHERENT ADDRESSING: In some cases the operator itself specifies where the operand is contained or where it should be stored. For example the instruction ABA (Add Accumulator B to Accumulator A) obviously states that the two operands involved will be found in the two accumulators and that the result should be stored in accumulator A. Within the 6800 instruction set there are 25 instructions which use inherent addressing.

b)　IMMEDIATE ADDRESSING: With the immediate mode of addressing the actual operand to be handled will occur in the second byte of the source statement, the first byte being the operation code, e.g. the instruction 860A would be decoded as Load ACC A (86) with the number 10 (0A). Note that for some instructions (CPX, LDS, LDX) a third byte of information is required.

c)　RELATIVE ADDRESSING: If this mode of addressing is used the second byte of the instruction will not contain the operand to be used but will effectively tell the microprocessor at what address in memory it may be found. Note that this form of addressing is only used by the "Branch" instructions. The relationship between the relative address and the absolute address is given by the formula: $A = (PC + 2) + R$ where A is the actual address of the destination of the branch instruction, PC is the actual address of the first byte of the branch instruction, and R is the 8 bit two's complement binary number stored in the second byte of the instruction.

d)　INDEXED ADDRESSING: With this mode of addressing the actual address of the operand is found by adding the contents of the index register to a number M which is contained as an 8 bit binary number in

BASIC CONCEPT OF MICROPROCESSORS

the second byte of the instruction. For example A607 would be decoded as "Load accumulator A" (A6) with the operand which will be found at the address given by whatever number is in the index register, plus 7.

Let us now look at a simple programme designed to add together two numbers held at two specific address in memory and to store the result at a third memory location.

1. After initialising the programme counter to tell the machine where to start, the contents of the PC will be put out onto the address bus. The programme counter will now be incremented by 1 (PC) + 1 → (PC). From the relevant memory location will come back an 8 bit word which will be routed to the data register and from there into the instruction de-coder which will decode it as a "Load" instruction. This will now set the internal logic to load into the accumulator the relevant data. But what is the relevant data? In order to find out,

2. The new contents of the (PC) will now be put onto the address bus. The (PC) will again be incremented by 1. if we assume that we are using immediate addressing then the 8 bit number which comes back from memory will be the operand that we which to load. When it arrives at the data register, the internal logic which had been set by the previous byte will route it into the accumulator.

3. The current contents of the PC will again be put out onto the address bus via the address register and will again be incremented in preparation for the next step. As before the byte which is selected from the memory will go to the data register and hence to the instruction decoder. This new byte will be decoded as an "Add" instruction.

4. The process will be repeated as in step (2) only this time the data will be added to that already in the accumulator.

5. When the next instruction is received it will be decoded as a "Store" which means that the contents of the accumulator will be put back out into memory.

6. The second byte of the store instruction, as we are using immediate addressing, will contain the address in memory where we wish to store the result. On receipt of this information the internal logic which was set by the "Store" instruction will transfer whatever is in the accumulator out on to the data bus and put it into memory at the location specified by the output of the address register.

7. The PC will now again address the memory and will go on doing so until it is told to stop. Thus the next instruction in the programme will be a "Halt" instruction, which will disconnect the clock and so halt the progress of the microprocessor.

Although the example given is a very trivial one, it demonstrates how a microprocessor handles data according to a given set of instruction. The programme used can be simple or complex but they are all handles in the same way, one instruction at a time. Microprocessors have made their presence felt in numerous and various application, and these range from domestic washing machines to VTR editors.

CHAPTER FOURTEEN

14.0 DIGITAL MEMORIES
14.1 INTRODUCTION:

A memory is a device for storing information. Many types of devices have existed in the past for storing information but it is within the last 50 years the need for storing binary electrical information arose.

Of the many devices developed for the above purpose the better known ones are the magnetic devices such as drums, discs, tapes and magnetic core. Other mechanical devices which originated earlier to these are the punched paper tape, punched cards etc. These are still in use.

The magnetic devices are extensively used now. The field of integrated circuit electronic memories are a later development and is now more popular than the magnetic core memories.

14.2 THE MAGNETIC CORE MEMORY

The storage medium of the magnetic core memory is a ring shaped piece of ferrite core, indicating its state (i.e.,/ logic state) by its permanent magnetism. One polarity represents logic '1' and reverse polarity represents logic '0'. The ferrite material used in the core has a square hysterisisloop as shown in fig. 14.2.1. This has a high remnant flux and requires a large current (= I) to change the magnetization. As can be seen from the figure a current equal to half this value (i.e., = ½) will have no effect on the magnetic core. The magnetic cores are arranged in a matrix form as shown in fig. 14.2.2. Each core stores one binary digit. The selection, reading and writing is done by the coincidence system of 4 wires. As shown in fig 14.2.2 each of the cores has one 'X' wire and 'Y' wire passing through it. Now if we want to write a logic '1' into a particular core, a current I in the correct direction is to be passed through the core. This is done by passing half the current through the X wire of the core, and half the current through the 'Y' wire; thus the full current will pass through only that core which is at the intersection of these wires. All the other cores on these wires will get only a small current and cause little disturbance to the remnant magnetism.

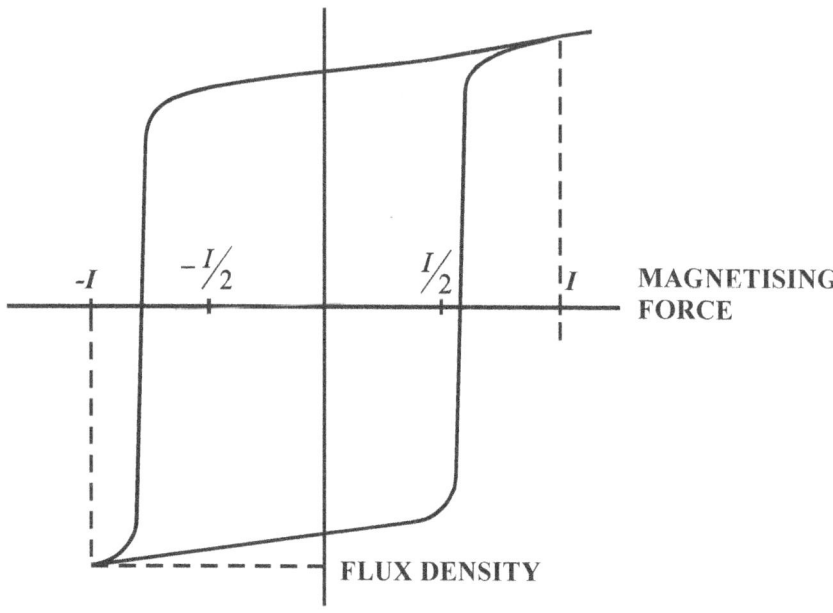

Fig. 14.2.1 Hyserisis Curve

When several matrix planes are operated simultaneously in a stack it is more economical to drive all the corresponding X and Y wires of each plane from the same drive circuits. To prevent all the cores in one location from being switched to the same state, an inhibit wire is threaded through every core on each plane.

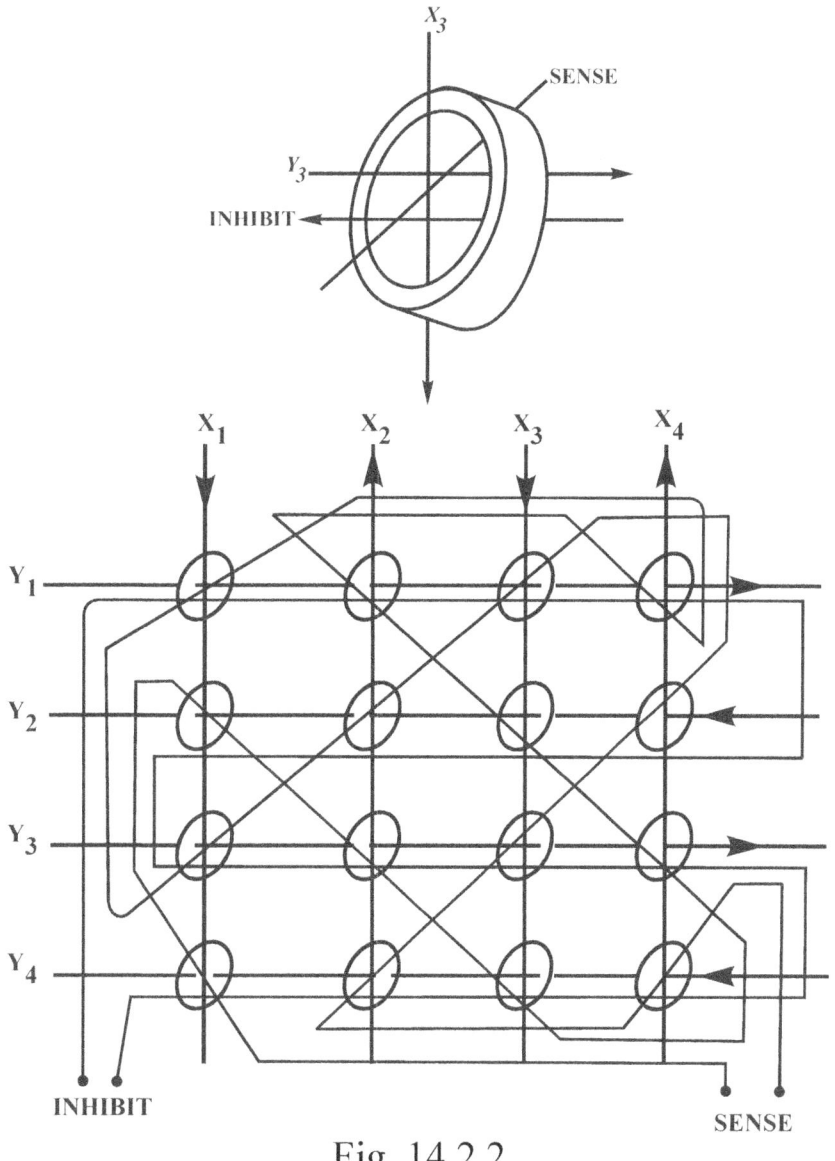

Fig. 14.2.2

The inhibit wire (one wire for each plane) will carry a current ½ and flows opposite to the normal 'Y' wire current. Hence whenever inhibit wire carries a current, that plane will not be affected by the X and Y currents. Thus the plane into which the "writing" is done is not given the inhibit current. In order to retrieve the stored data (i.e., t read) a fourth wire is included in the matrix plane, called the sense wire. This is also threaded through all the cores in fig. 14.2.2.

Reading from the memory is achieved by pulsing the appropriate X-Y wires with half current pulses and using the sense wire to detect whether the core changes state or not. If the read pulses are such that they will drive the core to the logic 0 state, then an output will only be produced by the sense wire when the original state of the core was logic 1. The sense is threaded in such a way as to eliminate the effects from the adjacent elements receiving the same half- current pulses.

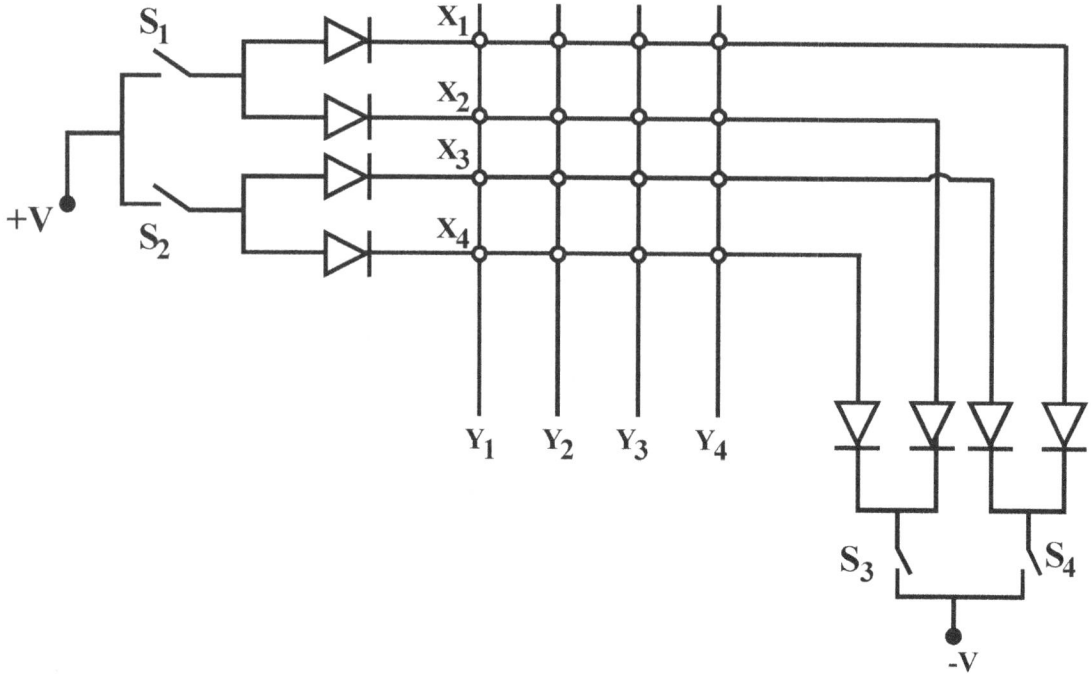

Fig. 14.2.3 Diode Selection Matrix

After the read process has occurred in all the cores in the stack, a particular location may be in the same logic state and so all those where a change occurred during the read process must have their state reversed. This is achieved by immediately following the read pulses by write pulses in the opposite direction and gating the inhibit wire pulse with the amplified output from the sense wire. The overall read-write process controls the cycle time of the memory.

DIGITAL MEMORIES

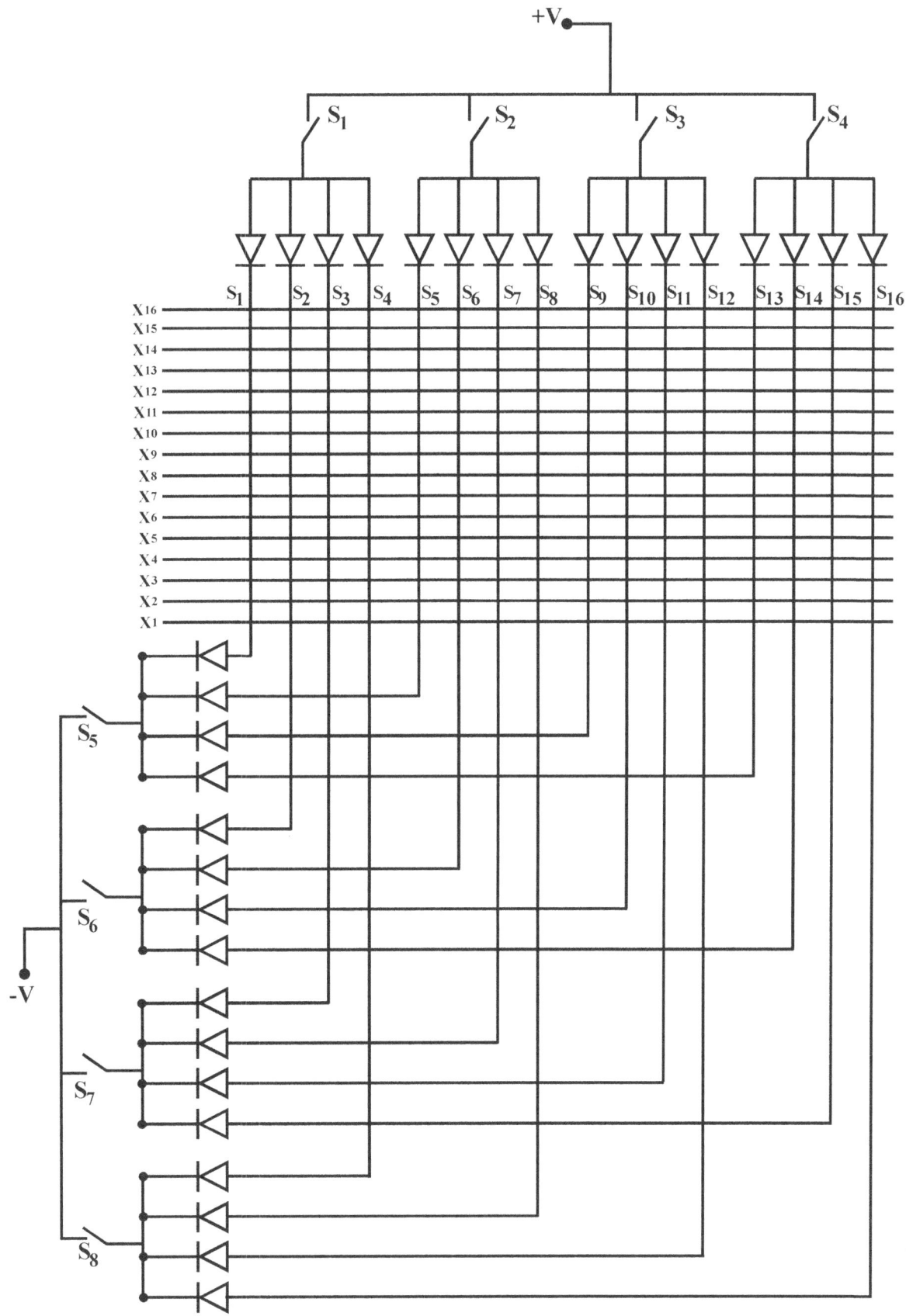

Fig 14.2.4 Diode Selection Matrix ($16 \times 16 = 2^8$)

DIGITAL MEMORIES

The various electromechanical systems have a wide of applications and are particularly valuable as they are all non-volatile systems. Modern technology has now produced semiconductor memories which will largely replace the ferrite core store in many of its applications.

The above methods help to reduce the number of leads which have to go to a memory system. A plane may have as many as 4096 calls ($64 \times 64 = 2^{12}$). Yet another method to reduce the leads is by using a diode selection matrix.

Fig. 14.2.3 shows a case where a 4×4 plane is controlled by four switches. If we want to select the line at X_3 switches S2 and S4 are to be closed. Then the current flows from +V through to S2 and X_3. Fig. 14.2.4 shows the circuit for a diode selection matrix ($16 \times 16 = 2^8$).

14.3 SEMI-CONDUCTOR MEMORIES

There are two basic of semi-conductor memories- ROM and the RAM.

ROM stands for READ ONLY MEMORIES.

Its name implies that the memory is pre-programmed-or data is already written in the memory and it cannot be changed once it is written.

There are two variations of the ROM which are of interest: the programmable read-only (PROM) and the electrically alterable read-only memory (EAROM).

In the PROM the matrix is fully wired with all the data links in place. These links are made from a fusible material and the PROM can be programmed using a special jig to fuse the links which are not required.

DIGITAL MEMORIES

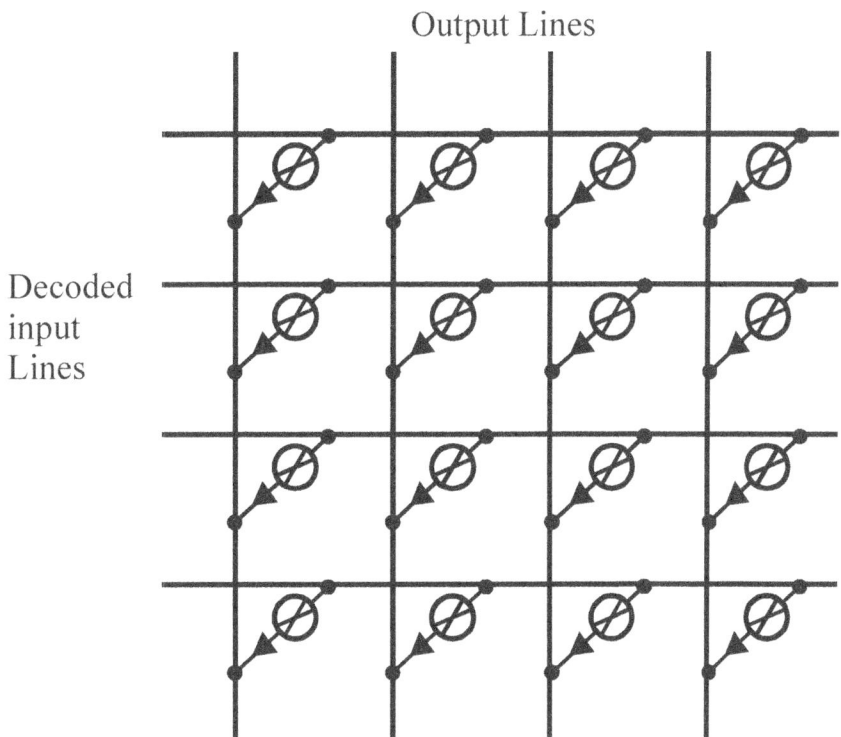

Fig. 14.3.1 EAROM structure.

The EAROM is a PROM in which the programming diodes have a series resistor as shown in fig. 14.3.1. The resistor consists of a glassy semiconductor which has both amorphous and polycrystalline phases. In its disordered amorphous phase it has a high coefficient of resistance (5×10^4 ohms cm) whereas the polycrystalline phase has a low resistively (0.3 ohms cm). The resistors thereafter are electrically alterable resistors that will remain in either resistance state indefinitely.

A typical structure of a ROM is shown in fig. 14.3.2. After the memory is programmed each of the calls will have either a 0 or a 1. The logic '1' is represented in the figure by a black '0' at the crossing point of lines. The input lines are decoded using gates. The figure 14.3.2 shows a memory capable of storing 8 words each of 8 digits.

The address is a 3 digit binary word. For example the address 101 (i.e., A = 1 B = 0 C = 1) gives the 3rd word from top which is 0 0 1 0 1 1 0 0 as the output. Similarly, the address 0 1 0 gives an output word 1 0 0 0 0 1 0 0. Thus the words are individually activated through the input gates which fully decode the appropriate binary addresses.

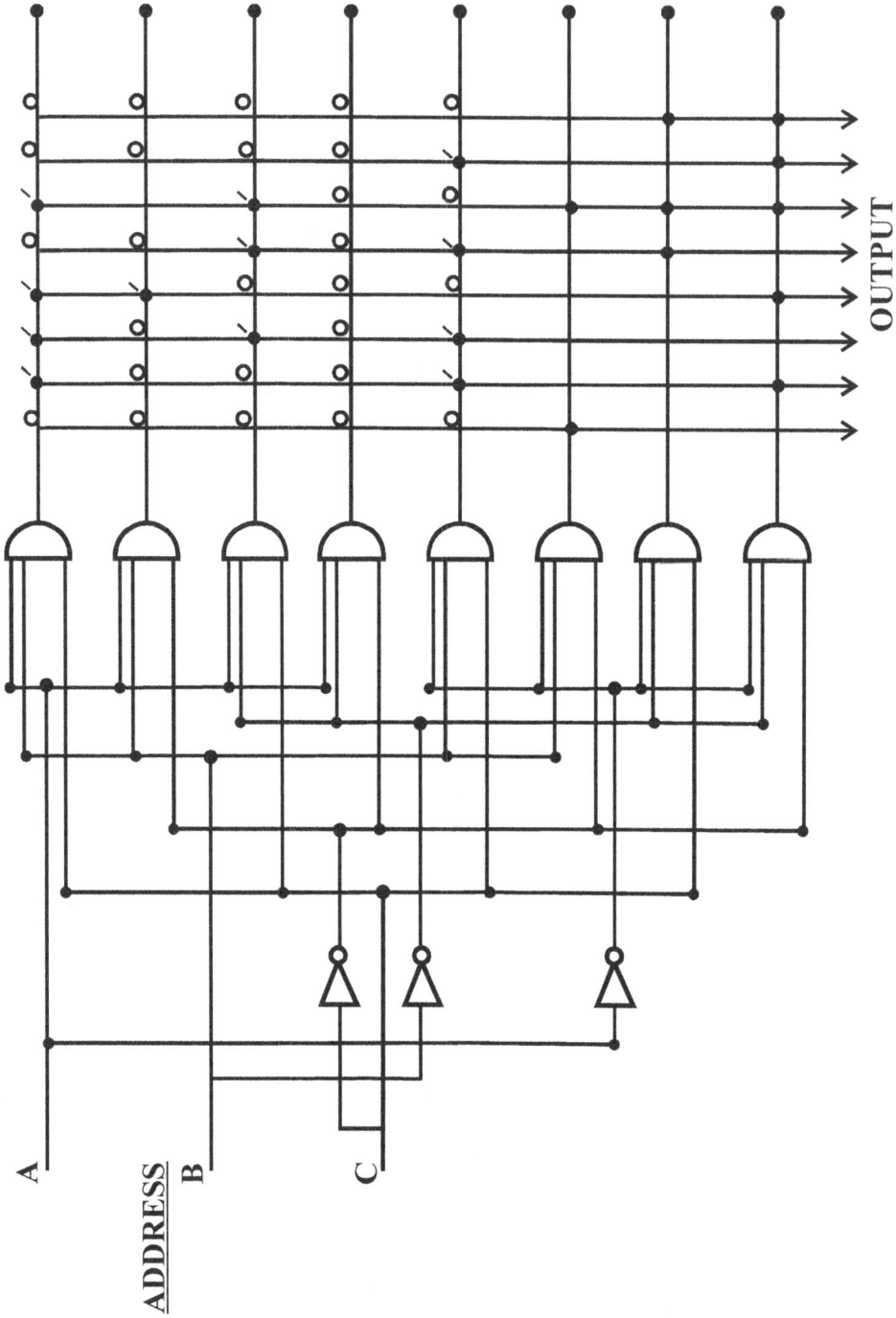

Fig. 14.3.2 ROM Structure

14.4 RAM (Random Access Memories)

The memory consists of a large number of cells each capable of storing one binary digit. There are two types of memory cells used- the static cells and dynamic cells.

The static cell consists of a bistable which has already been discussed earlier. This is normally a TTL device which has two stable conditions. The bistable can be set to any one of the two states and will remain in that condition indefinitely. Thus it can be used as a memory for storing binary data.

Fig. 14.4.1 shows a 4 transistor MOS cell

Fig. 14.4.2 gives circuit for two transistor TTL cell.

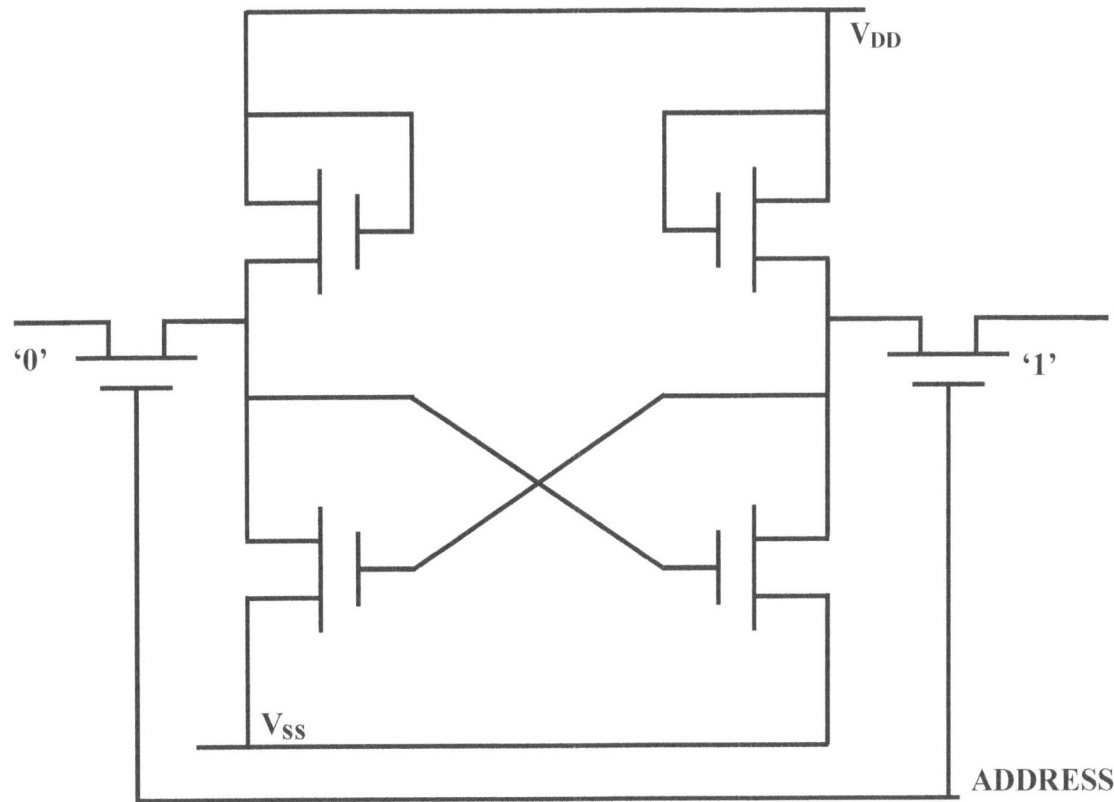

Fig. 14.4.1 Four transistor MOS cell.

In all the bistables one transistor will be conducting, In the ECL cells since the transistors are non-saturating, they are very fast, i.e.: the access time is very low. The TTL cells also can be made fast by using schottky diodes to clamp the collector base junctions out of saturation.

In the case of MOS cells 'N' channel devices have improved access time. (50 n sec. and less). The MOS cells need a different power supply and interfacing circuits for coupling to TTL devices. By using 'N' channel MOS, TTL compatible RAMS have been made.

In MOS memory cells (static) the load resistor is replaced by another MOS transistor with its gate connected to the drain to form a small area non linear resistor. This reduces the chip size. In CMOS techniques the load resistor is replaced by a complementary transistor since only one transistor of a complementary pair conduct at a time, the only current that flows is the leakage current through the top load resistors.

In non-complementary circuits power is continuously dissipated in the load. Thus the CMOS static cells have very low power consumption but in practice the transistors conduct together for an instant when the transistors are changing state. This causes a power pulse. These cells have a slow access time (500 n sec) because all the parasitic capacitances that exist between gate and substrate and in the reverse biased PN junctions have to be charge and discharged.

In the case of ordinary static cells the read-out is destructive because the sensing is done by forcing the bistable to a known state (say '1') and then sensing the change if any, in the bistable (a change indicates that the bistable was in '0' state before "reading")

Now we shall examine a static cell with a non-destructive read-out:

DIGITAL MEMORIES

The store element is a bistable. The circuit of the store element is shown fig 14.4.3. The bistable has got output gating transistors. This cell has two output lines-one gives output when the stored data is '1' and the other-gives the output if the stored data is 'O'. These output lines are called the digit lines. These lines are held at + 1.0V, and the X and Y lines (i.e., the address lines) held near zero volts (normally it will be about 0.3t)

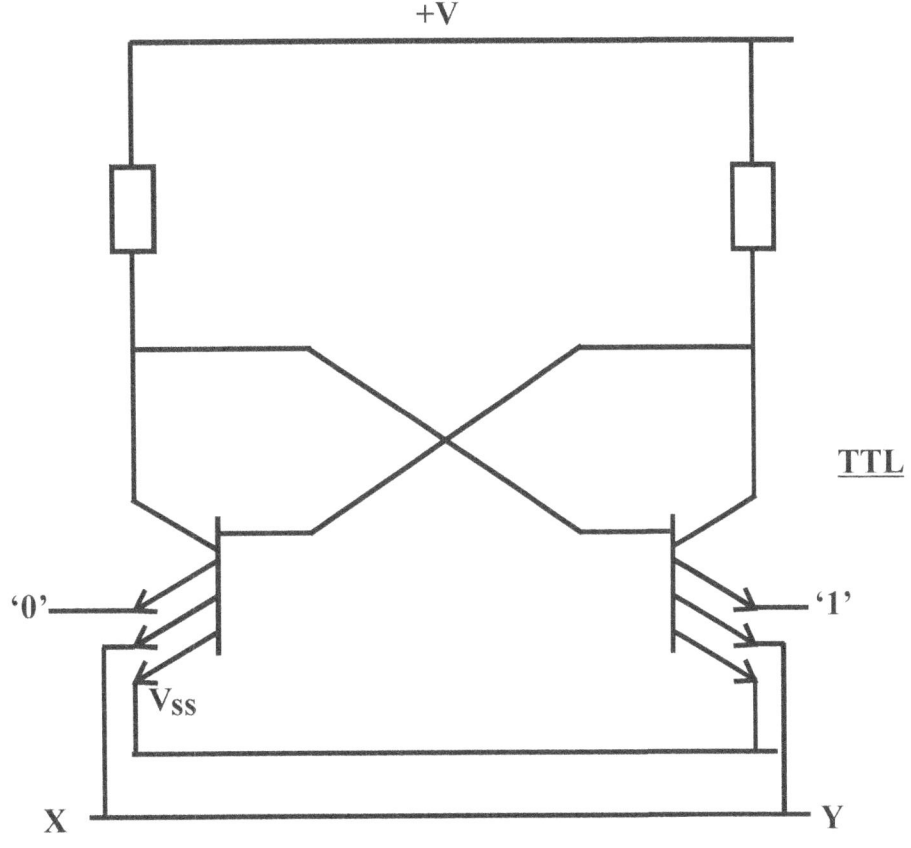

Fig. 14.4.2 Two Transistor TTL Cell

The double emitter transistors T2 and T3 form a bistable. The values shown in the diagram are for the state when the cell is not selected (i.e., X and Y lines are carrying nearly zero voltage). In the bistable one will be conducting (saturated) (and so low collector voltage of about + 0.6V) and the other transistor will be not conducting, (biased off). The total

DIGITAL MEMORIES

emitter current of the conducting transistor (say T2) is the sum of the collector and base currents which are nearly equal since all resistors have approximately the same value. The two emitters share the emitter current equally.

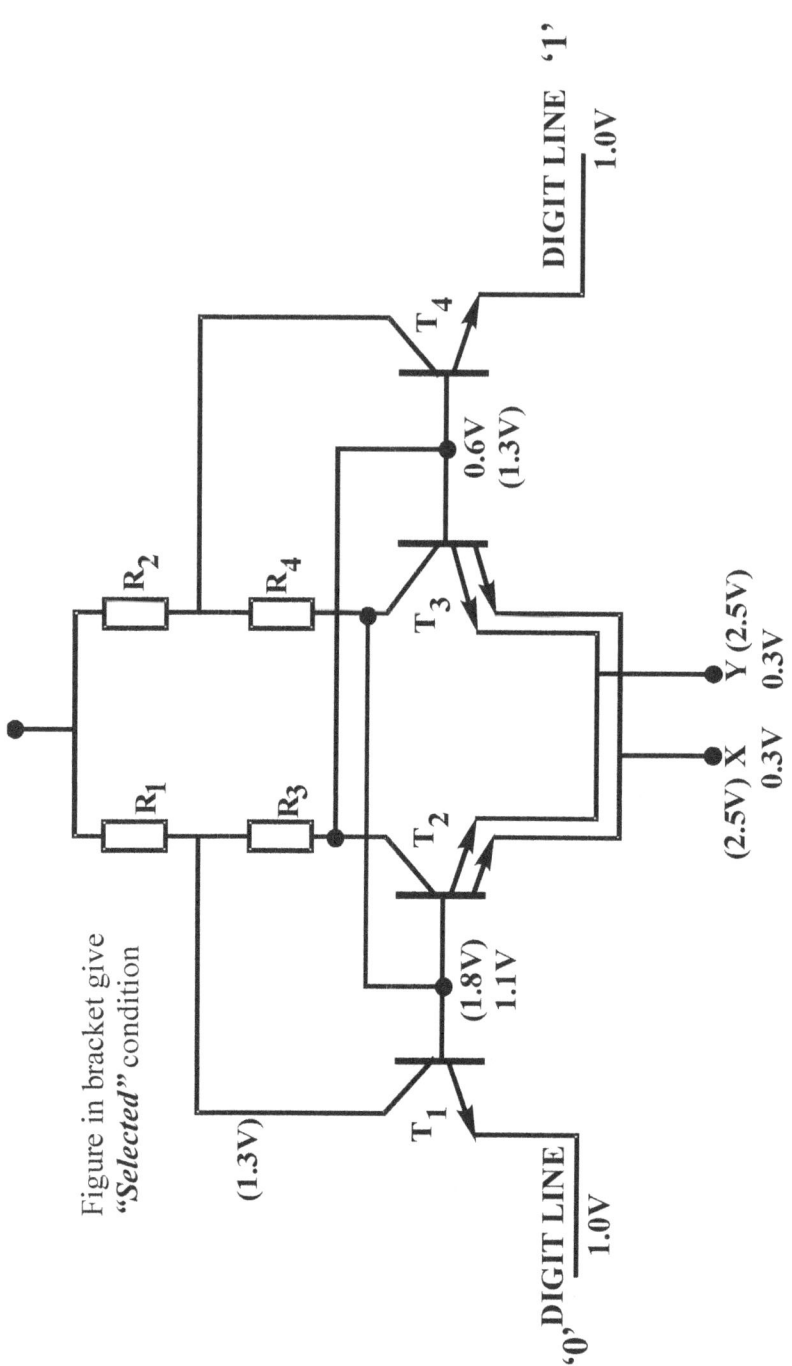

Fig. 14.4.3 TTL Static Cell (non-Destructive Readout)

DIGITAL MEMORIES

When one of the address lines is taken to +2.5V, total emitter current flows through the other emitter and address line and the internal voltages remain the same. Assuming T2 to be conducting and with one or both the address lines at 0.3V the base of T2 is approximately + 1.1V and so T1 has its base-emitter function forward biased by 0.1V. This is insufficient to cause any emitter current to flow (in T1). The other outer transistor T4 has its base at approximately 0.6V giving a reverse biased base-emitter junction and hence no emitter current.

When both the address lines X and Y are given + 2.5V (is. when the cell is selected) the base and collector voltages of T2 also move positively until the base-emitter junction of TI conducts - and then TI collector takes over the T2 collector current. When the X and Y lines reach +2.5 the base-emitter junction of T2 is reverse biased and TI is fully conducting (TI base is now at 1.8V and collector is at + 1.3V).

T4 base-emitter is now 0.3V forward biased which is not sufficient for it to start conduction. So now we have a bistable formed out of TI and T4. The T1 emitter current is decided by R1,R3 and R4.

The road-out consists of determining which digit line is carrying the current when the element is selected. At the end of selection when both the address lines are back at 0.3V the base-emitter of T2 conducts and original condition is restored (i.e., T2 conducting T1, T3, 14 not conducting).

Writing into the store element is accomplished by raising (or pulsing) the appropriate digit line above one volt after selecting (a, X + Y line at 2.5V) the element. For example, the '0' digit line (going to T1) is pulsed, the collector voltage of T1 rises causing T4 to conduct and takes over the base current from T1, T4 saturates, and T1 remains cut off when the pulse in the digit line is removed. When the address lines return to normal (0.3V) T3 will take over from T4. Thus T1, T2 and T4 will be none-conducting and T3 will be conducting. This completes "writing".

14.5 THE DYNAMIC CELL

The dynamic cell makes use of the gate capacitance of a M.O.S.F.E.T The capacitance stores data as an electrical charge. (Fig.14.5.1 the principle.). The "now select" line turns on the transistor allowing the capacitor either to charge up from the data line or discharge into it, depending on whether the operation is READ or WRITE.

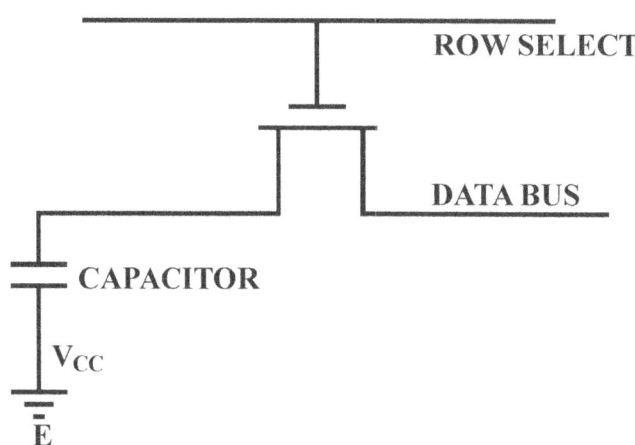

Fig. 14.5.1 Dynamic Transistor

This charge leaks away after some time and hence needs to be recharged periodically. Hence special "refreshing" circuits are provided to do this i.e., to charge each memory cell in the correct polarity periodically. Since no reading or writing can take place during refreshing period, the cycle time is increased and the operating speed reduced.

Fig. 14.5.2 shows a four transistor MOS cell . Fig. 14.5.3 shows a three transistor dynamic cell in which one transistor forms the cell.

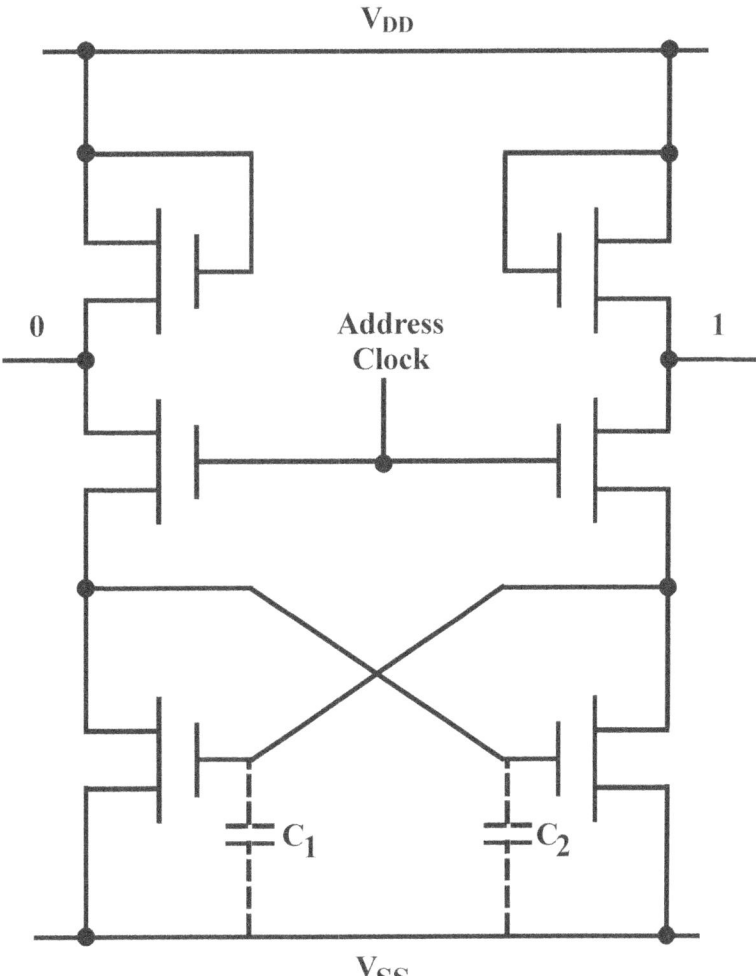

Fig. 14.5.2 Shows a 4 transistor MOSS cell

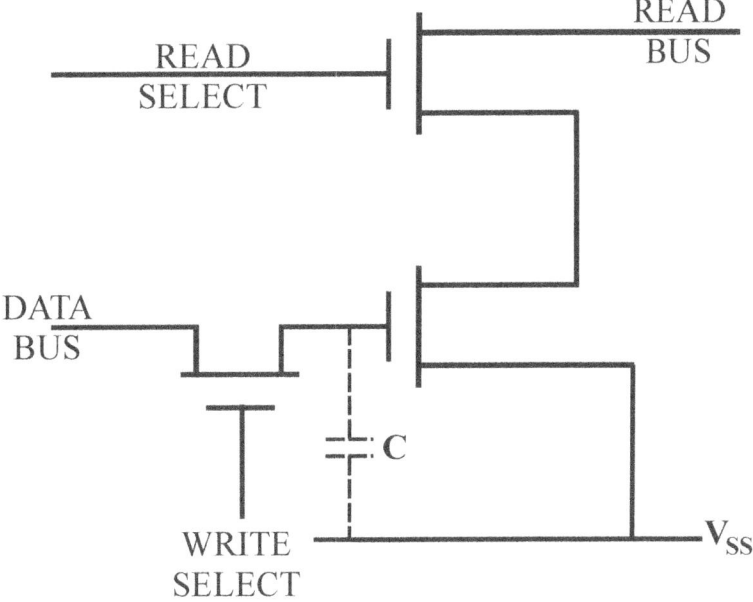

Fig. 14.5.3 Three Transistor Dynamic Cell

In fig. 14.5.4 the device uses resistors for load where most devices have another F.E.T. The row line turns on TR1 and TR4 allowing for the two column lines sense the state of the bistable-or during WRITE to change the state.

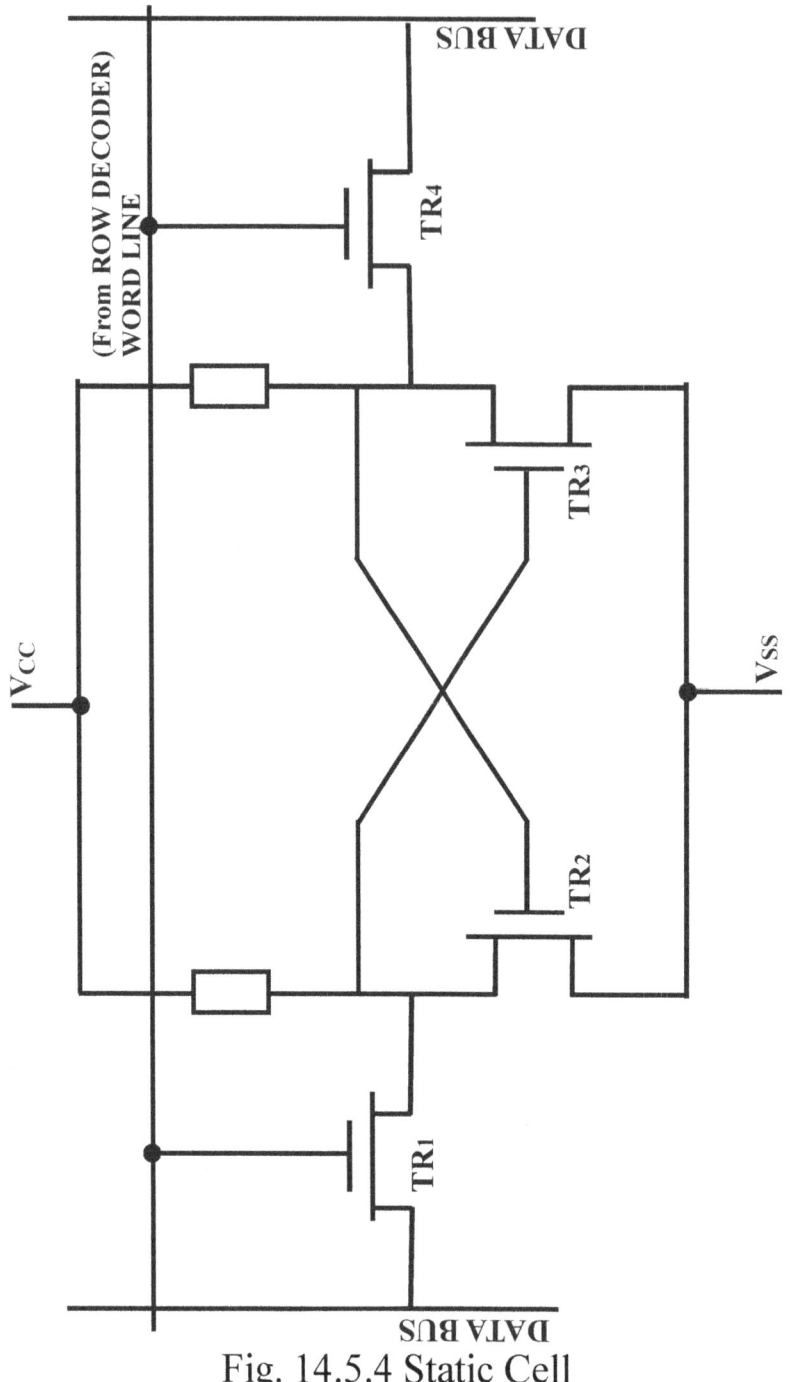

Fig. 14.5.4 Static Cell

14.6 ARRANGEMENT OF MEMORIES

Now we will briefly consider the arrangement of memories. The cells are arranged in a matrix form as in fig. 14.6.1

DIGITAL MEMORIES

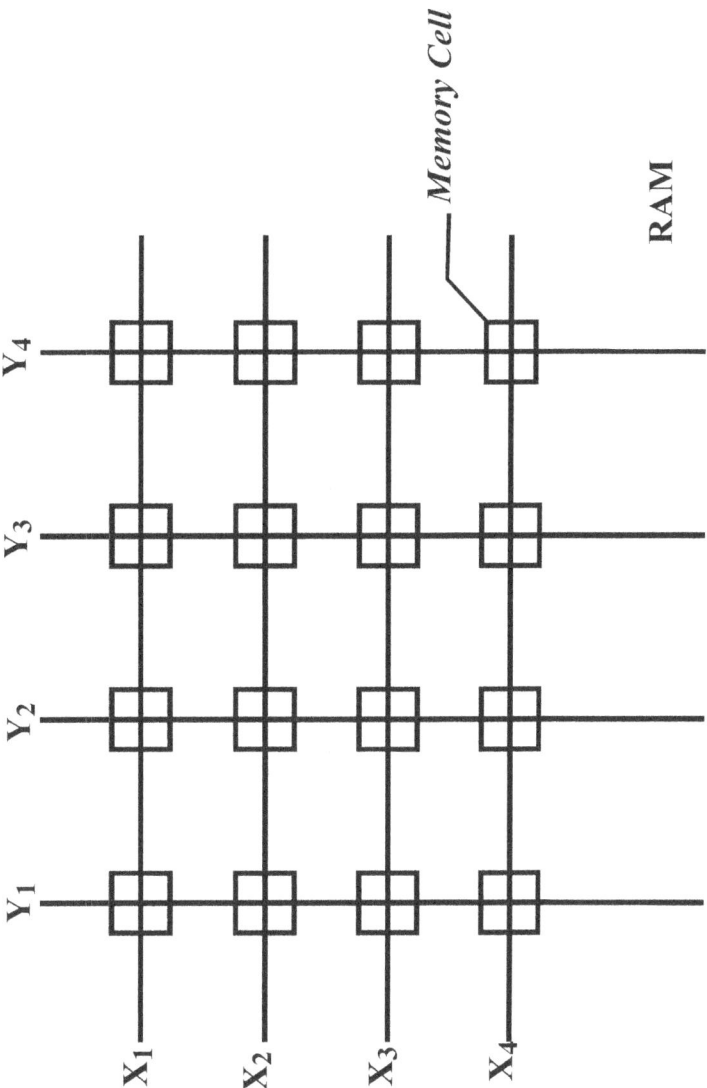

Fig. 14.6.1 RAM Matrix

The typical cell as we have already seen has an X and Y select line (address) and the data lines zero and 1. These have to be connected to the READ (or sense) amplifier and WRITE amplifier. A cell with the above connections are shown in fig. 14.6.2.

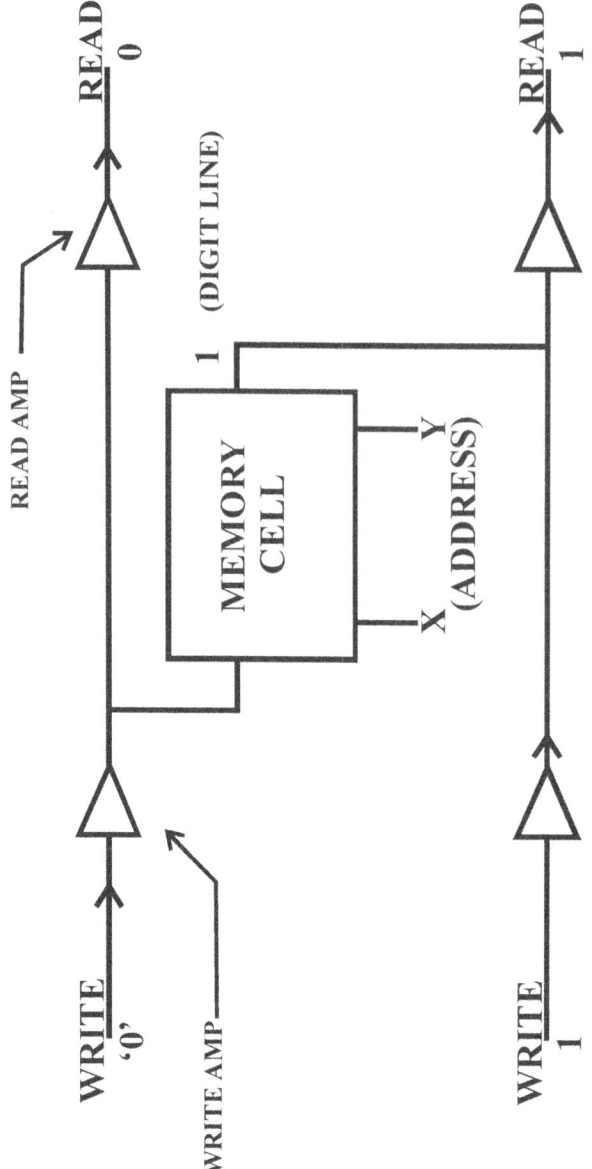

Fig. 14.6.2 Typical RAM Cell (with amplifiers)

Now if we substitute such cells in matrix plane as in fig. 13.5.3, we get the typical RAM layout as in fig. 14.6.3

DIGITAL MEMORIES

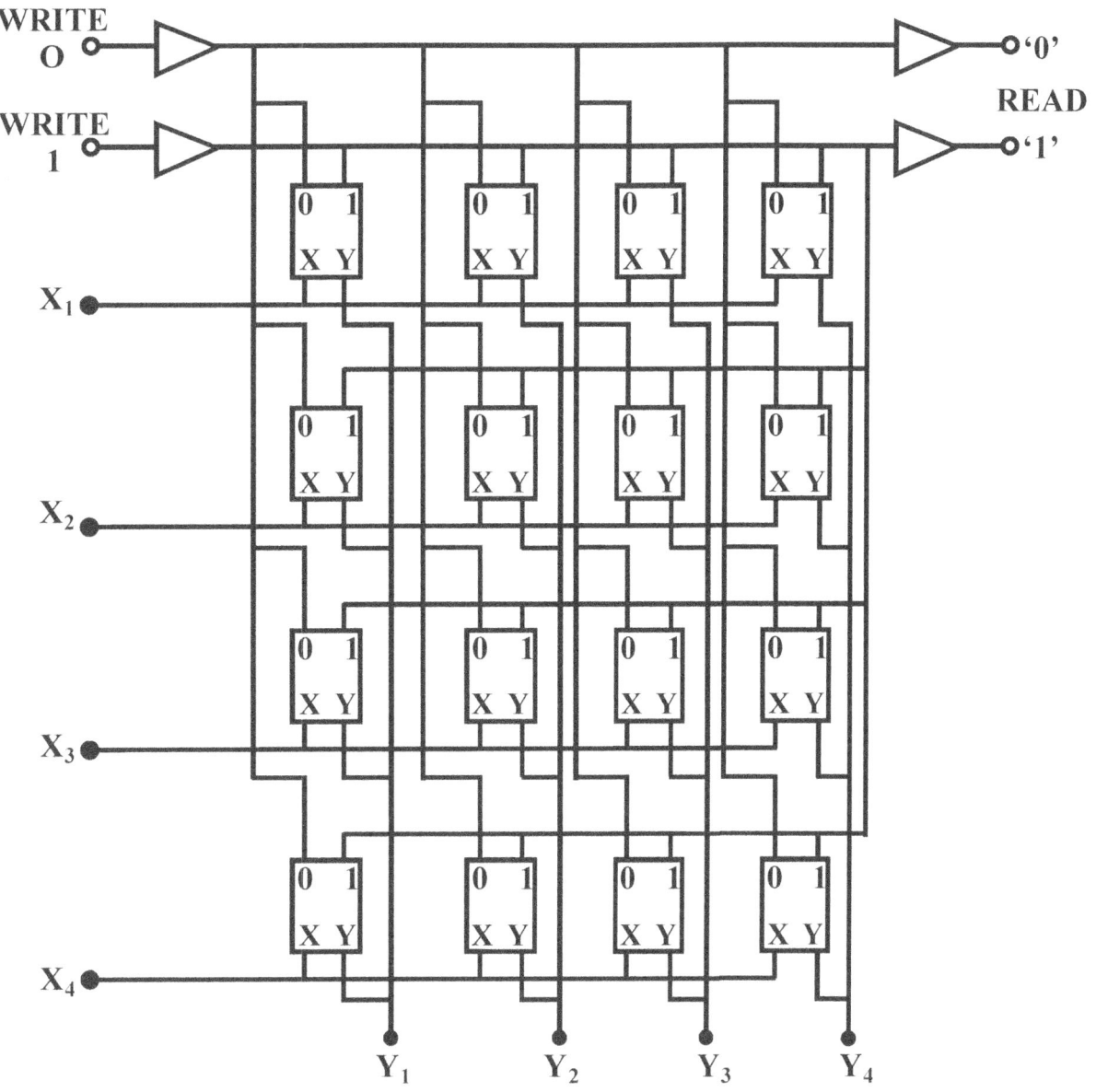

Fig. 14.6.3 RAM Matrix structure

All semi-conductor memories have the disadvantage of being volatile. Many techniques have been tried out to counter them. One method is to have a low power standby power supply. Another method is to have built in circuits in the chip which can sense a power failure and then transfer essential data to a battery operated part of the memory. Yet another method tried is to have a combination of RAM-ROM chip. On sensing a power failure about 1000 bit of data are transferred to an M.M.O.S FETS which can store it up to one year.

The field of electronic memories is fast growing both in terms of development and use. The newer concepts like CCD memories and magnetic bubble memories are being out and are likely to replace the present day memories.

APPENDIX

QUESTIONS

1. An amplifier have a gain of 500 and negligible phase shift at 500Hz. At 10KHz the gain has fallen to 350 but the phase shift is now 45^0. Feedback is applied to given an overall gain of 100 at 500Hz. Calculate the amount of feedback in dB and show what the effect is at 10KHz.

2. A common emitter amplifier with an unbypassed emitter resistors Re has a load of 5KΩ. Calculate the value of Re if the voltage gain is to be reduced by 14dB below. Its value in the absence of feedback due to Re, given that h_{ie} = 1KΩ, h_{re} = 2 × 10^{-4} h_{fe} = 70 h_{oe} = 50 μυ.

3. A tuned collector oscillator employs a transformer with a primary inductance of 5μH and resistance of 40Ω and the secondary is mutually coupled (closely) with M = 50μH. If the transistor has h_{ie} = 1KΩ with h_{re} and h_{oe} negligibly small calculate the minimum value of h_{fe} for oscillation and its frequency if the transformer is tuned by 50pf.

4. A Wein bridge oscillator uses two identical ganged resistor in conjunction with two equal fixed capacitors on four equal ranges varying from 200 to 2000Hz. If the minimum value of C is 100pF. Calculate the value of R and supplementary values of C for the other ranges.

5. A class A output stage is operated from a 6V supply with a transistor of 100mW collector dissipation to (i) A resistive load R_L

    (ii) A resistive load of 3Ω via a transformer.

    Calculate (a) maximum power output for each case

    (b) Quiescent operating conditions

(c) Value of R_L for (i) above and turns ratio in (ii) 50mW

6. A resistor loaded amplifier is to be used under maximum signal conditions only. Show how $P_{d.c}$ and P_o and P_{tr} vary with signal and calculate $P_{d.c}$ and P_o if $P_{tr} = 0.1$ watt.

7. A class B push pull amplifier uses transistors with a maximum dissipation of 2 watts. Calculate the maximum power output if the supply is 12V and the load is 5Ω. What is the required turns ratio?
If the input signal is halved after operating at its maximum value, calculate the new values of $P_{d.c}$, P_o and efficiency.

8. Repeat question 8 for class A conditions.

9. The circuit shows a CLASS A driver for a CLAA B stage whose approximate equivalent circuit is shown as 5Ω and diode. If $P_{tr(max)} = 500mW$ and $i_{c(max)} = 70mA$ and maximum C-E voltage is 40V, calculate (i) Turns ratio (ii) Quiescent operating conditions

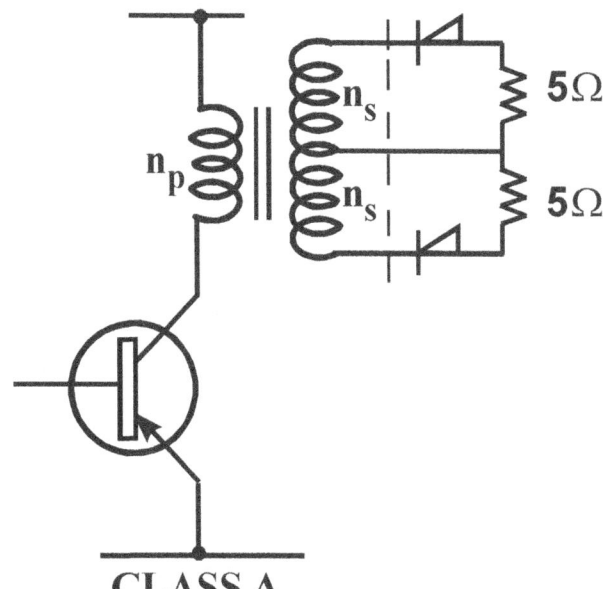

CLASS A

10. A single transistor is operated in CLASS B with a 1KΩ load. A moving coil meter in the collector circuit reads 10mA. How much sinusoidal signal power is delivered to the load?

11. Calculate the *h* parameters for the network.

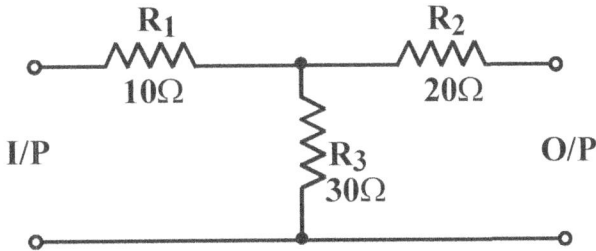

If the load on the network is 10Ω, calculate the input resistance, current gain, voltage gain and output resistance.

12. A transistor has the following **h** parameters $h_{ie} = 1.5K$, $h_{ie}=10^{-3}$ $h_{fe} = 50$ $h_{oe} = 50 \times 10^{-6} \mho$. Derive expressions and calculate the input resistance, output resistance, voltage gain of a common-emitter amplifying circuit using the above transistor and a load of 5KΩ

13. Two identical transistors are used in the circuit below.

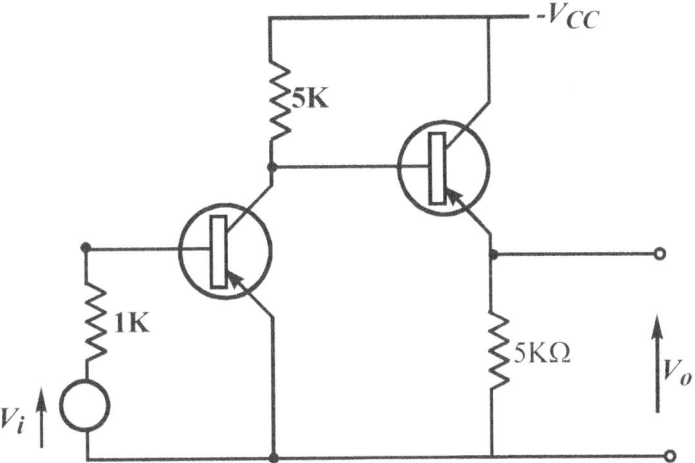

If the parameters are those given in question 12 calculate

   i. The overall voltage gain

   ii. Overall current gain

   iii. Input impedance

   iv. Output impedance

14. Design an emitter follower circuit to have an input resistance (in the absence of external load of 250KΩ with a transistor having small signal T parameters of $r_e = 10Ω$, $r_b = 500Ω$, $r_c = 750KΩ$ $\propto = 0.99$.

APPENDIX

If the collector-emitter voltage under the conditions of no a.c signal input is 5V and the collector current is 2.5mA calculate the required d.c supply voltage.

15. Calculate (i) midband current gain (ii) midband voltage gain (iii) lower cut off frequency (iv) gain-bandwidth product.

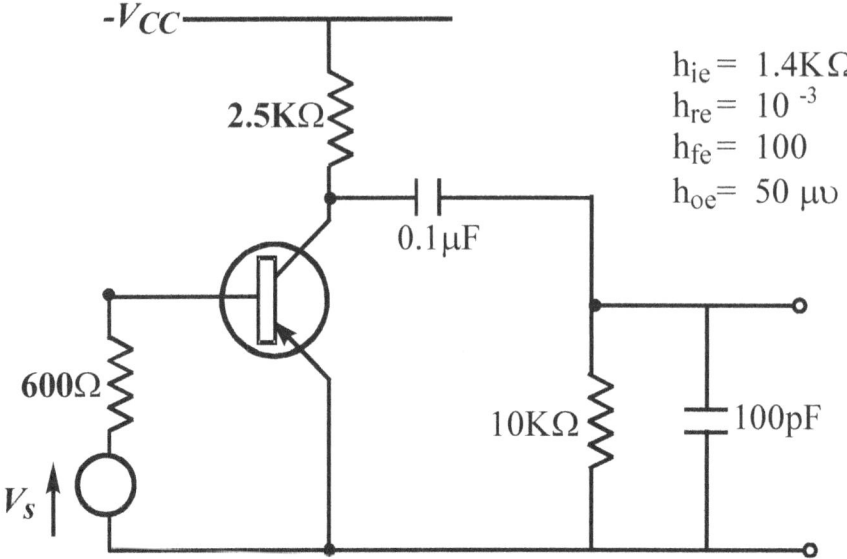

$h_{ie} = 1.4K\Omega$
$h_{re} = 10^{-3}$
$h_{fe} = 100$
$h_{oe} = 50\ \mu\mho$

16. Using reasonable approximations calculate the value of C if the lowest frequency to be amplified is 50Hz.

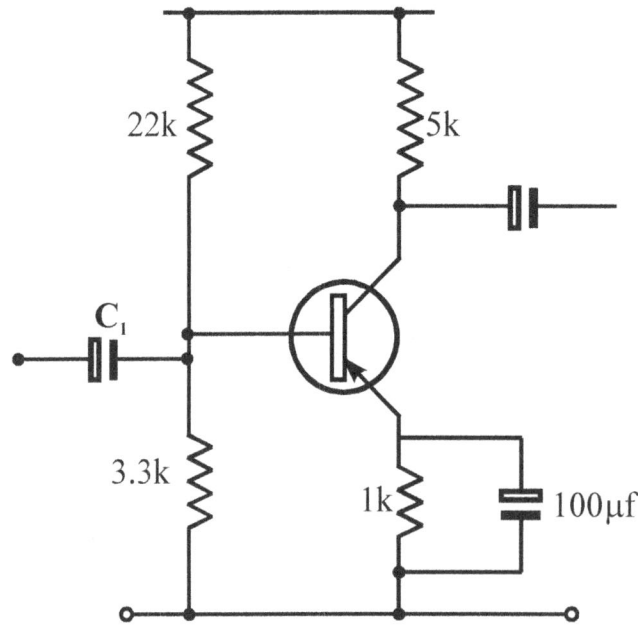

17. Calculate the input impedance of the above circuit at 500KHz if $C_{cb} = 2pF$, $C_{ce} = 5pF$, $C_{be} = 5pF$

APPENDIX

18. Derive expressions for the input resistance, output resistance, voltage gain of the following circuits.

    Does any configuration here particular advantages over the other two?

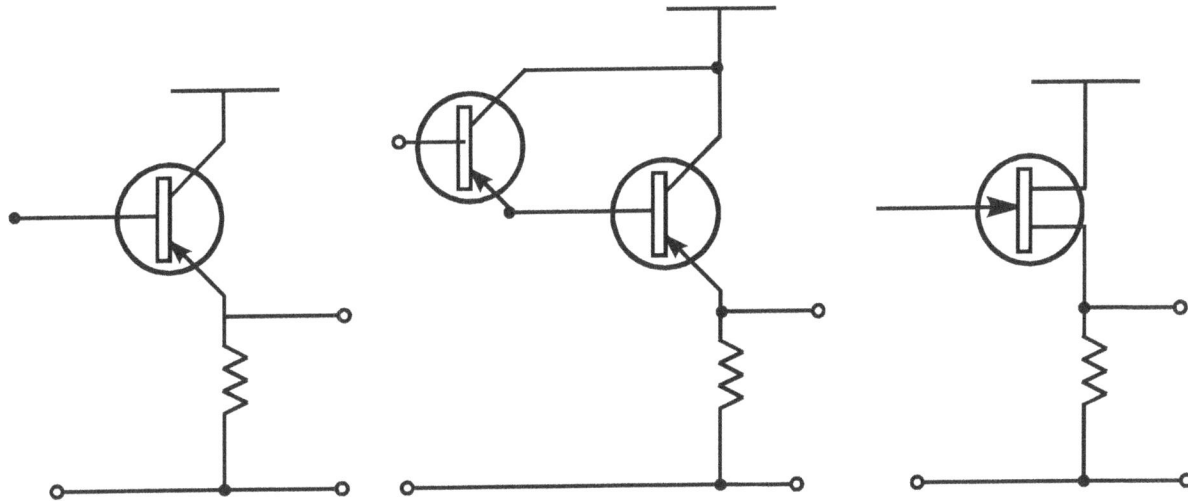

19. Define bandwidth. Calculate for circuit shown below.

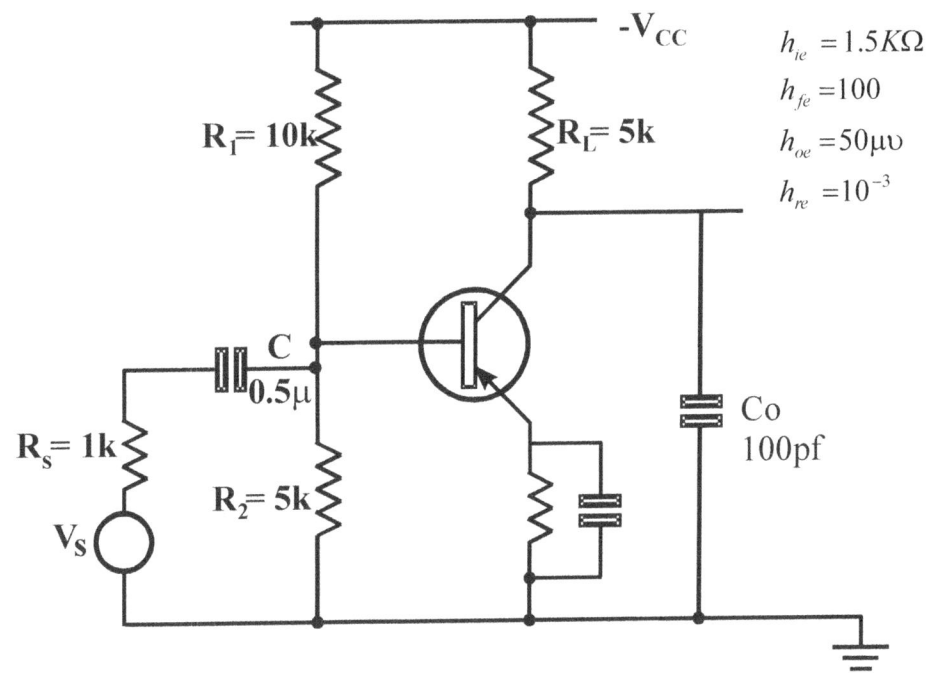

$h_{ie} = 1.5K\Omega$
$h_{fe} = 100$
$h_{oe} = 50\mu\upsilon$
$h_{re} = 10^{-3}$

   (i) lower break point frequency (ii) upper break point frequency

   (iii) midband volt gain.

20. Compare qualitatively common properties of emitter follower, source follower, emitter follower with Darlington pair.

APPENDIX

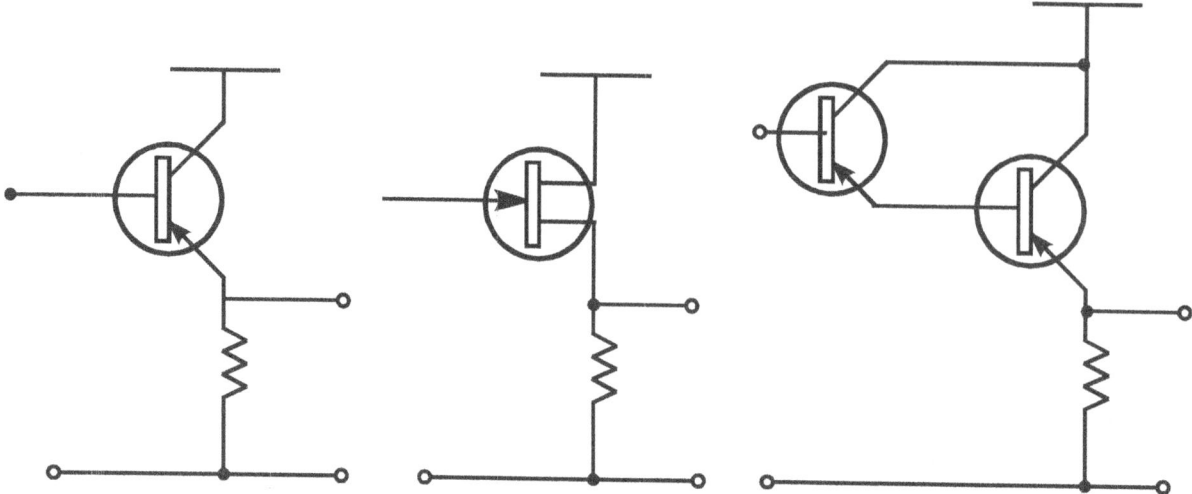

21. Phase shift oscillator employs two stage FET amplifier. Calculate

    (i) Transfer characteristics of phase shift network V_g/V_o

    (ii) Frequency of oscillation assuming network does not load amplifier.

    (iii) Smallest value of gain for sustained oscillation.

    (iv) If FET's are similar with $g_m = 1$ mA/V identical load resistors R_L, calculate value of R_L, calculate value of R_L to give minimum gain calculated above. Assume source resistors decoupled and R_g does not load FET1 with circuit having negligible reactance at frequency of oscillation

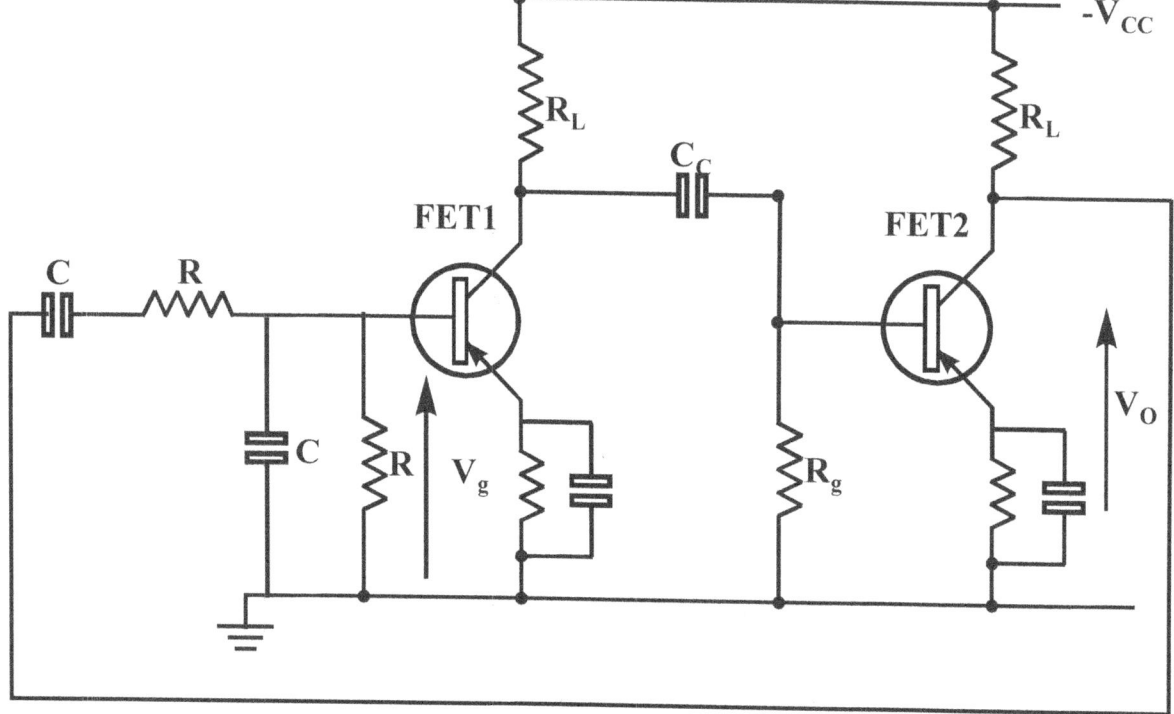

APPENDIX

22. Classify modes of operation for large signal (i.e., power) behaviour of transistor circuits for linear amplification. Compare collector circuit (or conversion) efficiency of CLASS A transistor amplifier driving a resistive load (i) directly (ii) via a transformer. Discuss relative merits or disadvantages of these configurations with other possible modes of operation.

23. Explain function of every component in figure with particular respect to its effect on (i) the d.c conditions (ii) the a.c or small signal behaviour (iii) both a.c and d.c conditions. What configurations are the transistors operating in and why are they couples as shown?

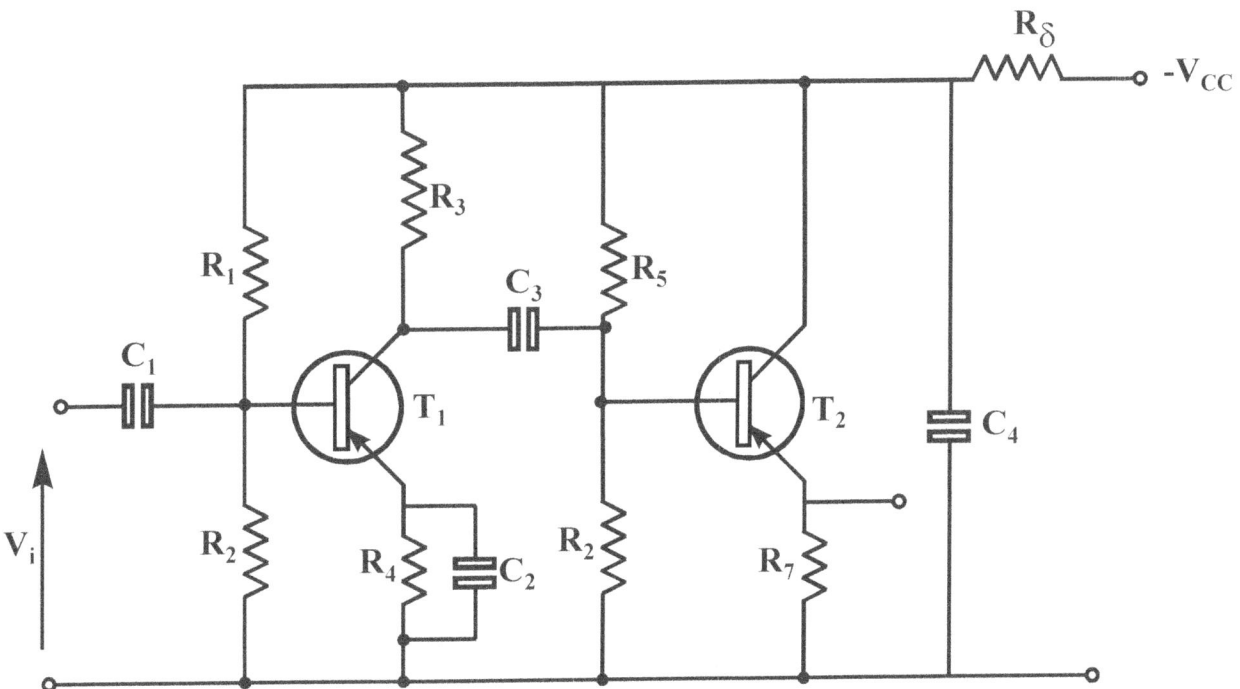

24. Derive expression for voltage gain of –ve forward biased amplifier stating assumptions. Amplifier has gain of 600, negligible phase shift at 400Hz and at 12KHz the gain has fallen to 300 with phase shift of -45^0. Feedback is applied to give an overall gain of 120 at 400Hz. Calculate amount of feedback and state what the effect is at 12KHz.

APPENDIX

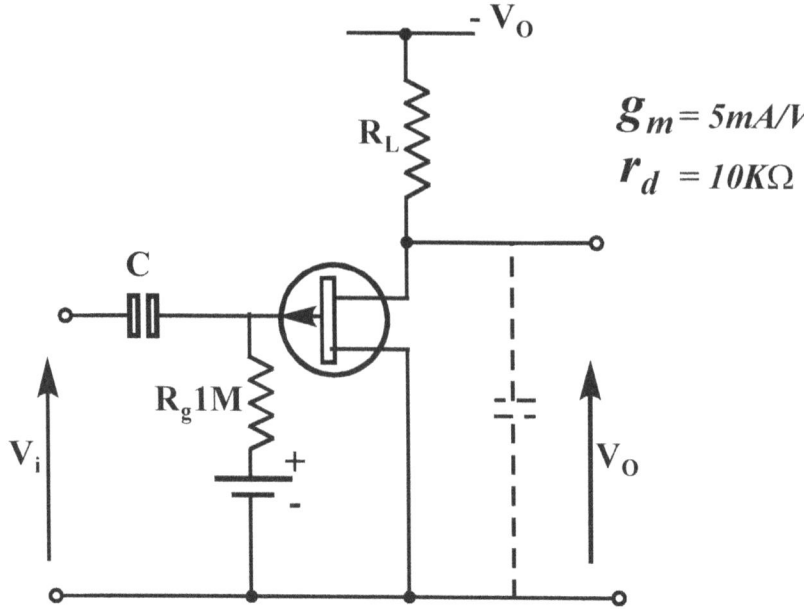

25. For common base amplifier draw low frequency equivalent circuit and calculate following at frequency of 1KHz:- voltage gain, current gain, input impedance if amplifier is feed from voltage source of 200Ω resistance.

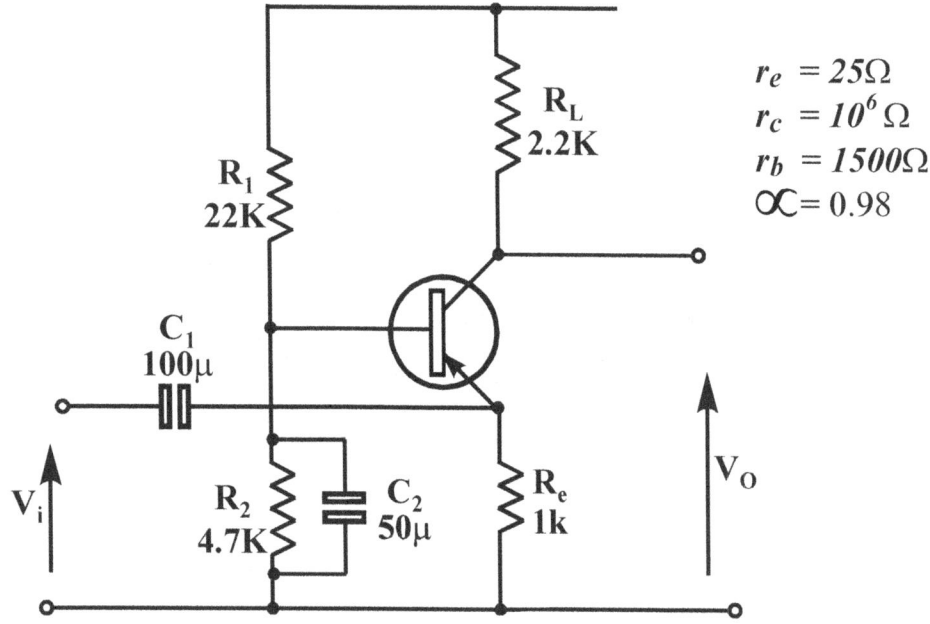

26. Using reasonable approximate calculate value of C if lowest frequency to be amplified is 50Hz. What value of Re connected between source and common line to reduce gain by, half at frequency where reactance

of C is negligible. What is its purpose and how can the reduction in gain be eliminated?

27. A PNP transistor is connected in common emitter configuration with 5KΩ load. Calculate current gain, input resistance and voltage gain of this simple amplifier, if

$h_{ie} = 1.5\text{K}\Omega \quad h_{re} = 10^{-3} \quad h_{fe} = 50 \quad h_{ie} = 50 \times 10^{-6}\text{υ}$.

28. Define Bandwidth and explain significance of gain bandwidth product in relation to RC couples amplifiers. Evaluate (i) bandwidth (ii) GB product of a CE amplifier with $R_L = 10\text{K}\Omega$ feed from source of 500Ω. The total stray capacitance, including that of transistor, between collector and common line is 20pF. $H_{ie} = 1.5\text{K}\Omega \quad h_{fe} = 50 \quad h_{ie}$ and h_{oe} negligible and effect of biasing may be ignored. If Bandwidth is to be increased to 2MHz. Calculate value of R_L and new value of midband gain.

i. Define Q factor explain what is meant by selectively a coil of 1mH and resist $10\sqrt{10}\,\Omega$ is tuned to resonance by 100pF. Calculated (i) resonant frequency (ii) Q factor for tuned circuit alone .

If is then used as collector load of tuned CE transistor amplifier. If effective trans conductance is $g_t = 1.5\text{mA/V}$ between amplifier input and output terminal in equivalent collector circuit. Shown calculate (iii) effective Q factor for tuned amplifier (iv) bandwidth of amplifier (iv) gain Vo/Vi of amplifier at resonance.

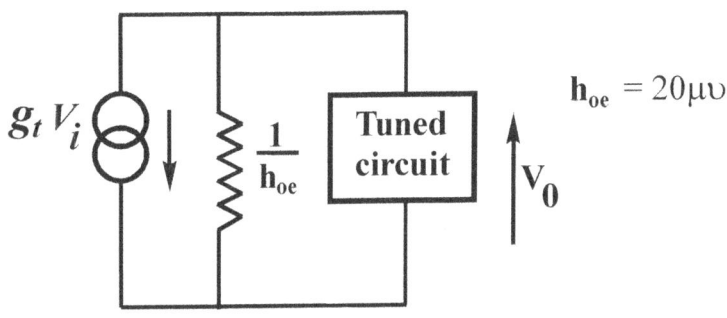

APPENDIX

30. Outline importance of –ve feedback in design of amplifier. Derive expression for volt gain of –ve feedback amplifier with amplifier gain *m* and feedback ratio β. Discuss validity of any assumptions made and how they may be approximately realised in practical amplifiers. Amplifier has gain (1000 + jo) at 800Hz and (300 – j 500) at 15KHz. If this is converter to –ve feedback amplifier by feeding back $\frac{1}{50}$th of output volt. Calculate gain at two frequencies stated. Flow is phase shift affected at 15KHz? What is % change at 800Hz if original gain reduced by 25%?

31. Discuss three methods of biasing applicable to CE amplifying stages. Without recourse to analysis point out the relative merits and defects of each configuration by considering the effect of temperature and change of transistor parameters on circuit behaviour.
State what influence changes in d.c operating conditions can have on the small signal (a.c) operation of the amplifier.

32. A CLASS B push pull amplifier uses transistors with maximum collector dissipation of 2 watts. Calculate (i) the maximum a.c power output which can be delivered to a 5 Ω load if d.c supply is 12 volts (ii) the require transformer turns ratio. If input signal is halved after operating at its maximum value calculate how values of (a) power from d.c supply (b) a.c power output (c) the conversion efficiency. Justify and expression need in calculate.

33. A FET is used in common source configuration for an R.C coupled amplifier with R_g = 100KΩ between gate and common line midband gain of amplifier is to be greater than 20dB. If g_m is 0.5mA/V, r_d = 40KΩ. Calculated (i) minimum value of R_L (ii) the output resistance

(iii) the current gain. If an identical circuit were cascaded with the above stage would the overall voltage gain be 40dB? Give reasons for answer.

34. State the Nyquist criterion of stability as applied to single loop feedback amplifiers maintains condition and frequency for a sinusoidal oscillator which uses a phase-inverting amplifier having very high input impedance and very low output impedance together with a phase shifting network. Draw a circuit configuration which uses a single transistor as the active element. Would the theoretical expressions derived above be valid for this circuit? Give reasons.

35. An a.c voltmeter of sensitivity $1000\Omega/v$ is connected across the output-terminals of a 1000Hz oscillator. On the 5-v range the meter reads 3.75V and on the 10-v range 4.09V. Find the voltage and the current equivalent circuits of the oscillator; assume that the oscillator output impedance is resistive. Evaluate the voltage which would be developed across;

    (a) 2000Ω resistor (b) $0.159\mu F$ capacitor

    (c) $2,000\Omega$ resistor in parallel with a $0.159\mu F$ capacitor when each is connected in place of the meter.

36. A CRO amplifier has a gain of X200 and the tube-sensitivity is 15V/cm. The input impedance of the amplifier is represented by $1M\Omega$ in parallel with 75pF. Find the screen deflection (in mm) when fed as shown.

37. A rectifier type AC voltmeter has full scale deflection (FSD) = 15V. Find the scale reading when connected to 10V d.c

APPENDIX

38. An emitter has $\phi = 4.54$ eV and runs at 2100^0C. Find the % change in emission for a change in temperature of 0.1% [take $Q_e/k = 11,600$]

39. Estimate the voltage increment per decade of current for a junction diode at (a) 20^0C (b) 100^0C. Estimate also the slope resistance for (a) and (b) at a current of 1 AMP.

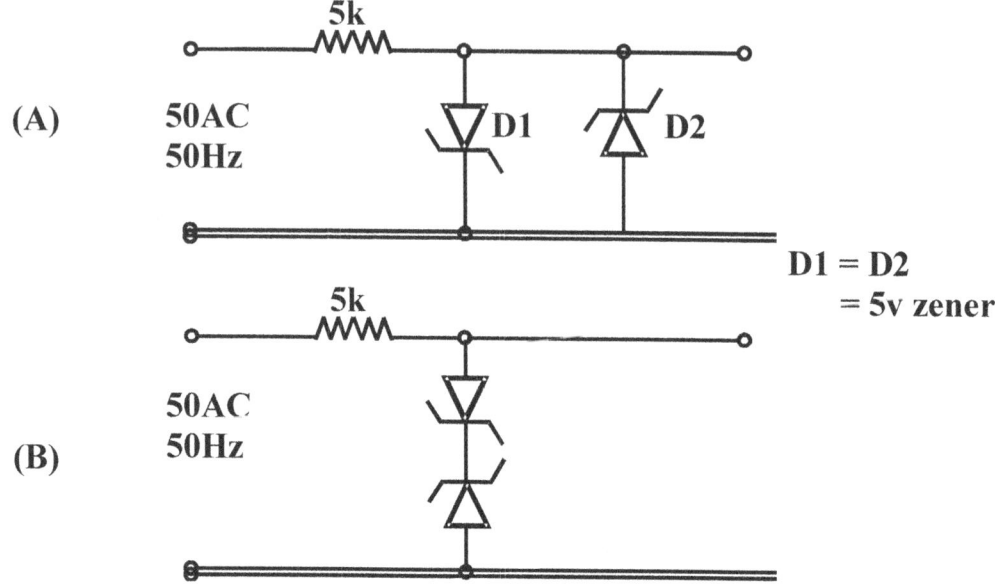

D1 = D2 = 5v zener

40. A junction diode at 20^0C has a reverse current of $2\mu A$. Estimate the p.d at a forward current of 10mA. What change in p.d (in μV) is caused by a 1% increase in the forward current of 10mA?

41. An electronic device has the following characteristics:-

V	120	140	160	180	volts
I	5.60	4.72	3.15	2.20	mA

    Comment on the device. A parallel tuned circuit of L= 1mH, C = 500pF is connected between the device and a +150V supply. Find the value of Q at which oscillation is just possible

42. Fully explain the term miller effect. Finds the input reactance at 100Hz of a triode having cgk=cga=5pF, given the voltage amplification, A= 10. Compare the reactance at 1MHz and comment on the difference.

APPENDIX

43. Explain how a P-type and N-type semiconductor can be realised from a pure germanium crystal. Sketch the characteristics of a PN junction when it is (i) Forward biased (ii) Reverse biased.

44. Sketch and label the symbols of the following electronic components
    (a) PN diode (b) NPN transistor (c) Zener diode (d) Varactor diode (e) photocell (f) Triode valve.
    Sketch a PNP transistor biased in the (a) common base mode
    (b) common emitter mode.

45. Define the following term. (a) Amplification factor for a common base circuit (b) Show the relationship between α and β. Solve for β if $\alpha=0.98$.

46. Distinguish between conductors, semiconductors and insulators. With the aid of sketched graph, show the relationship between resistance and temperature for conductors, semiconductors and insulators.

47. Write short notes on four of the following (a) thermistor (b) mutual conductance of the triode 9c) Fermi level (d) power amplification (e) emitter follower (f) extrinsic semiconductor.

48. Describe the action of a junction rectifier and discuss the main parameter which specifies the device.

49. Describe with appropriate diagrams the actions of a junction type field effect transistor. Sketch the characteristics and discuss the effect of temperature.

50. State how Nyquist's criterion may be used to determine the stability of a feedback amplifier from its open-loop characteristics. A phase inverting operational amplifier having an openloop voltage gain of magnitude 5000, input resistance of 20kΩ and output resistance of 1000Ω is to be used with series negative voltage feedback. The closed loop voltage

gain is to be -100. Calculate (i) The feedback ration β (ii) The input and output resistances of the feedback amplifier.

51. In the circuit shown in the diagram below, Calculate

(i) The current gain i_L/i_S

(ii) The input resistance measured at the point X-X

(iii) The voltage gain V_O/V_S

When the switch S is

   (a) Open

   (b) Closed

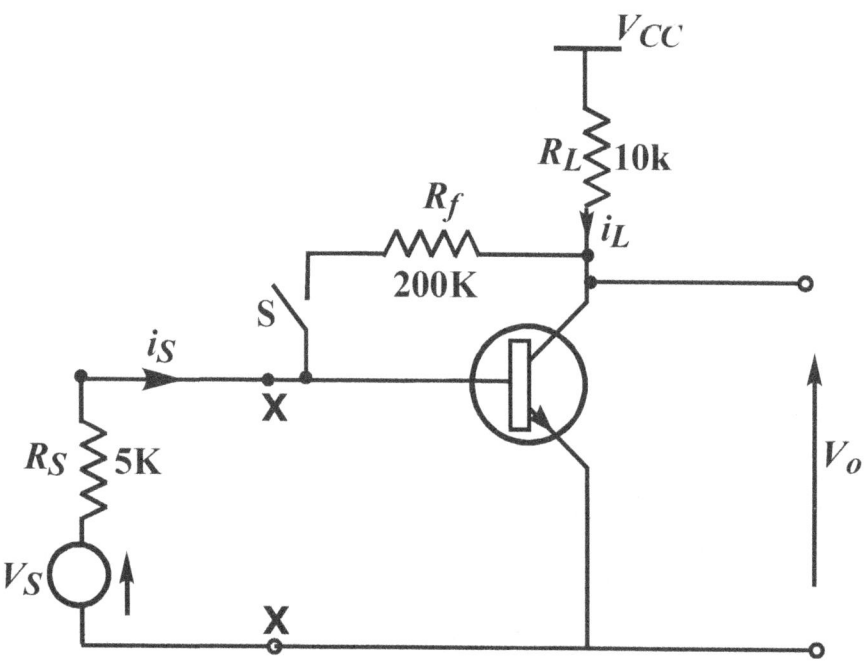

The relevant transistor parameters are

$h_{ie} = 1200\Omega$ $h_{fe} = 50$

$h_{re} = 2.5 \times 10^{-4}$ $h_{oe} = 25 \times 10^{-6} S$

52. Discuss the circuit parameters which in general limit the high frequency performance of any amplifier.

APPENDIX

A junction FEt is used in an amplifying stage shown in the diagram below.

If the FET parameters are $g_m = 1.5mA/V$, $r_d = 50K\Omega$, $C_{gd} = 2.5pF$, $C_{gs} = 3.0pF$ and $C_{ds} = 1.5pF$ calculate

1. The lower frequency break point
2. AT a frequency of 25KHz
   a. The input impedance
   b. The voltage gain in terms of both magnitude and phase

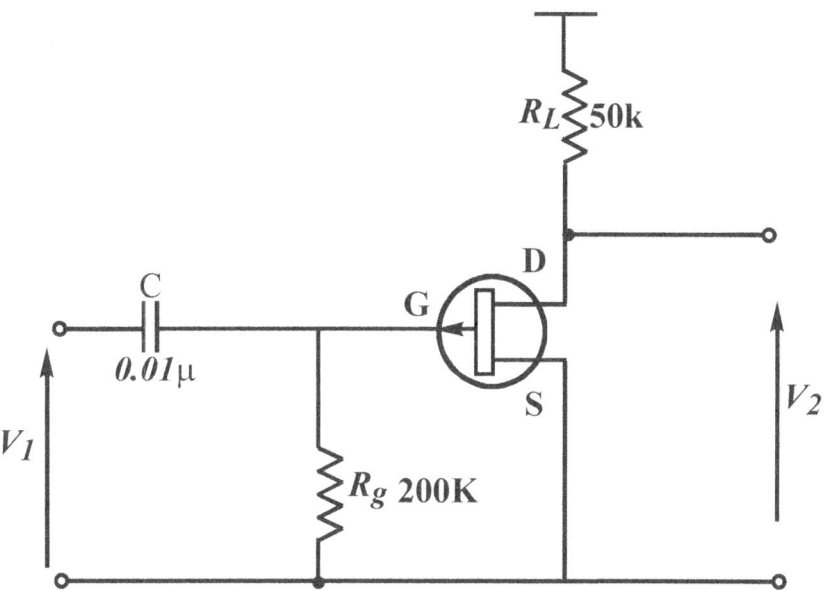

If an identical circuit was connected in cascade with the one shown below qualitatively discuss its effect on the voltage gain of the first stage.

28. How does the method of biasing affect the operation and classification of power amplifiers?

A transformer coupled CLASS B push pull amplifier is to be designed using transistors having a maximum dissipation of 15 watts. If the load resistance is 5Ω and the supply voltage is 50V calculate.

i. The maximum power output.
ii. The transformer turns ratio.

However, under normal working conditions the input signal is only half its maximum value. Thus determine for these conditions.

    a. The actual power output, P_O

    b. The power drawn from the supply, $P_{d.c}$

    c. The efficiency.

29. Describe the structure and functions of the cathode-ray oscilloscope stating how it may be used to measure voltage frequency and phase difference of two signals

30. What are semi-conductors? State fully the properties which make them useful in the construction of diodes and transistors.

31. Discuss two methods of biasing applicable to common emitter amplifying stages. What are the factors which affect the stability of the biasing? Derive the expression for the collector current in each case. If $R_L = 5K\Omega$, $R_f = 200K\Omega$ and $\beta = 49$, find the stability factor S for a feedback bias comment on the value of S.

32. A CLASS A output stage is operated from a 6V supply with a transistor of 100mW collector dissipation to a resistive load R_L of 3Ω via a transformer 11:1 turn ratio. Calculate the maximum power output.

33. Explain the function of every component in the circuit shown below with particular respect to its effect on (i) the d.c conditions (ii) the a.c or small signal behaviour.

APPENDIX

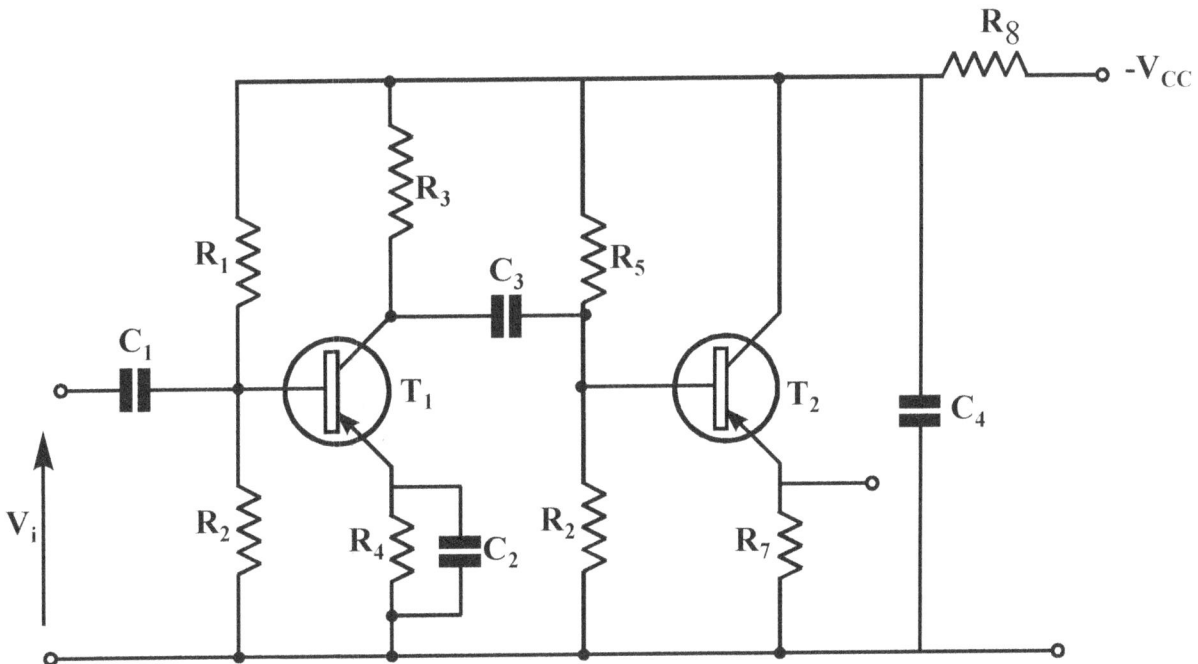

What configurations are the transistors operating in and why are they coupled as shown.

34. A coil of 1mH and resistance of $10\sqrt{10}\Omega$ is tuned to resonance by 100pF. Calculate (i) the resonant frequency (ii) Q-factor for the tuned circuit alone.

It is then used as collector load of tuned C.E transistor amplifier. If the effective transconductance is g_t= 1.5mA/V between amplifier input and output in the equivalent collector circuit shown.

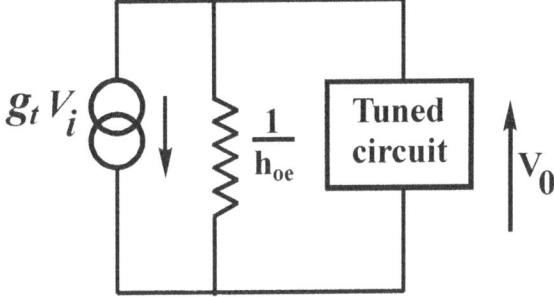

Calculate (iii) effective Q-factor for tuned amplifier (iv) Bandwith of amplifier (v) Gain $\frac{V_o}{V_i}$ of amplifier at resonance.

35. Discuss briefly three of the following topics

(i) A p-n junction photodiode (ii) Zener diode (iii) Capacitance of pn junction (iv) Voltampere characteristics of an ideal pn diode (v) Emitter follower.

35. What is thermoinic emission current? A thermionic cathode runs at 970^0K and consists of a cleave $1\frac{1}{8}$" long and $\frac{1}{8}$" diameter. If the constant A is $100 A/m^2$ and the work function is 1.1eV. Find the saturation emission current. (1in = 25.4mm)

36. A resistive loaded amplifier is to be used under maximum signal condition only show how the d.c power ($P_{d.c}$) a.c power (P_0) and transistor dissipation (P_{tr}) may be calculated. If P_{tr} is 0.1 watt. Calculate $P_{d.c}$ and P_o.

37. A varactor diode is run so that its capacitance is 12 pF and the
....

38. How does the method of biasing affect the operation and classification of power amplifiers?

A transformer coupled CLASS B push pull amplifier is to be designed using transistors having a maximum dissipation of 15 watts. If the load is 5Ω and the supply voltage is 50V. Calculate (i) the maximum power output (ii) the transformer turn ratio.

39. Define Q-factor and explain what is meant by selectively of the tuned circuit.

A coil of 1 mH and resistance of 10Ω tuned to resonance to 100pF capacitor. Calculate the resonance frequency and the Q-factor of the circuit.

Gradient of the characteristics is 1.2pF/volt. The diode is in parallel with an inductor giving a resonant frequency of 90 KHz. Find the response in KHz/millivolt.

APPENDIX

40 Derive an expression for the deflection 'D' of a beam of electrons on the screen of a cathode ray tube in terms of the deflecting voltage V_d. Show the relationship between the deflection and the sensitivity S of the tube.

41 Describe the volt-ampere characteristics of a p-n diode with particular reference to the reverse bias current.

What is the effect of temperature on the characteristics?

42. An electronic device has the following characteristic

V.   120   140   160   180   volts

I   5.60 4.72 3.15 2.20 mA

Comment on the device. A parallel tuned circuit of L = 1mH, C= 800F is connected between the device and a 150V supply. Find the value of Q at which oscillation is just possible.

43. A CLASS A output stage is operated from a 6V supply with a transistor or 100mW collector dissipation to

(i)   A resistor load R_L

(ii)   A resistive load of 3 Ω via a transformer.

Calculate the maximum power output for each case and the value of R_L for (i) above and the turns ratio in (ii).

44. An emitter has ϕ = 4.54eV and runs at 2100^0C. Find the % change in emission for a change in temperature of 0.1% ($Qe/K = 11{,}600$).

45. Find the maximum power dissipated in the transistor T shown in the diagram below

$e_1 = R_cI_1 + R_bI_2 + (I_1 + I_2) R_a$

$e_1 = R_1I_1 + R_2I_1 + R_3I_2$

$e_2 = (R_2 + R_3)I_2 + R_3 I_1$

$$e_1 = \frac{R_C(R_a + R_b)}{R_a + R_b + R_C} I_1 +$$

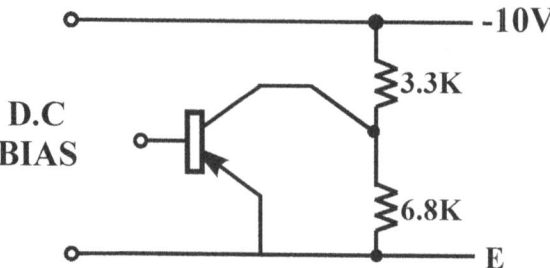

46. Define Q-factor and explain what is meant by selectivity.

    A coil of 1mH and resistance of $10\sqrt{10}\,\Omega$ is tuned to resonance by 100pF. Calculated (i) the resonant frequency (ii) Q- factor for the tuned circuit alone.

    It is then used as collector load of tuned C.E transistor amplifier. If the effective transconductance is $g_t = 1.5$ mA/V between amplifier input and output terminal in equivalent collector circuit shown. Calculate (iii) effective Q-factor for tuned amplifier (iv) bandwidth of amplifier (v) Gain V_o/V_i of amplifier at $V_c = i_c R_L$

REFERENCES

1) BROPHY, JAMES J. 1972 Basic Electronics for Scientists. Tokyo. McGraw-Hill Kogakusha Ltd.

2) COOK, NIGEL P. 1996 Introduction DC/AC Electronics (3rd Edition), New Jersey 07632 Prentice-Hall Inc.

3) HALKIAS, MILLMAN 1972 Integrated Electronics Tokyo McGraw-Hill Kogakusha Ltd.

4) HUGHES, E. 1970 Electrical and Electronic Technology, New Delhi

5) MEHTA, V. K 2003 Principles of Electronics New Delhi. S. Chand & Company Ltd.

6) RICHARD PEASSON 1979 Lecture Series 15F Glasgow. *The Thomson Foundation TV. Collage*

INDEX

A

a.c resistance, 6, 16, 27, 103
absolute temperature, 3
acceptor, 46
Acceptor, 44
Accumulator, 332, 334, 337
Address register, 334
Address Register, 333, 334
alternating component, 21, 26, 64
alternating voltage, 18, 19, 20, 21, 64, 298, 302
Amplification factor, 17, 18
amplifier bandwidth, 190
Amplitude Modulation, 237
amplitude stabilisation, 233
Angstrom Units, 33
Anode cathode capacitance, 22
Anode current, 3
anode slope resistance, 6, 16
anode transfer characteristics, 15
anode voltage, 3, 4, 5, 6, 13, 15, 16, 17, 25
antiphase, 21, 161, 194
Arithmetic Logic Unit, 333, 334
astable multivibrator, 289, 293, 297
avalanche, 52, 56, 175

B

Base, 91, 101
BASE BAND, 238
base-emitter junction, 109, 110, 112, 114, 116, 125, 132, 138, 296, 352
battery terminal, 10
Bessel functions, 248
binary digit, 273, 340, 348
BINARY system, 252
bipolar transistors, 307
bistable, 276, 277, 278, 289, 295, 348, 349, 350, 352, 355
Bohr, 31, 35
bombarding electron, 34
Boolean Algebra, 259
Botzmann's Constant, 51, 60
boundary, 1, 2, 47

C

breakdown voltage, 50, 51, 52, 192, 193, 196
buffer stage, 164

CARRIER, 238, 239
carrier signal, 237, 248
carrier wave, 236
characteristics of the triode, 12, 13, 14
chemical impurities, 41
circuit models, 150
clipper, 9
CLOUD, 2
CMOS techniques, 349
CODE system, 237
Collector, 101
collector leakage current, 120, 121, 122, 133, 283, 285
Collector Leakage Current, 283
comparators, 318
CONDUCTION BAND, 39
conservation of energy, 30
constituent atoms, 36
control grid, 23, 24, 27
control resistance, 55
conventional direction, 3
Coulomb's Law, 29, 37
coupling capacitor, 162
crystal structure, 39, 41
crystalline structure, 36

D

Darlington, 309, 364
Data Bus, 331, 332
data register, 333, 338
Data Register, 333, 334
De Morgan's Law, 264, 265
decimal system, 250, 251, 253, 254
de-coder, 338
depletion region, 174, 175
differentiating circuit, 297, 299, 300, 301, 302, 304, 305, 306
diffusion constant, 150
digital applications, 307
digital circuitry, 58
digital circuits, 106, 253, 257, 258
digital computers, 10
digital electronics, 327

digital techniques, 236
Diode Selection Matrix, 343, 344
direct electrical connection, 236
discrete components, 308
discrete orbits, 36
donor impurity, 44, 46
doping, 42, 44, 57
double tuned amplifiers, 211
dynamic characteristics, 8, 9, 20, 21
dynamic conductance, 58
dynamic impedance, 184, 187, 206
dynamic transfer, 19

E

EAROM, 345, 346
electrical conductivity, 42
electrical conductor, 1
electromagnetic radiation, 235, 236, 237
electromagnetic waves, 237
Electronic
switches, 279
electronic device, 279, 371, 378
Electrons, 38, 39, 40
electrostatic shield, 24
electrostatic shielding, 24, 27
Emitter, 101
energy level, 32, 33, 34, 35, 36, 37, 38
envelope, 10
Equivalent circuit, 22, 197, 198, 205, 206
excellent conductor, 58
extrinsic semiconductor, 41, 372

F

feedback capacitor, 318
feedback effect, 24
feedback voltage, 151, 217, 221, 312, 315
Fermi level, 39, 40
FERMI LEVEL, 39
filament temperature, 4
filament voltage, 4, 5
flip-flop multivibrator, 288, 289
fluctuation, 11, 80

INDEX

FORBIDDEN ENERGY GAP, 39
Fourier series, 195
free electron concentration, 45
free electrons, 1, 2, 48, 49, 52
Free running Multivibrator, 288
frequency compensation, 316
frequency modulation, 244
frequency of oscillation, 31, 228, 230, 232, 365
Full wave ratification, 10

G

Germanium, 43, 44
germanium atoms, 45
Gray code, 274
grid
  voltage, 13, 14, 15, 16, 18
Grid anode capacitance, 22
Grid cathode capacitance, 22
Grid voltage, 20
grid wires, 12
grid-anode capacitance, 23
grid-cathode voltage, 23
ground level, 34

H

half adder, 268, 269
half cycle, 10, 11, 62, 63, 64, 66, 111, 112, 299
Hartley circuit, 231
heater, 7, 10
HEXAGONAL, 329
high frequencies, 313
high impurity density, 58
high internal impedance, 23
high voltage winding, 10
HORSE code, 237
hybrid equivalent circuit, 148
hydrogen atom, 29, 33

I

impurity atoms, 41, 57
incandescent filament, 1
incident photons, 60
Index Register, 332
inherent thermal stability, 177
Input/Output devices, 332
instantaneous anode current, 7
Instruction Decoder, 333, 334

Instruction Set, 336
integrated circuit, 309, 318, 340
integrating circuit, 302, 303
integrator, 303, 317
inter-electrode capacitances, 22, 24
intrinsic semiconductor, 41, 44, 45
ionisation potential, 34, 35

J

JUGFET, 173

K

karnaughs map, 267
kinetic energy, 30, 35
knee voltage, 51, 52

L

Large power outputs, 193
laws of electromagnetism, 31
linear dynamic curve, 9
linear micro-circuits, 307
Load, 7
load line, 8, 99, 102
load resistor, 6, 19, 71, 92, 131, 310, 349
logarithmic amplifiers, 318

M

magnetic bubble memories, 359
majority carriers, 44, 51, 57, 172, 173, 174
Matrix structure, 358
memory location, 332, 335, 338
microprocessor, 327, 330, 332, 333, 335, 336, 337, 339
miller effect, 25, 371
Miller effect, 22, 164
minority carriers, 44, 57, 60, 283
modulating frequency, 247, 249
modulation factor, 245
modulation index, 239, 245
modulators, 180, 318
monolithic linear, 307
Monostable, 288, 289

monostable multivibrator, 289, 293
MOS F.E.T, 307
multistage amplifier, 211
multivibrators, 288, 300, 318
mutual characteristics, 13, 14
mutual conductance, 16, 18, 176, 372

N

NAND, 250, 264, 270, 318, 319, 327
Napierian base, 3
Narrowband, 247
negative bias, 18, 51, 60
Negative Feedback Amplifier, 211
negative resistance, 28, 58
neutralize, 12
noise, 177, 247, 311
N-type semiconductor, 44, 47, 49, 372
nyquist's criterion, 228

O

ohms law, 44
operation code, 337
operational amplifier, 311, 316, 372
oscillator, 23, 226, 228, 249, 288, 311, 312, 360, 365, 370
output characteristics, 13, 100, 110, 284
Output Waveform, 303

P

Parallel resonance, 183
parametric amplifiers, 54
peak current, 58, 103
peak forward voltage, 58
pentode, 23
Pentode, 27
permittivity of free space, 30
Phase angle, 216
phase angles, 232
Phase Modulation, 237
phase splitter stage, 203
physical behaviour, 143
Planks constant, 32
PN diodes, 90
polarity, 10, 14, 49, 217, 340, 353

INDEX

polycrystalline phases, 346
positive potential, 2, 12, 292, 294
potential barrier, 48, 52, 53, 57, 114
potential barrier voltage, 53
potential difference, 2, 3, 13, 16, 20, 25, 28, 220
potential energies, 30
power, 54, 55, 56, 61, 99, 103, 106, 108
power dissipation characteristics, 179
Programme Counter, 332, 333, 334
programming diodes, 346
PROM, 345, 346
P-type impurities, 44
push pull, 196, 198, 361, 369, 374, 377

Q

quantized, 32, 35
quantum mechanical behaviour, 58

R

radiated energy, 31
radiation resistance, 241
radio frequency, 10, 23
radio frequency signals, 10
RAM, 330, 331, 333, 335, 345, 348, 356, 357, 358
RAM Cell, 357
RAM-ROM chip, 358
reasonable amplifier, 316
Rectangular Wave Input, 304
repelled, 2, 13
requisite energy, 34
resistive load, 19, 21, 69, 180, 360, 366, 375, 378
resonance, 34
resonant frequency, 184, 368, 376, 377, 379
reverse direction, 3, 50, 58
ripple, 11, 55, 79, 80, 81, 82, 83, 87

S

saturation collector current, 284, 286
Saturation Collector Current, 283, 284
saturation current, 2, 3, 5, 51, 60
Saturation current, 51, 60
Saturation Current, 50
Schrödinger equation, 57
screen grid, 23, 24, 25, 26, 27, 28
screen-cathode voltage, 26
secondary emission, 25
secondary winding, 63, 64, 70, 71, 73, 76, 205
semiconductor photodiode, 59, 61
sense wire, 342, 343
servomotor, 179
short circuited, 158, 159, 164, 165
shunting, 54, 187
sine waves, 234
Single Tuned Stages, 205
sinusoidal, 9, 19, 62,
slope resistance, 26, 55, 371
SPACE CHARGE, 2
spectral lines, 31, 33
Square wave input, 303, 305
Stabilisation, 119, 139, 225, 226
stability factor, 122, 125, 127, 133, 134, 140, 375
Stability factor, 122, 129
STACK, 335
Stack Pointer, 332, 335
Static Cell, 351, 355
static characteristics, 4, 8, 9, 17, 25, 27, 28, 51, 56, 92, 93, 95
Status Register, 334
suppressor grid, 26, 27
synchronous detectors, 318
system memory, 330, 332

T

Tetrode, 24, 25
thermal runaway, 121, 125
thermal stability, 125, 133, 140

thermionic agitation, 49
thermionic valve, 3
thermistor, 46, 372
thermistors, 61, 233
transconductance, 165, 176, 376, 379
transient, 226, 286
transistor
  biased, 98, 134, 195, 198, 284, 285, 372
Triangular Wave Input, 300
triangular waveform, 316
trigger pulse, 289, 295, 296
triggering pulse, 289, 294
triode amplifier, 22
TRI-STATE, 329

U

unidirectional, 11
unmodulated carrier, 247

V

vacuum diode, 3, 4, 5, 10
Vacuum diode, 1
vacuum triode, 15, 17, 21, 22
valence electrons, 37, 39, 49, 52
valley current, 58
valve circuits, 177, 190
varactor diode, 54
VARACTOR DIODE, 54
voltage amplification, 21, 22, 24, 371
Voltage amplification, 99
voltage divider, 138, 140
voltage source, 152, 367
voltage stabilizer, 54, 55

W

waveform, 9, 10, 11, 19, 20, 83, 100, 195, 203, 238, 241, 244, 247, 298, 299, 300, 301, 303, 304, 316
wavelength, 34, 35, 57, 237

Z

zener diode, 50, 55
zero signal base current, 123, 132

Made in the USA
Columbia, SC
20 December 2019